GREEK THEORIES

OF

ELEMENTARY COGNITION

FROM ALCMAEON TO ARISTOTLE

GREEK THEORIES

OF

ELEMENTARY COGNITION

FROM ALCMAEON TO ARISTOTLE

BY

JOHN I. BEARE, M.A.

FELLOW OF TRINITY COLLEGE, DUBLIN
REGIUS PROFESSOR OF GREEK (SOMETIME PROFESSOR OF MORAL
PHILOSOPHY) IN THE UNIVERSITY OF DUBLIN

OXFORD

AT THE CLARENDON PRESS

1906

HENRY FROWDE, M.A.
PUBLISHER TO THE UNIVERSITY OF OXFORD
LONDON, EDINBURGH
NEW YORK AND TORONTO

PREFACE

MOST readers know the difficulty as well as importance of the *de Anima* and *Parva Naturalia* of Aristotle; and any genuine assistance would be welcomed by students who desire to master them. A great deal has been done by editors and others for the elucidation of the former of these works and, indirectly, of the latter, so far as they involve metaphysics, or psychology in its higher reaches. No one, however, has been at the pains to glean and put together systematically, from Aristotle himself and his predecessors, whatever may explain or illustrate the parts of his writings essentially concerned with empirical psychology. The results of this, it should seem, would be useful not only to students of ancient Greek psychology, but also to readers who, perhaps knowing and caring little about Greek, might yet desire a clear and objective, even if brief, account of what was achieved for the psychology of the senses by the ancient Greek philosophers. The purpose of this book, within the limits defined by its title, is to present such an account; and it will rightly be judged according to the degree in which it fulfils its purpose. Among its most competent critics will be the student who may test its usefulness in connexion with the many passages on the interpretation of which it directly or indirectly bears. To such critics and others its author leaves it; confiding less, however, in the merits of his work than in the fellow-feeling which all scholars, as well as students of philosophy, have for one who honestly grapples with their common foe, τὸ ἀσαφές, in whatever form this may present itself.

The books used or consulted are named in the list given

below ; but wherever even a hint has been borrowed, the writer to whom obligation has been thus incurred will always be found referred to in the notes.   There are many such references, especially to the publications of H. Diels ; but the mainstay of the whole work has been the actual text of Plato, Aristotle, and Theophrastus.   A list of the Greek passages explained or discussed has been added at the end.   In some—perhaps most—of these the points raised are of no great interest to scholars, but there is at least one exception ; and it is hoped that what has been said on Arist. 452^b 17–24 may be of some value.

The author wishes to thank the Delegates of the Clarendon Press for undertaking the publication of this work. His thanks are also due to the Press Reader and Staff for their great care and accuracy.   It remains for him, in conclusion, to express his deep gratitude to Mr. W. D. Ross, Fellow and Tutor of Oriel College, Oxford, for kindly reading the proofs, and making acute suggestions from which much profit has been derived.   He is indebted to Mr. Ross for having drawn his attention to Diels' palaeographical correction of Arist. 985^b 17, mentioned on p. 37, n. 2.

9 TRINITY COLLEGE, DUBLIN,
          January 10, 1906.

COMMENTARIES, MONOGRAPHS, &c.

USED FOR THE FOLLOWING WORK

Adam, J. Plato, *Republic* (Cambridge, 1902).
Alexander of Aphrodisias in Arist. *De Sensu*, Thurot (Paris, 1875).
" " " " Wendland (Berolini, 1901).
" " " *Metaph.* Hayduck (" 1891).
" " " *De Anima*, Bruns (" 1887).
" " " *Quaestiones etc.* Bruns (" 1892).
Archer-Hind, R. D. Plato, *Phaedo*, ed. 2 (London and New York, 1894).
" " *Timaeus* (London and New York, 1888).
Bacon, R. *Opus Maius*, Bridges (Oxford, 1899).
Bäumker, C. *Des Aristoteles Lehre von den äussern und innern Sinnesvermögen* (Leipzig, 1877).
" *Zu Aristot. 'De Sensu'* 2, 438ᵇ 16 ff. (*Zeitsch. f. d. öst. Gymn.*, Sept. 1877, 605 ff.).
Beck, H. *Aristoteles de Sensuum Actione* (Berlin, 1860).
Becker, Guil. Ad. *Aristoteles de somno et vigilia etc.* (Lipsiae, 1823).
Biehl, Guil. *Aristotelis Parva Naturalia* (Teubner, 1898).
Bitterauf, C. *Quaestiunculae Criticae ad Par. Nat.* (Monachii, 1900).
Boeckh, A. *Philolaos des Pythagoreers Lehren* (Berlin, 1819).
Bonitz, H. *Aristotelis Metaphysica* (Bonnae, 1848).
" *Aristotelische Studien* (Wien, 1862-7).
Brentano, F. *Die Psychologie des Arist.* (Mainz, 1867).
Burnet, J. *Early Greek Philosophy* (London and Edinburgh, 1892).
Bury, R. G. *Plato's Philebus* (Cambridge, 1897).
Bywater, I. *Heracliti Ephesii Reliquiae* (Oxford, 1877).
Campbell, L. Plato, *Sophistes* (Oxford, 1867).
" " *Theaetetus* (" 1883).
Chappell, W. *History of Music*, vol. i (London, 1874).
Dembowski, J. *Quaestiones Aristotelicae duae* (Regimonti Pr. (sic), 1881).
Diels, H. *Doxographi Graeci* (Berolini, 1879).
" *Die Fragmente der Vorsokratiker* (Berlin, 1903).
Eberhard, E. *Die aristotelische Definition der Seele etc.* (Berlin, 1868).
Freudenthal, J. *Zur Kritik und Exegese von Aristot.* περὶ τῶν κοινῶν σώματος καὶ ψυχῆς (Rhein. Mus. xxiv, pp. 81–93, 392–419).
" *Ueber den Begriff des Wortes φαντασία bei Arist.* (Göttingen, 1863).
" *Zu Aristot. 'De Mem.'* 452ᵃ 17 ff. (*Archiv f. Gesch. der Phil.*, II. Band, i. Heft, 1889).
Freytag, W. *Die Entwickelung der griechischen Erkenntnistheorie bis Aristoteles* (Halle, 1905).

vi   COMMENTARIES, MONOGRAPHS, &c.

Galenus, Claudius.   *De Placitis Hippocratis et Platonis*, I. Müller (Lipsiae, 1874).
Goclenius, R.   *Libelli Aristotelis de Sensu et Sensilibus castigata versio et analysis logica* (Francofurti, 1596).
Gomperz, T.   *Greek Thinkers* (E. Tr.) (London, 1905).
Görland, A.   *Aristoteles und die Mathematik* (Marburg, 1899).
Grote, G.   *Aristotle*, 3rd ed. (London, 1883).
„   *Plato and the other Companions of Socrates* (London, 1875).
Hammond, W. A.   *Aristotle's Psychology: De Anima and Parva Naturalia*, transl. with introduction and notes (London and New York, 1902).
Hayduck, M.   *Emendationes Aristoteleae* (Meldorf, 1877).
„   *Observationes criticae in aliquot locos Arist.* (Greifswald, 1873).
Hippocratis *Opera*.   E. Littré (Paris, 1839-61).
Ideler, J. L.   *Aristot. Meteorologica* (Lipsiae, 1834-6).
von Jan, C.   *Musici Scriptores Graeci* (Lipsiae, 1895).
Jourdain, C.   *Recherches critiques sur l'âge et l'origine des traductions latines d'Aristote*, Nouv. éd. (Paris, 1843).
Kampe, F. F.   *Die Erkenntnisstheorie des Aristoteles* (Leipzig, 1870).
Karsten, S.   *Empedoclis Agrig. Carm. reliquiae* (Amstelodami, 1838).
Marchl, P.   *Des Arist. Lehre von der Tierseele*, i. Teil (Beilage zum *Jahresberichte des humanistischen Gymnasiums* (Metten, 1896-7)).
Michaelis Ephesius.   *In Arist. Parva Naturalia*, Wendland (Berolini, 1903).
Mullach, F. W. A.   *Democriti Abderitae Fragmenta* (Berolini, 1843).
„   *Fragmenta Philosophorum Graecorum* (Parisiis, 1857-79).
Neuhäuser, J.   *Aristoteles Lehre von dem sinnlichen Erkenntnissvermögen und seinen Organen* (Leipzig, 1878).
Ogle, Dr. W.   *Aristotle on the Parts of Animals*, trans. and annot. (London, 1882).
„   *Aristotle on Youth and Age etc.*, trans. and annot. (London, 1897).
Pacius, J.   *Aristotelis Parva (ut vocant) Naturalia* (Francofurti, 1601).
Panzerbieter, F.   *Diogenes Apolloniates* (Lipsiae, 1830).
Philippson, L.   ὕλη ἀνθρωπίνη (Berlin, 1831).
Poschenrieder, F.   *Die naturwissenschaftlichen Schriften des Arist. in ihrem Verhältnis zu den Büchern der hippokratischen Sammlung* (Bamberg, 1887).
Prantl, C.   *Aristoteles über die Farben* (München, 1849).
Ritter and Preller.   *Fontes Philosophiae*, Ed. 7 (Gothae, 1886).
Rohde, E.   *Psyche*, Ed. 3. (Tübingen u. Leipzig, 1903).
Schaubach, E.   *Anaxagorae Clazomenii Fragmenta* (Lipsiae, 1827).
Schieboldt, F. O.   *De Imaginatione Disquisitio ex Arist. Libris repetita* (Lipsiae, 1882).

Schmidt, J. *Aristotelis et Herbarti Praecepta, quae ad Psychologiam spectant, inter se comparantur* (Wien, 1887).

Siebeck, H. *Geschichte der Psychologie,* 1. Teil, 1. Abt. (Gotha, 1880).

„   *Aristotelis et Herbarti doctrinae psychologicae etc.* (Halis Sax. 1872).

Simonius, S. *Arist. de Sensu et de Memoria* (Genevae, 1566).

Sperling, K. *Aristoteles' Ansicht von der psychologischen Bedeutung der Zeit als 'Zahl der Bewegung'* (Marburg, 1888).

Stein, H. *Empedoclis Agrigentini Fragmenta* (Bonnae, 1852).

Stewart, J. A. *Notes on Aristotle's ' Nicomachean Ethics'* (Oxford, 1892).

Sturz, F. G. *Empedocles Agrigentinus* (Lipsiae, 1805).

Susemihl, F. Various 'Scholae' on passages in Aristotle (Greifswald).

Themistius (Sophonias). *In Arist. Parva Naturalia,* Wendland (Berolini, 1903).

Theophrastus Eresius. *Op. Omn.* Wimmer (Parisiis, 1866).

Thurot, C. *Études sur Aristote* (Paris, 1860).

Torstrik, Ad. *Arist. de Anima* (Berolini, 1862).

Trendelenburg-Belger. *Arist. de Anima* (Berolini, 1877).

Usener, H. *Epicurea* (Lipsiae, 1887).

Volprecht, A. *Die physiologischen Anschauungen des Aristoteles* (Greifswald, 1895).

Wachtler, J. *De Alcmaeone Crotoniata* (Lipsiae, 1896).

Waitz, T. *Aristotelis Organon* (Lipsiae, 1844).

Wallace, E. *Aristotle's Psychology in Greek and English* (Cambridge, 1882).

„   *Outlines of the Philosophy of Aristotle* (Cambridge, 1898).

Zeller, E. *Aristotle and the Earlier Peripatetics* (E. Tr.) (London, 1897).

„   *Pre-Socratic Philosophy* (E. Tr.) (London, 1881).

„   *Plato and the Older Academy* (E. Tr.) (London, 1876).

Ziaja, J. *Aristoteles, de Sensu* 1–3 *übersetzt etc.* (Breslau, 1887).

„   *Die aristot. Lehre vom Gedächtniss etc.* (Leobschütz, 1879).

„   *Die aristot. Anschauung von den Wesen und der Bewegung des Lichtes* (Breslau, 1896).

„   *Zu Aristoteles' Lehre vom Lichte* (Leipzig, 1901).

THE FOLLOWING TRANSLATIONS HAVE BEEN CONSULTED :—

(*a*) Those contained in the Berlin and Didot editions of Aristotle.

(*b*) That of the *Parva Naturalia* by H. Bender (Stuttgart, not dated).

(*c*)   „   „   „   „ F. A. Kreutz (Stuttgart, 1847).

(*d*) Saint-Hilaire, J. Barthélemy, *Arist. opuscules, trad. en français* (Paris, 1847).

Also, of course, the translations of Plato by Jowett, of Plato's *Timaeus* and *Phaedo* by Archer-Hind, and of Aristotle's *de Anima* by E. Wallace.

GREEK THEORIES OF ELEMENTARY COGNITION FROM ALCMAEON TO ARISTOTLE

INTRODUCTION

§ 1. THE aim of the following pages is to give a close General historical account of the various theories, partly physio- Plan. logical and partly psychological, by which the Greek philosophers from Alcmaeon to Aristotle endeavoured to explain the elementary phenomena of cognition. The pre-Aristotelean writers who applied themselves to this subject, and of whose writings we possess any consider- able information, are Alcmaeon of Crotona, Empedocles, Democritus, Anaxagoras, Diogenes of Apollonia, and Plato. We propose to set forth here their speculations, together with those of Aristotle, as to the so-called Five Senses, Sensation in general, and the psychical processes, such as Imagination and Memory, which involve the syn- thetic function referred by Aristotle to Sense, and named by his Latin commentators the *Sensus Communis*. We shall concern ourselves as little as possible with metaphysical or epistemological questions, attending rather to what the writers above mentioned, together with Aristotle, did, or tried to do, for *empirical* psychology, to the extent which we have defined. Aristotle in his psychological teaching sums up for us the results of the work of his predecessors, whose doctrines he sifted and compared. Accepting, reject- ing, or modifying these, he developed a scheme of psycho- logy which in minuteness and comprehensiveness transcends anything of the same kind achieved before. But if this is to be thoroughly understood, it must be considered in due connexion with preceding schemes ; and to place it in this connexion we have here brought together all that can be positively ascertained of what earlier philosophers had bequeathed to him. This information we have arranged under three heads—I. The Five Senses; II. Sensation in general; III. The *Sensus Communis*. The subject of each heading is dealt with in such a way as to exhibit the teachings of the successive writers from Alcmaeon to Aristotle respecting it; and with regard to each of the five senses, also, the same order and division have been adopted.

Psychology without meta- physics.
§ 2. All the philosophers above named held certain meta- physical theories which to some extent, no doubt, ruled their psychological thinking [1]. But though they were meta- physicians first and psychologists afterwards, the effect of their metaphysics upon their psychology was by no means as great as might be supposed. The extreme generality of their philosophic views in nearly all cases rendered it im- possible, or at least difficult, for them to effect a real junction between these and the particular phenomena of mind with which psychology deals. As regards the latter, all had before them the same concrete facts; and even those whose fundamental principles differed most widely may sometimes be found giving similar explanations of the elementary phenomena of perception. Hence no grave injury to the practical value of an account of their psychology need be apprehended from the fact that our study of the latter does not connect itself organically with a study of their respective philosophical theories. Theoretically, no doubt, such a connexion is not only desirable, but necessary. A philosophical history of psychology could not be complete without it. But psychology *as a science* may, and must, stand without metaphysics. Whether the psychologist is a materialist or an idealist (or if the antithesis be preferred, a spiritualist), he will, so far as he is true to the conception of science, deal with the elementary phenomena of percep- tion according to ascertained natural laws. If he touches

[1] No one who reads this will be ignorant of what these theories were; therefore it would be superfluous as well as tedious to give a detailed statement of them here.

upon questions which exceed the bounds of phenomena, e. g. as to the nature of mind out of relation to the living organism, he passes the limits of science and therefore of psychology, as this term is here employed. As regards the study of mind, empirical psychology, assisted by physiology, will and ought to have the *first* word, though it cannot have the *last.*

§ 3. The ancient Greek psychologists endeavoured to give observation its due weight in determining such psychological questions as they raised. For this reason they deserve to be called the founders of psychological science. Their honest differences from one another, as well as from their better informed successors, and their helpless ignorance of much which is now familiarly known and fundamental for psychology, contribute to the curious interest which a history of their efforts has for a modern reader. This history is, of course, largely a history of failure. Those, however, who know how far empirical psychology is still from the achievement of its aims will not hastily disparage the Greeks on this account. It was not so much the defectiveness of their psychological methods—defective as these were no doubt—as that of their physical and physiological science that rendered fruitless their best attempts to comprehend the elementary facts of sense-perception, and to place them in an intelligible connexion with their conditions. The most ancient Greek psychologists treated psychology as an integral part of physics or of physiology. With the possible exception of Anaxagoras, they looked upon 'knowing,' for example, as one of the many properties of matter. Problems as to the nature of space, critically considered, lay beyond their horizon. They never asked how it comes to pass that we 'project' our percepts in an extra-organic space, and fall into the habit of speaking of them as outside *ourselves.* Questions of the objective existence of things whose qualities are perceived or known only in virtue of our faculty of cognition did not come up for discussion until some centuries after Thales. Before the Sophists—or 'die Sophistik'—all agreed that there is on one hand such a thing as *truth* (however difficult to discover sometimes), and, on the other, such a thing as its opposite, *error* or *falsehood*. The spirit of the Sophistic age, however, dissolved the barrier which divided Truth from Error, making a new departure neces-sary if philosophy and science alike were not to cease utterly among men. For want of positive knowledge and of method, science and philosophy alike were ultimately endangered in the confusion to which undisciplined specula-tion led the followers of Heraclitus.

As regards scientific method, it was not to be expected that it could exist at a period when logic—deductive and induc-tive—was as yet unknown, and when the provinces of the various departments of thinking had as yet no boundaries assigned to them. As regards positive knowledge, again, the disadvantages under which the Greek psychologists laboured were insuperable. Pure mathematics had advanced to an important degree of attainment, but empirical sciences, e.g. physics and physiology, were in their infancy. Even Aristotle, like his predecessors, with whom he so often places himself in controversy, possessed only the scantiest means of physical observation. In fact, observation did not go beyond what could be accomplished by the naked eye. Physical experiments only of the most rudimentary kind were possible at a time when, of all our varied mathe-matical and physical implements, inquirers had to content themselves with what they could achieve by the aid of the rule and the compasses. ' Chemical analysis, correct measurements and weights, and a thorough application of mathematics to physics were unknown. The attractive force of matter, the law of gravitation, electrical phenomena, the conditions of chemical combination, pressure of air and its effects, the nature of light, heat, combustion, &c., in short all the facts on which the physical theories of modern science are based, were wholly, or almost wholly, undiscovered [1].' In their attempts at psychology under such circumstances it is not to be wondered at if they met with but little success. They had, for example, to arrive at

[1] *Vide* Zeller, *Aristotle*, i. p. 443, E. Tr.

a theory of vision without a settled notion of the nature
of light, or of the anatomical structure of eye or brain.
They had to explain the operation of hearing without
accurate knowledge of the structure of the inner ear, or
of the facts and laws of sound, or at least with only some
few mathematical ideas gleaned from the study of harmonics.   Physiology and anatomy, chemistry and physics,
as yet undifferentiated, lay within the body of vague floating possibilities of knowledge studied by them under the
name of Nature.   For want of a microscope their examination of the parts of the sensory organs remained barren.
They had no conception of the minuteness of the scale on
which nature works in the accomplishment of sensory
processes and in the formation of sensory organs.   The
retina, as well as the structure of the auditory apparatus,
was wholly unknown to them.   The nerve-system had not
been discovered, and the notions formed of the mechanism
of sensation and motion [1] were hopelessly astray.   The
veins, with the blood or (as some thought) the air coursing
through them, were looked upon as discharging the functions
now attributed to the sensory and motor nerves.   Even
Aristotle did not know the difference between veins and
arteries.   When this difference was first perceived, it was
for a time still supposed that the veins conducted the blood,
the arteries the air.   Perhaps the climax of our surprise is
reached when we find Plato of opinion that not only air,
but also drink, passed into the lungs [2].   Yet in this opinion
Plato was at one with the best, or some of the best,
medical teaching of his time.   As early as Alcmaeon of
Crotona the brain had been thought of as the central organ
of sentiency, and, in short, of mind ; and Plato held that it
was so.   But Aristotle, again, declares this to be untrue,
and holds that the heart is the great organ of perception

---

[1] *Vide* Galen. *de Placit. Hipp. et Plat.* §§ 644 seqq.; especially
'Ερασίστρατος [294 B. C.] μὲν οὖν, εἰ καὶ μὴ πρόσθεν, ἀλλ᾽ ἐπὶ γήρως γε τὴν
ἀληθῆ τῶν νεύρων ἀρχὴν κατενόησεν· 'Αριστοτέλης δὲ μέχρι παντὸς ἀγνοήσας
εἰκότως ἀπορεῖ χρείαν εἰπεῖν ἐγκεφάλου.

[2] *Timaeus* 70 c.

and of mind so far as this has a bodily seat. Empedocles had supposed the blood, especially that in the region of the heart, to be the locus or habitation of mind. Thus ignorant of, and therefore free to differ about, cardinal facts and laws of anatomy, physiology, and physics, the ancient Greeks were unable to make real advances towards explaining the conditions of the most obscure of all pheno-mena—those of Mind.

Dialectical psycho-logy. § 4. Under these circumstances many of the Greeks, perhaps feeling the hopelessness of such attempts at em-pirical psychology, occupied themselves for the most part with discursive speculations which really aimed at little more than the clearing up of common ideas or words. Thus Plato's *Theaetetus* is largely occupied with an endeavour to determine the meaning of ἐπιστήμη, or *knowledge*. Dis-quisitions on methodology, too, came to receive much attention from Plato as well as Aristotle ; but the scientific experimental work itself, on which real advance depends, was lacking. Laborious efforts of genius like Plato's ended, too often, for the time in the production of categories, which, however they may have enriched philosophy, left empirical psychology no better off than it had been before. But in place of empirical there came a sort of dialectical or 'rational' psychology, studying, or professing to study, the soul and its faculties *per se*, apart from experience and from organic life in this physical world. With this form of psychology, whether it shows itself in Aristotle or in his predecessors, we shall here have as little as possible to do.

Sources of our know-ledge of ancient Greek psy-chology. § 5. In order that we may most conveniently illustrate the progress of psychological speculations, we shall allow the authors of these speculations to a great extent to speak for themselves through the medium of a translation. Some commentary will be, occasionally, necessary not only to explain particular *dicta* but to exhibit special doctrines in their due relationship to others.

Our first and greatest authorities for the history of psychology, as of so much else in philosophy and science, are of course Plato and Aristotle, especially the latter. We

shall avail ourselves also of the valuable fragment of Theophrastus *de Sensu.* The information derived from these
writers as to the tenets of previous thinkers has always
to be scanned closely in order to discover whether it is
objectively true, or whether allowance has to be made for
differences of standpoint, or for misrepresentation due to
antagonistic attitudes. Still we are most favourably
situated when we have Plato, Aristotle, and Theophrastus
as our guides. Records such as are preserved in the pages
of incompetent historians of philosophy or compilers of
philosophic dogmas who may have lived several centuries
after Christ, when the works of some of the authors with
whom they deal were no longer extant or only survived in
doubtful tradition, must be received with steady scepticism
and tested by every means in one's power. In many cases
these records contain intrinsic proof of untrustworthiness ;
and they are nearly always tinged with the colour of later
theories which had superseded in the popular mind those
promulgated by the earlier psychologists. Thus much of
what is ascribed in Stobaeus or the Pseudo-Plutarch to
Democritus is, from the terms in which it is couched,
evidently contaminated with the teaching of Epicurus ;
much that is ascribed to Plato or Aristotle is expressed
in the terminology of Stoicism.

§ 6. We shall commence by giving a detailed account of Method of
what the writers already named in § 1 each had to say following exposition.
of the particular functions, organs, &c. of *seeing, hearing,
smelling, tasting, touching*; of the objects of these senses
as such, and of the media through which the objects were
supposed to operate. Next we shall present such theories
as they have left on record of *sensation in general,* and
of the faculty (referred by Plato to intelligence, by Aristotle
to sense) which compares and distinguishes the data of the
particular senses, and to which such activities as those of
imagination and *memory* belong. Finally, we might be
expected to discuss the connexion between the faculty of
sense and that of *reason.* With this subject, however, we
shall at present have nothing to do. To discuss it would at

once take us beyond the limits which we have prescribed for ourselves. The nature of the process, if process it can be called, which leads from the elementary phenomena of cognition to the higher functions of thinking, cannot be scientifically in any real sense explained, but must long remain obscure in a sort of metaphysical twilight. The same is true of the process which leads from purely physical to psychical functions; if indeed we are within our rights in thus contrasting them. We have chosen to restrict ourselves to the more positively intelligible subject of empirical psychology, and to the contributions made to the advancement of this by the ancient Greeks.

Greek conception of psychological problem, as regards perception.
§ 7. The conception which the Greeks formed of the conditions of psychology was not lacking in comprehensiveness. They saw that it demanded for its successful prosecution a thorough knowledge (*a*) of the stimulus of perception; (*b*) of the organ of perception as well as of the whole organism; and (*c*) of the medium which somehow connects the object with the organ, and by the help of which the stimulus takes effect in quickening sensation so as to bring the object home 'to consciousness.' Thus a psychological interest not only excited them to physical inquiries but aroused them to investigations which have since culminated in anatomy, physiology, and histology. But they had only vague anticipatory conceptions, such as enabled them to put questions which they were utterly unable to answer, although upon the answers depended the progress of psychological knowledge. Thus for centuries this subject remained totally unprogressive. Any useful progress made by it in modern times has resulted chiefly from advances made in physiological and physical knowledge. If with all that biology, chemistry, and physics can do to help it forward, the most interesting questions of psychology are still unanswerable, or at least unanswered, it is easy to see how fruitless the most intelligent attempts of the ancients were doomed to be in dealing with such questions before these auxiliary sciences existed.

PART I. THE FIVE SENSES

THE ANCIENT GREEK PSYCHOLOGY
OF VISION

§ 1. THE speculations of the ancient Greeks as to the conditions of seeing, and the nature of the proper object of vision, may be chosen to illustrate the strength or weakness of their whole position in elementary psychology. The capital of knowledge which they possessed respecting the facts of seeing was of the scantiest and most superficial kind. They knew (as the most ignorant person knows) that the eye is the organ of sight, and that without light the eye cannot see; that, besides light and the eye, an object is also necessary for vision; and that, moreover, the relationship of the eye to the organism, or certain parts of it, requires to be considered before seeing can be fully explained. Of most of the clear and fine distinctions marked by modern anatomy and physiology between the various parts of the visual apparatus the Greeks, from the time of Alcmaeon to that of Aristotle, were totally ignorant. They had not noticed the retina; they knew of the crystalline lens as an anatomical fact, but had not any notion of its refractive properties, or of the eye as an optical system. They were hopelessly ignorant of the mechanism and need of optical adjustment or accommodation. Such were their shortcomings in physiology, and consequently in the empirical psychology of vision.

§ 2. I. Almost all the early attempts at a theory of vision agree in regarding the 'pupil' of the eye as a matter of primary importance for visual function[1].

Ancient Greek speculation as to the sense of sight.

The chief data of ancient Greek

---

[1] The Greek κόρη and Latin *pupula*, or *pupilla*, as meaning 'pupil,' are both named originally from the circumstance that an observer looking into a person's eye can see in the dark central spot an image of himself

II. Another fact which greatly influenced this branch of study was that when the eyeball is pressed, or moved hastily, in darkness, a flash of light[1] is seen within the eye. From this was drawn the conclusion that the eye has within it a native fire, and that on this native fire, not less than upon the image in the pupil, its faculty of vision somehow depends.

III. A third fact which formed a basis of visual theory was that the interior of the eye is found to contain aqueous humours—roughly called ' water ' by the Greeks. The functions of the retina being altogether unknown, and the optic nerves being perhaps known, but certainly not known in their true character, the primary business of the early psychologists who treated of vision seemed to be, to determine the parts played in vision by the *image*, the *fire*, and the *water*, respectively. As regards the assumed intra-ocular fire, the question was frequently agitated, whether its rays went forth from the eye as from a luminary, and (either by themselves or in combination with a column of light proceeding from the object) as it were *apprehended* the object of vision, and brought it within the purview of ' the soul '; or whether the fire merely lurked within the periphery of the eye, and there seized the image which, coming to it from outside, was reflected in the aqueous interior, as if in a mirror. The seat of the inner fire was the pupil, which, at least from the time of Empedocles, was identified generally with the ' lens.' With these facts before us we shall be better prepared to understand the purport of the extracts which are to follow. We

reflected there. This is dwelt upon by Plato (?) *Alcib.* i. 132 E καὶ τῷ ὀφθαλμῷ ᾧ ὁρῶμεν ἔνεστί τι τῶν τοιούτων (sc. τῶν κατόπτρων); . . . ἐννενόηκας οὖν ὅτι τοῦ ἐμβλέποντος εἰς τὸν ὀφθαλμὸν τὸ πρόσωπον ἐμφαίνεται ἐν τῇ τοῦ κατ' ἀντικρὺ ὄψει ὥσπερ ἐν κατόπτρῳ, ὃ δὴ καὶ κόρην καλοῦμεν, εἴδωλον ὄν τι τοῦ ἐμβλέποντος. This image of B mirrored in A's eye and seen there by B, was by many regarded as the essential objective equivalent of the psychic fact that A sees B, just as if it were an image on A's *retina*, not in the *pupil* of A's eye. This early view of κόρη was, however, soon modified. It came to represent what is now called the ' lens.' Cf. Theophr. *de Sens.* § 36.

[1] The Greeks knew nothing of pressure of the eyeball serving as *retinal stimulus*, and so causing this sensation of light.

shall consider these according as they bear upon the *organ* (or *function*), the *medium*, or the *object* of vision. It is to be noticed that Alcmaeon, with whom we begin, has left us no information on what he conceived to be the nature of the *medium* or the *object*. His recorded views are concerned only with the visual *organ*, its functions, and its relationship to the organism as a whole.

Alcmaeon of Crotona.

§ 3. 'Seeing takes place,' says Alcmaeon [1], 'by reflexion in the diaphanous element.' 'Alcmaeon of Crotona [2] held that the eyes see through the environing water. That [each eye] contains fire is, indeed, manifest, for a flash takes place within it when it receives a stroke. It is with the glittering and diaphanous element, however, that it sees, whenever this reflects an image (ἀντιφαίνῃ), and it sees better in proportion to the purity of this element [3].'

Chalcidius [4] tells us that Alcmaeon was the first to practise dissection, and that to him, as well as (long afterwards) to Callisthenes and Herophilus, many important

[margin note: Alcmaeon of Crotona on the sense of seeing.]

[1] Stob. *Ecl. Phys.* i. 52 (Diels, *Dox.*, p. 404, *Vors.* p. 104). I have translated Diels' (*Dox.* proll. p. 223) suggestion ἀντίλαμψιν for MS. ἀντίληψιν = 'apprehension' by the diaphanous element, which still brings us to the idea of reflexion. 'Αντίλαμψιν = reflexion, corresponds to the ἀντιφαίνῃ of Theophr. § 26; see next extract. To ascribe 'apprehensive' power to the διαφανές within the eye is quite out of keeping with the doctrine of Alcmaeon, nor is he likely to have employed the term ἀντίληψις. Indeed it surprises one to find even τὸ διαφανές—a distinctively Aristotelean word in this connexion—ascribed to him.

[2] Theophr. *de Sens.* § 26 (Diels, *Vors.*, p. 104).

[3] Wachtler, *de Alc. Crot.* (Teubner, 1896), p. 49, refers τῷ στίλβοντι here to the fire and τῷ διαφανεῖ to the water within the eye. But στίλβειν is not often found used of the gleam of fire (which would rather be λάμπειν), whereas it is regularly used of lustre, and of the glittering of water. Cf. Arist. 370ᵃ 18 φαίνεται τὸ ὕδωρ στίλβειν, and 561ᵃ 32 ὑγρὸν ἔνεστι λευκὸν καὶ ψυχρόν, σφόδρα στίλβον. Both participles should, notwithstanding the repetition of the article, be referred to the same thing, viz. the ' diaphanous' element in which the image is said to be reflected. C. Bäumker (*Arist. Lehre von den äussern und innern Sinnesvermögen*, p. 49) notices that in the passage above translated, the words ὁρᾶν δὲ τῷ στίλβοντι καὶ τῷ διαφανεῖ form an iambic trimeter.

[4] *In Plat. Tim.*, p. 279, ed. Wrobel, pp. 340-1, ed. Meursius.

discoveries respecting the anatomy of the eye and the optic nerves are due. It is not possible, however, to determine from the words of Chalcidius how much of the anatomical knowledge of which he speaks was discovered by Alcmaeon, and how much by the others; nor can much weight be assigned to the authority of this commentator on such matters. But, according to the Hippocratean treatise Περὶ Σαρκῶν (or 'Αρχῶν), the connexion between eye and brain is formed by a 'vein' passing from the membrane which covers the latter to each of the two eyes. Through this 'vein' the viscous substance of the brain is said to prolong itself into the eyes, where it forms the transparent membranes which cover the eyes. In this the light and all bright objects are reflected, and by this reflexion we *see*. Things, again, are seen because they have brightness, and can therefore be reflected by the transparent membrane of the eye. This fact of reflexion, according to the Pythagorean theory[1], is accomplished by 'a visual ray' from eye to object, which reaching the object doubles back again to the eye, like a forearm outstretched and then bent back again to the shoulder[2]. The above pseudo-Hippocratean tract may (as Siebeck says) really present us with an account of Alcmaeon's theory of vision. 'The membranes, of which there are many protecting the visual organ, are diaphanous like the organ itself. By means of this quality of diaphanousness it reflects (ἀντανγεῖ) the light and all illuminated objects; accordingly it is by means of this, which so reflects, that the visual organ (τὸ ὁρέον) *sees*[3].'

The intra-ocular fire and the image reflected in the water co-operate

§ 4. According to Alcmaeon, therefore, it would seem that vision is effected by the 'image,' *and* by rays which issue from within and pass outwards through the water; that these rays emanate from a fire within the eye; as if the glistening and diaphanous element in the eye were merely

---

[1] It is not improbable that Alcmaeon was to some extent influenced by the Pythagorean teaching : *vide* Arist. *Met.* i. 5. 986ª 29; Siebeck, *Geschichte der Psychologie*, i. 1, pp. 103–106.

[2] Cf. Plut. *Epit.* iv. 14; Diels, *Dox.*, p. 405.

[3] Cf. Hippocr. viii. 606 L.; Diels, *Vors.*, p. 104. For ἀντανγεῖ cf. Eur. *Or.* 1519, and ἀντηύγει σέλας, Stob. *Flor.* ii. p. 392 (Teub.).

instrumental. If, as is probable, Alcmaeon, with the Pythagoreans and other mathematical philosophers, held that seeing is accomplished by means of such rays issuing from the eye, we may suppose that the reflexion in the eye, which is instrumental or subsidiary to vision, is the result of this process : that the visual image is collected somehow by the energy of the internal fire, going out to the object and thence returning to the eye with its impression, which is there mirrored in the diaphanous element[1]. Thus the fire would represent the 'active' force of vision, while the water would serve to bring the object seen home to the eye itself. The fact of the fire-flash was regarded as demonstrating the presence of fire in the eye, and a function had to be assumed for this fire in connexion with seeing. The presence of the watery element was manifest, and it, too, required to have its visual function explained, which was most simply done, as it appeared, by making the water the mirror in which the image in the 'pupil' (also manifest to observation) is reflected. Considering the natural obscurity of the act of vision on its psychical side, we need not look for greater accuracy or consistency of view than this on Alcmaeon's part. But there is a popular confusion lurking in the position thus described. The 'visual ray' hypothesis, which makes seeing an 'act' of the mind or of the eye, cannot be really harmonized with the other hypothesis by which the eye with its aqueous humour is regarded as a mere mirror reflecting objects as is done by a standing pool[2].

[1] Though διαφανές strictly means 'transparent,' and a *purely* transparent substance would reflect no image, this does not prevent the use of the word in such connexion as the present by all writers including Aristotle. Water and air were held to be diaphanous and yet the great instruments of reflexion. Of course when they do 'reflect' images there are present conditions which modify their mere 'transparency' and render such reflexion possible.

[2] It is hard to agree with Prantl, *Arist.* Περὶ Χρωμάτων, p. 37, that Alcmaeon's statement regarding vision and its organ are in harmony with and anticipate those of Aristotle. Aristotle distinctly denies that the eye contains fire, and explains the 'flash' differently from Alcmaeon.

Empedocles.

Empe-
docles :
general
view of his
system of
thought in
its bearing
on the
questions of
psycho-
logy of
sense.
Does not
refer to
pupillar
image.

§ 5. According to the doctrine first enunciated by Empedocles, *like perceives like.* All bodies are formed of the four elements, *earth, air, fire, water.* All have passages (πόροι) or 'pores' in them, and from all emanations or effluences (ἀπόρροιαι) come, and enter into the said pores or passages. Thus all bodies are in a state of physical communion, and all interaction whatever between bodies depends upon the facts thus stated. On this basis it is that Empedocles founds his theory of perception. Emanations from what we may call the *percipiendum*, or object, enter into the pores of the *percipiens*, or percipient organ. These emanations, to result in perception, must be 'symmetrical' with the pores : if they are either too small or too large for these, no perception takes place. Hence it is with the eye only that we see, although emanations of colour pass into the pores of other organs also ; for these emanations are symmetrical with the pores of the eye, not with those of the other parts. In the same way, the eye is incapable of perceiving odour, as the emanations of this, which are symmetrical with the pores of the olfactory organ, are not so with the pores of the eye. The specific differences of the sensations and of their objects are thus the result of differences in the pores of their respective organs which restrict them to the reception of certain kinds of emanations, thus destined to be characteristic of them. Different organs, or organs with different pores, take different impressions of the same object. Thus Empedocles thinks he explains sense-perception when he shows how the objects of the extra-organic world enter into the bodily organs. In general his explanation of seeing is the following :—The eye, like all other things, is constituted of the four elements. In its interior is fire ; next outside this comes water ; both being again enclosed by air and earth. The whole eye is compared by him to a lantern in the centre of which (corresponding to the crystalline lens) is the fire. Between this and the earthy cornea comes the water, which is separated from the fire by a fine, delicate membrane. The fire can penetrate these outwards, as light passes through the sides of a lantern, while emanations from objects also can come in, so that according as they proceed from bright or from dark objects they may enter into and pass through the corresponding pores of the fire or of the water. 'By like we know like.' With the intra-ocular fire we perceive the emanations of fire, i. e. *white*; with the water we perceive those of water, i. e. *black*; and so on. The pores of the fire and those of the water alternate in the eye ; and the fire being able to pierce the water, we may suppose them thus arranged at the outer surface of the eye, so that both meet the emanations from objects at this outer surface. Empedocles, who never mentions the pupillar image, does not explain any colours in detail save white and black, as above. Stobaeus [1] tells us that he looked upon four colours as primary : white, black, red, green, corresponding to the four elements. Normal vision he considered to depend on the due proportion in the eye of fire and water—the ocular elements essential to vision. As will be seen below, it is not easy to ascertain *how far* the rays of fire passed outwards : whether (*a*) merely through the water to the outer surface of the eye [2], or (*b*) all the way to the object, however distant [3]. The third possibility, that the inner fire formed a junction with the emanations from the object at some point intermediate between this and the eye, cannot, on any positive authority, be ascribed to Empedocles, but would seem to constitute the distinguishing feature of Plato's visual theory.

§ 6. Diels [4] suggests that Empedocles may have derived his knowledge of the structure and functions of the eye from Alcmaeon. But, like Alcmaeon, he was himself a physician, nor does he speak on these subjects like one who took his information at second hand. The most interesting passage of Empedocles on the constitution of the eye is one contained in the verses of his poem Περὶ Φύσεως, quoted by Aristotle in the tract *de Sensu* [5]. It is as follows : 'As when

Organ and function of vision, according to Empedocles.

[1] *Ecl.* i. 16; Diels, *Vors.*, p. 181, *Dox.* proll. p. 222.
[2] So Siebeck, *Gesch. der Psych.* i. 1, p. 271, thinks.
[3] μέχρι τῶν ἄστρων, Arist. 438ª 26.
[4] *Vide* Wachtler, *Alcm.*, p. 49.   [5] Arist. 437ᵇ 23 seqq.

one who purposes going abroad on a stormy night maketh
him ready a light, a gleam of blazing fire, adjusting thereto,
to screen it from all sorts of winds, a lantern which scatters
the breath of the winds as they blow, while the fire—that
is, the more subtile part thereof—leaping forth shines along
the threshold with unfailing beams: thus then did Nature
embed the primordial fire pent within the coatings of the
eye, *videlicet* the round pupil, in its delicate tissues, which
had been pierced throughout with pores of wondrous
fineness, and, while they fenced off the deep surrounding
flood, allowed the fire—i. e. the more subtile part thereof—
to issue forth (διέσκον) . . .' Empedocles here describes
either Φύσις, or perhaps more especially 'Αφροδίτη, as having
stationed the primeval fire in the lens of the eye, like the
light in the centre of a lantern, the capsule of the lens
corresponding to the transparent sides of the lantern.
Μήνιγξιν, which Alexander refers to the capsule of the
lens (ὁ τὴν κόρην περιέχων χιτών), may, however, refer to the
outer coatings of the eye, while λεπτῇσιν ὀθόνῃσι refers to
the capsule of the lens itself. At all events, the finer part
of the fire darts forth through these membranes and through
the water, as the light does through the sides of the lantern [1].

'And the flame innocuous gat for itself a small portion

---

[1] See Prof. Burnet's *Early Greek Philosophy*, p. 231, and Diels, *Vors.*,
p. 206. The latter renders ὡς δὲ τότ' ἐν μήνιγξιν κτλ. 'so barg sich
das urewige Feuer damals (*bei der Bildung des Auges*) hinter der
runden Pupille in Häute und dünne Gewänder eingeschlossen.' If,
with Diels, giving up the play on κούρη, we make πῦρ subject of λοχάζετο,
we may explain that the 'primordial fire *ensconced* (or *ambushed*) itself
in the round pupil.' There is no need of τ' in v. 8. In fact it injures the
sense, as ὀθόνῃσι λοχ. seems to refer to a further process, not co-ordinate
with ἐεργμένον. He translates ὅσον ταναώτερον ἦεν in vv. 5 and 11 'weil
es soviel feiner war,' but the ὅσον is limitative, indicating the precise
amount of the fire which was capable of leaping forth, the same to which
Plato, *Tim.* 45 B–C, refers in the words τοῦ πυρὸς ὅσον τὸ μὲν καίειν οὐκ
ἔσχε, τὸ δὲ παρέχειν φῶς ἥμερον. The expression κατὰ βηλόν seems to
favour Siebeck's view (*op. cit.*, p. 271) that Empedocles contemplates a
co-operation between the fire from within and the ἀπόρροιαι from without
at the surface of the eye. There seems to be no sufficient reason for
following Alexander in rendering these words by κατὰ τὸν οὐρανόν, as
Diels does in his 'zum Firmament.'

of earth (in the formation of the eye)¹.' The eye was formed of the elements, for Empedocles further says : ' Of these (elements) divine Aphrodite made up the fabric of the tireless eyes².'

§ 7. In these passages we notice that no reference is made by Empedocles to his doctrine of pores and emanations, so fundamental for perception. Aristotle, too, observes³ that Empedocles, while at one time explaining vision, as we have seen, by means of fire issuing from the lens, at other times explains it by ἀπόρροιαι, as if imputing inconsistency to his theory of vision⁴. It is not easy to assent to the suggestion of mere inconsistency; yet on the other hand it is difficult to reconcile the two standpoints here contrasted. There is indeed another record which seems to bear upon the matter. ' Empedocles mixed the rays with the images, calling their joint-product by the compound term *ray-image*⁵.' But this passage is intrinsically suspicious. By the εἴδωλα would seem to be intended something between the ἀπόρροιαι of Empedocles and the εἴδωλα of Democritus and Epicurus; and the theory here ascribed to Empedocles, of the mixture of the rays with the ἀπόρροιαι to form the ἀκτινείδωλον, reminds one too much of the distinctively Platonic theory known later as the συναύγεια⁶. Empedocles and Plato both accept the existence

Empedocles' doctrine of 'pores' and 'emanations': its bearing upon visual function.

¹ Simpl. *ad Arist. Phys.* (Diels), p. 331. 3 (Diels, *Vors.*, p. 206). Simplicius instances this, because of the use of the word τύχε here, as illustrating the fortuitousness of the formation of things according to Empedocles ; in which he overstrains the meaning of this word. The position of the adjective is noticeable in the words ἡ δὲ φλὸξ ἰλάειρα : it seems to give it conditional force, like that given by ὅσον ταναώτερον, reducing the φλόξ referred to to what Plato calls φῶς ἥμερον.

² Simpl. *ad Arist. de Caelo* (Diels), p. 529. 21 (Diels, *Vors.*, p. 206). From this we conjecture that in the passage quoted by Aristotle the subject of λοχάζετο was also Ἀφροδίτη.   ³ *De Sens.* l. c.

⁴ The words of Stob. *Ecl.* i. 52 (Diels, *Dox.*, p. 403) πρὸς τὸ διὰ τῶν ἀκτίνων καὶ πρὸς τὸ διὰ τῶν εἰδώλων (Ἐμπεδοκλῆς) ἐκδοχὰς παρέχεται merely repeat what Aristotle here says.

⁵ Plut. *Epit.* iv. 13 (Diels, *Dox.*, p. 403) Ἐμπεδοκλῆς τοῖς εἰδώλοις τὰς ἀκτῖνας ἀνέμειξε προσαγορεύσας τὸ γιγνόμενον ἀκτινείδωλον (Diels' correction of ἀκτῖνας εἰδώλου) συνθέτως, Gal. *H. P.* 94.

⁶ *Timaeus* 45 B seqq.

and agency of the intra-ocular fire; but the former, at least in his own verses, has nothing to show that he held, as Plato did, the theory of a confluence of the rays from the eye with the emanations from objects. The notion of an εἴδωλον, too, i. e. an image pictorially resembling the object, is quite foreign to the visual theory of Empedocles and of Plato [1], though proper to that of Epicurus, and (if we can trust the references in Aristotle and Theophrastus) used also by Democritus for the immediate object of vision. From Aristotle's argument against Empedocles, in which he urges that vision is not, as the latter thought, due to fire issuing from the eye, and from the words of Empedocles himself φῶς (or πῦρ) δ' ἔξω διαθρῷσκον κτέ., it is certain that, according to the opinion of the latter, the essential constituent of the eye —the ὠγύγιον πῦρ—was a principal factor of vision [2], which is effected by visual rays proceeding outwards. From the statements of Theophrastus (§ 9 infra), again, it is equally certain that according to Empedocles vision, like the other senses, is effected by ἀπόρροιαι. How are we to harmonize the two positions? They must be regarded as complementary parts of one theory. We really do not know how far outwards Empedocles regarded the rays as proceeding. If we assume that they merely went so far as to meet the ἀπόρροιαι, this will to some extent help us to a reconciliation of the views attributed to Empedocles by Aristotle. The assumption would [3], however, bring the theories of Plato and Empedocles into very close connexion, and tend, at least, to justify Zeller's view of their affinity or identity [4].

The doctrine that 'like perceives like' and the doctrine of 'emana- § 8. Empedocles, holding that *like perceives like*, connects his doctrine of visual perception with that of the four elements, thus: 'With earth we see (ὀπώπαμεν) earth, with water we see water; with air we see the bright air; with fire we see destroying fire; just as with love we [perceive] love,

---

[1] In *Soph.* 266 B–C, *Alc.* i. 132 E &c. visual theory is not discussed.

[2] In this point Empedocles is at one with Goethe in his *Farbenlehre*, though the German writer does not observe the agreement.

[3] Notwithstanding what Mr. Archer-Hind says Plato, *Tim.*, p. 156.

[4] Zeller, *Pre-Socratics* (E. Tr.), ii. 166–7 n.

and with hate, baleful hate[1].' 'Some hold that each and tions' both
every affection results from the agent in its ultimately combined for his
simplest and most essential form entering through certain theory of vision.
pores of the patient ; and they say it is in this manner that
we see and hear and exercise all the other senses ; and,
moreover, that vision takes place through air and water
and other transparent bodies, inasmuch as all these have
pores, invisible from their smallness but close together and
arranged in rows, and all the more so arranged in proportion
to their greater transparency. Some writers have laid down
this doctrine in certain instances without confining it to
cases of agency and patiency : they go further, and say that
mixture takes place only between bodies which have pores
mutually symmetrical[2].' Thus it was recognized by Aristotle,
and doubtless by others, that Empedocles did endeavou to
make his theory of seeing, and of perception in general,
conform to his physical (or metaphysical) theory of the
communion of all substances by pores and ἀπόρροιαι[3].

§ 9. 'Empedocles, explaining the nature of the eye as Different
organ of vision, states[4] that its inner part consists of fire constitu- tion of
and water[5], while the environment of this consists of earth different
and air, through which it (the internal fire) being of a subtile eyes, and consequent
nature passes, as the light in a lantern passes through the differences of visual
sides. The pores of the fire and water alternate in position power.
with one another. By those of fire we cognize white
objects, by those of water, black objects ; for these two
sorts of objects fit into these two sets of pores respectively.

---

[1] Arist. 404b 13–16.   [2] Arist. 324b 26 seqq.
[3] If in the verses above referred to, containing the lantern-simile,
the line αἱ χοάνῃσι δίαντα τετρήατο θεσπεσίῃσι finds its proper place (as is
assumed by Diels, *Vors.*, p. 206, and Blass, *Fleckeisens Jahrb.*, 1883,
p. 19), we can believe that there too he was thinking of the doctrine of
pores and ἀπόρροιαι, and would perhaps be found to mention and
harmonize it with the visual ray theory if we had his poem complete.
The membranes of the pupil are in this verse spoken of as 'pierced
right through with pores (χοάνῃσι) divinely formed': 'die mit göttlich
eingerichteten, gerade hindurchgehenden Poren durchbohrt waren'
is Diels' version.
[4] Theophr. *de Sens.* §§ 7–8.
[5] Adopting καὶ ὕδωρ, from Diels after Karsten.

C 2

Colours are carried to the eye by emanation.' In these sentences Theophrastus introduces us to the two main but unharmonized doctrines already spoken of: vision by means of emanations entering the pores of the eye, and vision by means of fire issuing forth (from the *eye*, or from the *pupil* to the outer surface of the eye); but he seems not to feel the difficulty or necessity of reconciling them. He goes on: 'All eyes are not constituted alike of the contrary elements; some have in them more fire and less water than others; some less fire and more water; some again have the fire in the centre and others at a point outside this [1], which affords the reason why some animals see more keenly in the daytime, others by night. Those which have less fire than water in the eye see better by day, for in them the defect of internal light is repaired by the excess of external; while those that have less of the contrary see more keenly by night, since to these also that element which they lack is supplied by compensation; and under opposite conditions they are keen-sighted in opposite ways. For those which have the fire in excess are dim-sighted (by day) since the further augmentation of this fire in the daylight fills [2] and obstructs the pores of the water; while those which have the water in excess suffer the corresponding result by night, as the fire then has its pores obstructed by the water. These states continue until, in the one case, the obstructing water has been separated (from the pores) by the light from without, and, in the other, the obstructing fire has been cleared away by the air [3]. The eye is best in temperament, and therefore in visual power, which consists of both (fire and water) in equal quantities.' Thus the eye in its constitution

---

[1] Does ἐκτός, sc. τοῦ μέσου, here imply a divergence from the view stated in the Empedoclean verses that the primeval fire is in the crystalline lens? or simply that (according to Empedocles) the lens itself need not always be in the centre? For the text, cf. Diels, *Dox.*, p. 500 n., *Vors.*, p. 177; Karsten, *Emp.*, pp. 484–5; Prantl, Περὶ Χρωμ., p. 47.

[2] ἐπιπλάττειν Schneider: ἐπιλάμπειν is suggested by Prantl = 'shine upon,' and so obstruct.

[3] ἀήρ is to ὕδωρ what τὸ ἔξωθεν φῶς is to πῦρ. The light of day corrects the excess of water in the eye; so the dampness of night corrects the excess of fire. ἀήρ as usual = '*damp* air.'

contains the opposites, viz. the fiery and watery elements, in definite relationship to light and shade, or white and black. A passage of Aristotle[1] corroborates the information contained in the foregoing extract from Theophrastus. 'To suppose that, as Empedocles says, gleaming eyes (γλαυκὰ ὄμματα) are fiery, while black contain more of water than of fire, and that on this account the former, the gleaming, see dimly by day owing to lack of water, and the latter by night owing to lack of fire, is an error; since we must assume that the visive part of the eye in all cases consists not of fire but of water[2].'

§ 10. Plato in the *Menon*[3] tells us that Gorgias, as a follower of Empedocles, held the doctrine of pores and emanations; and that by means of this doctrine he furnished an explanation of colour as object of vision. According to this, colour is an emanation consisting of figures symmetrical with the pores of the visual organ and for this reason capable of being seen. We read elsewhere also[4] that Empedocles regards colour as 'that which fits into the pores of the eye.' To this Stobaeus[5] adds the statement already referred to (§ 5 *supra*) that 'Empedocles regarded *white, black, red, green* (or, with ὠχρόν for χλωρόν, *yellow*) as the primary colours[6], being equal in number with the

Object of vision: Colour.

[1] 779[b] 15 seqq.

[2] Philoponus (in Arist. *de Gen. An.* v. 1, Hayduck, p. 217, 15), in his remarks on this passage, says that 'Empedocles makes the organ of sight to consist of the four elements ... and asserts (but H. reads φημί) that vision itself is the power of the soul in virtue whereof we see, inasmuch as it (vision) is the form (εἶδος) of the eye.' This (if φησι be kept) well illustrates the untrustworthiness of late commentators on early philosophers whose views they looked at only through the medium of their successors. Here Philoponus represents Empedocles as an Aristotelean. The opinion of Empedocles about gleaming and black eyes is referred to also in the Pseudo-Arist. *Problems*, 910. 13. We find similar views held on this point by Anaxagoras and Diogenes.

[3] *Men.* 76 C–D.   [4] Plut. *Epit.* i. 15. 3 (Diels, *Dox.*, p. 313).

[5] *Ecl.* i. 16. 3 (Diels, *Dox.*, p. 313).

[6] For MSS. ὠχρόν, χλωρόν has been adopted; yet the change may be not worth while making, if the suspicion mentioned below be well founded. ὠχρός is used by Arist. 559[a] 18 to denote the colour of the yolk of an egg; i. e. it means *yellow*. Cf. Diels, *Dox.*, Prol. p. 50; and Mullach, *Democritus*, p. 353. Curiously enough, the same error of ὠχρόν for

elements[1].' This is perhaps supported by the fact that in *Fragment* 71, Empedocles teaches that colours are produced by the mixture of the four elements[2]. The following criticism of Empedocles' colour-theory by Theophrastus[3] will help to place this theory itself in a clearer view.

Theophrastus criticizes Empedocles' theory of vision.

§ 11. 'Empedocles teaches that like is perceived by like,' but this gives rise to difficulties as regards his own theory of the particular senses. 'When he makes the visual organ to consist of fire and its contrary, we may observe that it could indeed perceive *white* and *black* by the operation of similars; but how could it perceive *grey* and the other composite colours[4]? For he does not explain such perception (of grey, &c.) as taking place either by the 'pores' of the fire or by those of the water, or by others formed of both together[5]; yet we see these just as well as we see the simple colours. It is, moreover, a strange doctrine that some eyes see better by day, others by night. For the smaller fire is destroyed by the greater[6], which is the reason why we cannot gaze directly at the sun or at any excessively bright

χλωρόν affects the statement of Stob. (*Ecl.* i. 16. 8; Diels, *Dox.*, p. 314) attributing the same 'four-colour' theory to Democritus. That χλωρός is the true word in Democritus we know from Theophrastus (§ 75). As regards Empedocles, however, we have not this assurance, Theophrastus (§ 59) merely telling us that Empedocles held two primary colours white and black, while the remaining colours are formed by mixtures of these. It has been suspected (Diels, *Dox.*, p. 222) that the compiler of the *Placita* erroneously ascribed to Empedocles the four colours of Democritus.

[1] For the ancient and traditional conception (cf. Prantl, *Arist.* Περὶ Χρώμ. p. 30) of white and black, as the primary colours from which the other colours can be obtained by mixing them in various proportions, cf. Aristotle, §§ 41-2 *infra*.

[2] Diels, *Vors.*, p. 203

Πῶς ὕδατος γαίης τε καὶ αἰθέρος ἠελίου τε
Κιρναμένων εἴδη τε γενοίατο χροιά τε θνητῶν.

[3] *De Sens.* §§ 17-19.

[4] Here we seem to find an echo of Arist. *de An.* i. 5. 409[b] 23 seqq. when criticizing Empedocles' general theory of cognition.

[5] As Diels, *Dox.*, p. 504 n. remarks, according to the critic 'μικτοὶ πόροι μικτοῖς χρώμασι conveniunt.'

[6] This notion which we so often find referred to probably arose in the popular mind from the disappearance of the stars when the sun rises.

object[1]: so that those in whom the light within the eye
is defective should see worse by day[2]. Or if (as Empedocles
thinks) its like augments the visual fire in the daytime[3],
while its opposite destroys or thwarts it, then *all* should see
white objects better by day, both those whose internal
light is less and those whose internal light is greater;
while again *all* should see black objects better by night.
The fact is, however, that all animals except a very few see
all objects better by day than in the night-time. It is
natural to suppose that in these few their native fire has
this peculiar power, just as there are animals whose eyes
in virtue of their colour are luminous at night[4]. Again,
as regards the eyes in which the fire and water are mixed
in equal proportions, it must follow that either is in turn
unduly augmented by day or by night: hence, if water
or fire thwarts vision by being in excess, the disposition
(διάθεσις) of all eyes would be pretty nearly alike[5].'

Democritus.

§ 12. For Democritus, as for Empedocles, the most General
obvious explanation of perception seemed to be that which view of the
physical
showed how particles of external things come into the theory of
pores of the sensory organs. He differed from Empedocles Democritus in its
in his doctrine of the existence of void, which Empedocles bearing on
visual
did not allow. They agreed, however, in the belief that function:

¹ This is perhaps—though see note 4 *infra*—an *arg. ad hominem*
against Empedocles: Theophrastus, as a disciple of Aristotle, would
not hold that the eyes contain a 'small fire,' to be quenched by the
greater fire of the sun.
² Instead of better, as Empedocles asserts
³ i. e. if (instead of the greater fire without destroying the less within
the eye) the daylight augments the intra-ocular fire.
⁴ Not '*cutis* noctu magis splendet,' as in Wimmer's Latin version.
There would seem to be here on the critic's part an admission which
is contrary to the teaching of Aristotle. Theophrastus seems to
attribute the capacity of some animals to see by night to the possession
of a peculiar fire in their eyes.
⁵ i. e. the so-called best class of eyes, having water and fire in equal
proportions, would both by day and by night, in one or the other way,
be out of keeping with the conditions of perfect vision, and would
therefore not have the superiority claimed for them by Empedocles:
they would be no better than the eyes already referred to.

'like is perceived by like [1].' Instead of holding, like Empedocles, that there are four elements qualitatively distinct, Democritus with Leucippus (of whom so little is known separately that we can neglect him or merge him in his pupil) taught that the elements of things are homogeneous atoms, infinitely numerous, moving eternally in void. The introduction of atoms in certain ways through the organs 'to the soul' was for him (as the introduction of ἀπορροαί was for Empedocles also) the essence of perception. We perceive an external thing by its being thus introduced into the soul ; but the soul, for him as for Empedocles, is itself material, so as to be capable of being affected in the way perception implies. It consists of atoms of a certain shape endowed with a certain order and movement. The impression made by the atoms of the object on the soul must be of a certain initial strength, in order to be noticeable. For Democritus (as for Empedocles [2], to some extent) the organs are thus essentially *passages*—thoroughfares for instreaming atoms. All the senses are modes of one, viz. Touching [3]. The essential feature of the eye is, for Democritus, its moist and porous nature, while the ear is a mere channel for the conveyance of sonant particles inwards 'to the soul.' To reach the soul, the particles conveyed inwards require to be disseminated through the body. It is impossible for us, he thought, to receive wholly exact impressions of external things through the organs of sense. For example, in seeing,

[1] As against the doubt of Theophr. *de Sens.* § 49 see Arist. 405ᵇ 12–16; Sext. Emp. *adv. Math.* vii. § 116 ; Mullach, *Democr.*, pp. 206, 401, and Theophr. himself § 50. Indeed, Democritus also held that 'like is affected by like'—a *physical* principle—while according to Aristotle (323ᵇ 3 seqq.) most philosophers with one accord assert that like cannot be affected by like (τὸ ὅμοιον ὑπὸ τοῦ ὁμοίου πᾶν ἀπαθές ἐστι). It is hard to see how the acceptance of the latter physical principle could be, as Mr. Archer-Hind (Plato, *Tim.*, p. 205) says, compatible with that of the psychological axiom 'like is known by like.'

[2] In his account of the formation of the ear, which he compares to a κώδων, Empedocles seems to have regarded this sense-organ, at least, as something more than a mere passage, and as having a determining power over the *quality* of the sensation to be produced by the ἀπόρροιαι.

[3] Cf. Arist. 442ᵃ 29 Δημόκριτος καὶ . . . ἀτοπώτατόν τι ποιοῦσι· πάντα γὰρ τὰ αἰσθητὰ ἁπτὰ ποιοῦσι.

the air intervening between us and the object interferes with our obtaining a correct impression or image of this, as is evidenced by the blurred look of distant things. Democritus first laid down the distinction of the qualities of body [1] into the so-called primary and secondary qualities, to which, however, he did not himself remain always consistent. As Theophrastus (*de Sens.* § 80; see p. 35 *infra*) points out, we cannot quite follow his doctrine of the formation of colours unless we assume a φύσις χρώματος—an objective existence of colour. He held that vision is the result of the image of the object mirrored in the eye. But when we ask—what exactly is mirrored? the answer for him is not easy ; since between object and eye come what he called δείκελα (generally spoken of by Aristotle and Theophrastus as εἴδωλα), things which in the case of this sense are also referred to as ἀπόρροιαι τῆς μορφῆς. These δείκελα, not the object, are therefore the immediate and proper *data* of sense.

§ 13. Democritus regarded the pupillar image as the essential factor of vision. ' Democritus,' says Aristotle [2], ' is right in his opinion that the organ of vision proper consists of water, but not when he goes on to explain vision as the mirroring (ἔμφασιν) of objects in this water. For this mirroring is due to the fact that the surface of the eye is smooth, and the image exists really not in the mirroring eye but in the eye that beholds this [3], inasmuch as the case is merely one of reflexion [4]. But on

Aristotle's criticism of Democritus' visual theory. The latter misunderstood the function of the water in the eye.

[1] The non-objectivity of colour is stated as a doctrine of his by Arist. 316ᵃ 1 Δημόκριτος . . . χροιὰν οὔ φησιν εἶναι, τροπῇ γὰρ χρωματίζεσθαι. Cf. Theophr. *de Sens.* § 64; also Galen. *de Elem. sec. Hipp.* i. 2 νόμῳ γὰρ χροιὴ . . . ἐτεῇ δ᾽ ἄτομον καὶ κενὸν ὁ Δημόκριτός φησιν. He is alluded to by Arist. 426ᵃ 20 οἱ πρότεροι φυσιολόγοι οὐδὲν ᾤοντο οὔτε λευκὸν οὔτε μέλαν εἶναι ἄνευ ὄψεως κτέ. [2] Arist. 438ᵃ 5-16.

[3] The subject of ἔστιν is ἡ ἔμφασις derived from τοῦτο, sc. τὸ ἐμφαίνεσθαι ἐκείνῳ = τῷ ἐκεῖ ὁρωμένῳ. Here Aristotle's argument does not require the seeming admission of the Platonic view, viz. that vision is effected by an ὄψις, or ray, which goes forth from the beholder's eye and returns to this from the object. This view, rejected by him 435ᵃ 5, and *de Sens.* ii, is one which Aristotle himself, provisionally at least, adopts *Meteor.* iii. 2. 373ᵇ seqq.; *vide* Ideler, *Meteor.* ii. pp. 273 seqq.; Galen. *de Placit. Hipp. et Plat.* § 640.

[4] Ἔμφασις in the eye, like all other ἔμφασις, is to be explained by

the whole it would seem that in his day no scientific knowledge yet existed of the way in which images are formed in mirrors, or of the reflexion of light in general. It is strange, too, that Democritus should never have asked himself why, if his theory of vision be true, the eye alone sees, while none of the other things, in which images are also mirrored, do so.' 'Democritus holds [1] that we see by the reflexion of images, but describes this latter process in a way peculiar to himself. It does not, he says, take place directly in the pupil from the object; but the air between object and eye is impressed with a sort of stamp while being dispatched in a compact form from the object to the organ [2]; for emanation is always taking place from everything. This air, then, being solid, and of different colour [3], reflects itself in the eyes, which are moist. A dense body does not admit (this air-impression), but one that is moist, like the eye, gives it free passage. Hence moist eyes see better than those that are (dry and) hard, provided that their outer membrane is as thin and dense as possible, and that the inner parts are spongy and free from dense and solid tissue [4], as well as from such moisture as is thick and glutinous; and that the veins of (or, connected with) the eyes are straight and free from moisture, so as to conform in shape to the images moulded by, and thrown off from, the object [5].'

This intermediate effect of the object in moulding the

Marginal note: Peculiarity of Democritus' theory: the object of vision impresses the air, and this impressed air is what affects the eye.

ἀνάκλασις, i. e. the bending back of the ὄψις from the reflecting surface. The image, supposed to be *in the mirror*, is a set of rays reflected *to* this from the object, and *from* it to the beholder's eye, in which therefore it really is. Thus the image 'seen in the eye' of A cannot explain how A sees. Cf. R. Bacon, *O. M. Persp.* III, *Dis.* i. *cap.* 2, 'nihil est in speculo . . . ut vulgus aestimat.'

[1] Theophr. *de Sens.* § 50 (Diels, *Dox.*, p. 513 n.).

[2] The reading suggested by Diels κατά—for καὶ—τοῦ ὁρῶντος has been translated, but συστελλόμενον has been preferred to his στελλόμενον: the preposition is defended by the words of Theophrastus, § 52 ὠθούμενος καὶ πυκνούμενος.

[3] From the eye: see *infra* Anaxag. § 20, Diogenes of A. § 23.

[4] Adopting Usener's στιφρᾶς for ἰσχυρᾶς.

[5] ὡς (= ὥστε) ὁμοιοσχημονεῖν τοῖς ἀποτυπουμένοις.

air into definite visible forms (ἀποτύπωσις) is the peculiar characteristic of Democritus' theory of vision. He held that if there were pure vacuum, and not air, around us, the emanations or images sent from the visible objects would reach the eye unblurred: that is to say, they would then report the exact form of an object, no matter how great the distance from which they might come. ' Democritus,' says Aristotle [1], 'is not correct in his view that, if the space between object and eye were pure void, an ant could be seen clearly in the sky.' As it is, however, the air takes the first copy of the object, and the eye receives it only at second hand, while the likeness of this copy to the original becomes more and more imperfect in proportion to the distance it has to travel.

§ 14. Theophrastus [2] criticizes these tenets of Democritus: ' His notion of modelling (ἀποτύπωσις) in air is quite absurd. Whatever is capable of being moulded into shape must have density, and must not be liable to dispersion ; this he implies when he illustrates the process, and compares it with the stamping of impressions on wax. In the next place, such modelling might take place more successfully in water than in air, water being more dense ; hence we should see better in water. As a fact, however, we see worse. In the third place, why should one who (as Democritus in his treatise περὶ εἰδῶν does) believes in the emanation of the shape of an object [3], hold this further belief in the modelling of the air ? For the actual images (εἴδωλα αὐτά) of the objects are represented in the eye, according to the former belief. But, again, if we grant that, as Democritus says, the air is moulded into shape, being like wax impressed and condensed, how does the reflexion of an image take place, and of what nature is it ? If there is really such an image, i.e. an impression taken by the air from the object seen, it must be, in this as in other instances, on the side facing the latter. Such being the case, the image cannot come opposite to the eye unless the moulded portion of air is first

(margin note: Theophrastus criticizes Democritus' theory of vision.)

[1] 419ᵃ 15.   [2] De Sens. §§ 51 seqq. (Diels, Dox., pp. 513-15).
[3] ἡ ἀπορροὴ τῆς μορφῆς.

turned round[1]. Now it was for Democritus to show by what and how this turning process was to be effected, without which seeing would still be impossible. A further point is this. When several objects are seen together, how can we understand the presence of a plurality of impressions at the same time in the same air? And how do two persons see *one another* at the same time? The two impressions must meet as they travel in opposite directions from one to the other, each of them facing the object from which it came. Therefore this again is a point which requires further inquiry and elucidation. But we may add another point. How is it, on Democritus' hypothesis, that each person does not see himself in the course of the process? As the impressions of one's body reflect themselves from the air in the eyes of others, so they should reflect themselves back in one's own eye, especially if they directly face the latter, and if the phenomenon of reflexion is one which takes place in the same way as the repercussion of sound in an echo; in which case, according to Democritus, the voice is reflected back (ἀνακλᾶσθαι) also to the very person who gave it utterance. But this theory of air-modelling, taken all round, is absurd. From what Democritus says, it should follow that the air is continually having formed in it models of all kinds of objects, of which many would cross one another's paths, thus causing an impediment to vision, and being generally improbable. And, moreover, if the impressions made in the air are permanent, one should, even when the bodies from which they come are no longer in view or are far distant, be able to see them still, if not at night, at all events in the day-time; though, indeed, it would be even more credible that the impressions should remain in the air at night, as the atmosphere is at that time more endowed with animation[2].

---

[1] The image will come to the eye 'wrong side on.'

[2] ἐμψυχότερος, which at first seems strange, suits the argument and the theory of Democritus better than Wimmer's conjecture ἐμψυχρότερος. Democritus held that ψυχή consists of atoms of a certain sort (i. e. exceedingly small and round), which exist in countless myriads in the air, and from which the ψυχή within the living body is constantly being recruited through the respiratory process. Cold tends to expel them

Perhaps one might say that in the daytime the sun causes the reflexion of images in the pupil by bringing the light[1] to the eye, and this is what Democritus seems to have meant; since that the sun should, as he says, *condense* the air, pushing and striking it off from itself, is an absurd notion. The sun naturally rarefies air instead of condensing it. It is to be remarked also, as an anomaly in Democritus' theory, that he gives not the eye alone, but also the remainder of the body its part in visual perception. This he implies when he states that the eye must contain void and moisture for the purpose of receiving impressions more freely and then *transmitting these to the rest of the body*[2]. A still further anomaly is involved in Democritus' assertion that cognate things best see their kindred, while nevertheless he also asserts that reflexion is due to difference of colour, which would imply that like things are not reflected in their likes. Besides this: how are *magnitudes* and *distances* reflected in the eye? this is a question which he undertakes but fails to answer. Thus Democritus, in enunciating his peculiar theory of vision, instead of settling the old problems, bequeaths them to us in a more difficult form than before.'

§ 15. 'Leucippus, Democritus, and Epicurus, hold that the visual affection (τὸ ὁρατικὸν πάθος) takes place by the entrance of images (κατὰ εἰδώλων εἴσκρισιν)[3]. (Democritus' term for the visual

from the body; and, as at night and in sleep the body is colder than by day, the quantity of soul-atoms in the air at night is greater than by day. Cf. Arist. 471[b] 30 seqq. Diels, *Vors.*, p. 391, now defends ἐμψυχότερος.

[1] The text here translated is corrupt and obscure.

[2] ἵν' ἐπὶ πλέον δέχηται καὶ τῷ ἄλλῳ σώματι παραδιδῷ. These words suggest the answer which Democritus would have made to Aristotle's question (§ 13 *supra*)—'Why on Democritus' theory does not every other mirror, as well as the eye, *see*?' 'Mirrors,' Democritus would reply, 'are not connected with a bodily organism.'

[3] Plut. *Epit.* iv. 13; Stob. *Ecl.* i. 52 (Diels, *Dox.*, p. 403). Theophrastus, as we have seen, and Aristotle, 438[a] 16, both use this word εἴδωλον with reference to Democritus' object of vision. Cicero, too, *ad Fam.* xv. 16. 1, implies that Democritus himself so used it : ' quae ille Gargettius et iam ante Democritus εἴδωλα, hic "spectra" nominat.' Yet nowhere do we find the word thus used in the remains of Democritus himself. The term which he employed usually, if not always, was δείκελον (or δείκηλον), which

image):
further
authorities
for his
visual
theory.

'They assigned as cause of vision certain images (εἴδωλα) which emanate (ἀπορρέοντα) continually from the objects seen, of like form with (ὁμοιόμορφα) the latter, and impinge upon the eye. Such was the theory of Leucippus and Democritus [1].' 'Democritus asserts that *seeing* is the reception of an image reflected from the object seen. This word image (ἔμφασις) here means the form (εἶδος) reflected in the pupil. The case is like that of all other transparent surfaces which show an image reflected in them. He holds that certain images (εἴδωλα), similar in shape to the things from which they come, streaming off from all the things which are visible, impinge upon the eyes of those who see them, and that thus *seeing* takes place; in proof whereof he adduces the fact that in the pupil of the eye of those who see any object there is invariably the image or likeness of the object seen. This is the whole account of seeing according to Democritus [2].'

Democritus'
theory of
the object
of vision—
Colour, its

§ 16. Democritus is the earliest philosopher in whose recorded writings we find an attempt at a detailed theory of colour. The *white* and the *black* he refers immediately to affections of touch: the former to the *smooth*, the latter

seems to have been, by its derivation, fitted to express generally the ἀπορροή from an object of whatever sense. It properly signifies not a 'spectrum' but what we mean by (the *English* word) specimen : i. e. an emanation qualitatively like the thing from which it comes. This, in reference to the sense of sight, would be no doubt a ' specimen ' (in the *Latin* signification) of the object *qua* visible : a copy of its figure and colour. In reference to other senses it would denote the qualities respectively which these are fitted to perceive, whether odour, or sound, or taste. Only in reference to the sense of seeing could it coincide in meaning with εἴδωλον, but as this, which Aristotle calls the sense *par excellence*, tends to absorb the attention of psychologists, either the word δείκελον was narrowed to the idea of εἴδωλον (= ἡ ἀπορροὴ τῆς μορφῆς), or else the latter was extended to cover all the meanings of the more general term. That δείκελον was *capable* of expressing εἴδωλον, appears from the phrase of Parthenius δείκελον 'Ιφιγένης, the *image*, or *effigy*, of Iphigenia. In Laconian δεικελίσται was = Attic μιμηταί (*Etym. Magn.* 260, 48).

[1] Alexander, in Arist. de Sens. p. 56 (Wendland), and Arist. de Sens. 440ᵃ 15–18.

[2] Alexander ad Arist. de Sensu 438ᵃ 5, p. 24 (Wendland). This reproduces the theory of Democritus in the simpler aspect in which Aristotle criticizes it, 438ᵃ 5–16.

to the *rough*[1]. He asserts that the simple (ἁπλᾶ) colours[2] physical
are four: *white, black, red,* and *green* (χλωρόν). *White* is the production: four
smooth[3]. For if anything is not rough, and neither throws primary colours.
shadows nor is difficult of penetration, it is, in every case,
bright (λαμπρόν). The things that are bright must be straight-
bored (εὐθύτρυπα), and hence translucent (διαυγῆ). Of white
objects, those which are hard—as, for example, the flat inner
surfaces of bivalve shells—consist of such atomic shapes[4], for
thus they would be shadowless and luminous (εὐαγῆ) and
straight-pored (εὐθύπορα). Those, on the other hand, which
are friable (ψαθυρά)[5] and brittle (εὔθρυπτα) consist of atoms
which are spherical but obliquely situated in position with
regard to one another, and in their mode of combination
in pairs[6], and their whole atomic structure is as far as
possible uniform. This being so, such bodies must be
friable, because the amount of conjunction between each pair
among their atoms is slight; and they must be brittle,
because the disposition of the atoms is uniform; while they
must be free from shadow, because they are smooth and
flat. Things are whiter one than another in proportion
as the figures aforesaid are more exact and less mixed
with others, and possess the aforesaid order and disposition more perfectly. Such, then, are the atomic figures of
which white is composed. *Black* consists of figures of the
contrary kind, those which are rough, uneven (σκαληνῶν),

[1] Arist. *de Sens.* 442ᵇ 10.

[2] For what follows in this paragraph see Theophr. *de Sens.* §§ 73–5
(Diels, *Vors.*, p. 394).   Distinguish χλωρόν from πράσινον.

[3] Plato, *Tim.* 60 A, regards τὸ λεῖον as διακριτικὸν τῆς ὄψεως which is
the characteristic quality of white.

[4] σχημάτων, the most noticeable of the intrinsic differences of the
atom—its figure—serving for the general name, as often in Democritus
himself.

[5] ψαθυρός here is opposed to σκληρός, not (as in Arist. 441ᵃ 25) to
γλίσχρος.

[6] ἐκ περιφερῶν μὲν λοξῶν δὲ τῇ θέσει πρὸς ἄλληλα καὶ κατὰ δύο συζεύξει:
which seems to mean that a cross-section of the structure would exhibit
the atoms in a quincuncial arrangement.   Prantl (Περὶ Χρωμ., p. 52)
keeping the older text τὰς δύο συζεύξεις τήν θ' ὅλην τάξιν ἔχειν ὁμοίαν
translates—'aber in der ganzen Ausdehnung jedenfalls in σχῆμα θέσις
und τάξις einander gleich.'

and dissimilar; for thus they would cast shadows, nor would their pores be straight or easily permeable. Their emanations, moreover, must be slow and confused [1]; for the emanation makes a difference, by its quality, in the nature of the sense-presentation: and its quality is liable to change owing to the intervention of the air. *Red* is formed of the same kind of atomic figures as the hot [2], only that those of red are larger; for a hot thing is redder the larger the aggregations of its atomic figures are, when these figures are similar in kind [3]. A proof that red is composed of such atoms as those which form the hot, is that we ourselves are red when heated, just as other things are when ignited, as long as they continue to have the character of 'the igneous'; but ignited things are redder in proportion as they are formed of large figures; such are flame, coals of wood whether green or dry, and also iron and other metals which are subject to ignition. Those are brightest [4] which contain the most and finest fire; while those are more red in which the fire is coarser and in less quantity. Whence it is that things at a more red heat are less hot (sc. than those at a white heat); for (in the world of atoms) the fine, which is the essence of the bright, is also that which constitutes the hot [5]. *Green* (χλωρόν), again, is formed of the solid and the void, being compounded of both, but the colour varies in tint (διαλλάττειν) according to their position and arrangement [6].

[1] We cannot guess what this new factor—the *speed* of the ἀποῤῥοαί—has to do with colour according to Democritus. There is no thought here of 'rapidity of vibrations.' Mullach (*Dem.*, p. 221) punctuates so as to separate διαφέρειν from πρός, wrongly.

[2] The atoms of fire are spherical, Arist. 303ᵃ 14. By 'larger (μειζόνων)' here must be meant 'in larger aggregates,' as in next clause.

[3] Diels (*Dox.*, p. 521) compares Arist. 329ᵇ 26 θερμὸν γάρ ἐστι τὸ συγκρῖνον τὰ ὁμογενῆ· τὸ γὰρ διακρίνειν, ὅπερ φασὶ ποιεῖν τὸ πῦρ, συγκρίνειν ἐστὶ τὰ ὁμόφυλα.

[4] i. e. show the *whitest* heat.   [5] θερμὸν γὰρ τὸ λεπτόν.

[6] It is remarkable and noticed afterwards by Theophrastus (§ 18 *infra*) that Democritus explains green by the solid and the void, not by the *shape* of the atoms, like the other colours. Prantl supposes that Democritus in explaining green thought of this as the colour of plants and of

§ 17. Thus, then, Democritus accounts for his four Formation of other colours by mixture of the four primary colours. primary colours. 'Each colour[1] is purer the more the figures of which it is properly composed are free from admixture of others. The other colours are generated by mixtures of these four. *Gold* and *bronze* and such colours are formed by a mixture of white and red. They derive their brightness (τὸ λαμπρόν) from the white, and their reddishness (τὸ ὑπέρυθρον) from the red. The red falls, in the process of mixture, into the void interstices of the white. If to these be added pale-green (χλωροῦ), the most beautiful colour is produced; but the proportion of green so added must be small; it cannot be great when the white and the red are thus compounded. The resulting colours will differ according as the amount of admixture in every such case is greater or less. *Purple* is formed of white, black, and red, the red being in largest quantity and the black in small[2], the white coming midway in amount, which is the reason why it appears pleasant to sense. That the black and the red are in it appears from mere inspection; that it contains white is shown by its brightness and lustre, since it is white that produces these. *Woad*-colour[3] arises from a mixture of the very black with green, but with a preponderance of black. *Leek-green*[4] arises from purple and woad-blue, or from pale-green and purplish (πορφυρο-ειδοῦς). For sulphur[5] is of this colour, and shares the quality of brightness. *Deep-blue*[6] is formed of woad-colour and fire-colour (πυρώδους), but of figures round and needle-

vegetation generally, and from its great extent and abundance in nature, conceived it as resulting *directly* from the two primordial causes of things. [1] Theophr. *de Sens.* §§ 76-8.

[2] This adopts μικράν, which Mullach and Diels, *Vors.* read. Diels (*Dox.*, p. 522 n.) prefers the better attested, though seemingly less probable, μακράν, with the remark 'at atri permultum inesse elucet ex v. 11.'

[3] ἴσατις, the plant woad, used here for woad-blue.

[4] τὸ πράσινον, a colour which like φοινικοῦν and ἀλουργόν, according to Arist. 372ᵃ 5, is not capable of being produced artificially. *Vide* Plato, § 31 *infra*.

[5] Diels (*Dox.*, p. 522 n.), agreeing with Burchard that this example is inappropriate, conjectures τὸν ἰόν, sc. 'aeruginem, in quam splendor certe cadit.' [6] τὸ κυανοῦν.

like, so that the black should contain the quality which makes it brilliant [1]. The *nut-brown* colour (καρύινον) is formed of green and purplish. If bright be mixed therein [2], *flame-colour* arises, since this is shadowless and the dark is excluded. Red mixed with white renders green lustrous [3], not black. Hence growing fruits are at first green, before they become heated, and so diffused [4]. So many are the colours described by Democritus. But he asserts that colours, like tastes, are really infinitely numerous [5] according to the ways of mixing them ; i. e. according as one removes some of this, or adds some of that, ingredient, or mixes less of this or more of that. The colour resulting in the one case will never be like that in the other.'

Theophrastus' criticism of Democritus' theory of colour and its varieties.

§ 18. Theophrastus criticizes the above account of colour and its varieties. Democritus, he says [6], creates a difficulty by suggesting *four* primary colours, instead of the *two*, black and white. 'His assigning different atomic shapes to explain the whiteness of objects according as these are hard or friable is unsatisfactory. For though (εἰ) it would be natural to explain these two classes of objects differently regarded simply as tangibles, one surely must not go on to suppose the *figure* of the atoms to be the cause of their difference in colour; the *position* of the atoms is rather what would account for this. Round figures, and indeed all figures, may overshadow one another. For example, the very argument which Democritus himself employs, when discussing smooth things which appear black, shows this to be so. He asserts that their appearance is due to their

---

[1] The 'figures' have heads shaped like conical bullets on a small scale.

[2] Adopting λαμπρόν for χλωρόν, and (τοῦτο γὰρ ἄσκιον) with Diels, *Dox.*, p. 522 n.   [3] εὐαγές.

[4] διαχεῖσθαι, rendered by Mullach 'antequam maturescant.' This is better than Diels' διακαίεσθαι. The διάχυσις referred to is a process resulting from heat (the opposite of πῆξις, which results from cold), denoting the softening of ripe fruit—a sort of *concoctio* of its tissues. Cf. Arist. 380ª 11, 382ª 29.

[5] So Plato, *Tim.* 68 D (§ 30 *ad fin. infra*), declares that God alone could create or explain their infinite variety. Aristotle denies the infinity of varieties of colour.   [6] *De Sens.* §§ 79–82.

atomic conjunction (σύμφυσιν) and arrangement, this being
in them the same as in the black. And, again, he implies
it when explaining the colour of rough things which are
white. For these, he says, are formed of large figures of
which the commissures are not indeed round but serrated [1],
while the outlines of the figures are broken like stair-steps,
or the tops of vallated mounds [2] erected before a city wall.
This feature in the edge of the atom renders it shadowless, so
that there is nothing in it to hinder brightness from appearing [3]. . . . In general Democritus here explains not so much
the whiteness as the transparency or brightness of bodies ;
since that it should be transparent, and that its pores should
not zigzag, is the essential characteristic, or condition, of
the structure of the *diaphanous* body. Again, that the pores
of white things should be in straight lines, while those
of black should be in zigzag lines, is a condition which
can explain these colours only on one assumption, viz. that
colour is *an objective thing*, which enters into and passes
through the pores [4]; but Democritus does not assume this.
He asserts that seeing is due to the emanation and the
image reflected in the eye [5]. But if seeing is due to this (sc.

---

[1] οὐ περιφερεῖς, ἀλλὰ προκρόσσας. ‘Democrito πρόκροσσος latius patet,
ut pinnae in hanc figuram ⌇⌇⌇⌇ continuatae significentur,’ Diels, *Dox.*,
p. 323 n.

[2] I follow Diels' text (*Dox.*, p. 523).

[3] The conception referred to here seems to be this, that in white
objects, which are formed of smooth atoms, the atoms are always
so disposed that there are straight passages, through the bodies
which they compose, for the uninterrupted transmission of light ;
while in black or dark-coloured objects, formed of rough atoms, the
passages are crooked or darkened by the overlapping of atoms which
stand as it were in one another's light. Yet the smooth atoms may
be so arranged as to throw shadows and produce black; and the
rough may have their angularities so matched and arranged as not to
obstruct light, and so may produce white.

[4] ὡς εἰσιούσης τῆς φύσεως ὑπολαβεῖν ἔστιν. As Diels (*Dox.*, p. 523)
observes, ‘ opponuntur φύσις χρωμάτων et ἀπορροή.’

[5] διὰ τὴν ἀπορροὴν καὶ τὴν ἔμφασιν τὴν εἰς τὴν ὄψιν. Colour was for
Democritus a purely subjective thing : hence, as Theophrastus remarks,
the explanation which treats it as something objective passing into
and through atomic interstices involves him in a contradiction of his
own theory.

the entrance of χρῶμα), what difference does it make whether the pores lie in straight lines over against one another, or in zigzag lines? Nor is it easy to see how an emanation comes from void, and an explanation is due from him on this point also [1]. For he makes white to arise from light or some *positive* thing. Nor is it easy to understand his account of black. For a shadow is something black, a sort of eclipse of the white [2], hence white as a colour has a positive natural primacy. He assigns, too, as cause of black, not merely shadows, but also the density of the air, and therefore of the emanation that enters the eye, and the disturbance or confusion in the eye itself. But he does not make it clear whether these things are due to want of transparency [3], or may arise from some other cause, and, if so, from what sort of cause. It is curious, too, that he does not assign some atomic shape as the cause of green, but explains it only by the solid and the void. These last, however, enter into all things whatever, no matter what atomic shapes things consist of. He should have assigned some characteristic cause in the case of this as of all other colours; and if it be opposed to red, as black is to white, he should have assigned it the opposite atomic shape as its base; while if it be not opposite, this fact in itself might make one wonder, viz. that he does not represent the primary colours as opposites, such opposition being assumed by all writers [4]. He should, in particular, have explained in detail what sort of colours are simple; why some are, and some are not, composite; since it is regarding the first elements that uncertainty is greatest. But this he found, no doubt, a difficult problem.'

Colour,
according
to Demo-
critus, not
a ' primary
quality of
body.' The § 19. Democritus teaches that colour *per se* is nothing objective, for the ultimate elements—the *plenum* and *vacuum* —are destitute of all sensible qualities, while the things composed of them possess colour (as they do every sensible quality) owing merely to the *order, figure,* and *position* of

---

[1] Here (as below) Theophrastus hits at a difficulty in Democritus' account of green.　　　　　[2] ἐπιπρόσθησις τοῦ λευκοῦ.

[3] διὰ τὸ μὴ εὐδίοπτον.　　　　[4] Read ἅπασιν with Diels, *Dox.,* p. 524.

the atoms, i.e. (*a*) to their order relatively to one another, _{way in which the} (*b*) to their several shapes, and (*c*) to the position of each in _{sensible} its place. The subjective aspects—the qualities—of sensible _{qualities are genera-} objects are all due to these three things[1]. Colour has no _{ted from} objective existence, since the colours of bodies are due to the _{the atoms} position of the atoms in them[2]. (Cf. TOUCHING, § 2, p. 182.) _{and void.}

Anaxagoras.

§ 20. Following Heraclitus, Anaxagoras is sharply op- Difference posed to his contemporaries and predecessors in holding, _{of principle between} as he did, that perception is effected not by the operation Anaxa- of like upon like, but of contrary upon contrary. This _{goras and his} accords, on the one hand, with his metaphysical doctrine contem- of νοῦς ἀμιγής, and, on the other, with the empirical fact _{poraries respecting} that many perceptions, e. g. that of temperature, seem to _{the theory of percep-} rest upon a contrast between the condition of the perceiving tion.Unlike organ and the object it perceives. If the temperature of _{perceives unlike. Ap-} water is exactly that of the hand, this may be thrust into plication of it without perception of it as either cold or hot. The _{this to the theory} contrariety required by the doctrine of Anaxagoras as one _{of vision.} of the conditions of perception exists for all possible cases ; since, according to the Anaxagorean doctrine πᾶν ἐν παντί, we have within us the contraries of all possible external objects. Our information as to the psychological teaching of Anaxagoras is scanty, yet contains evidence of his being influenced by these principles.

---

[1] Stob. *Ecl.* i. 16 (Diels, *Dox.*, p. 314).

[2] Arist. *de Gen. et Corr.* 316ᵃ 1 τροπῇ γὰρ χρωματίζεσθαι. The terms for *order, figure*, and *position* are, in ordinary Greek, τάξις, σχῆμα, and θέσις, but the terms used by Democritus for these respectively were διαθιγή, ῥυσμός, and τροπή. Cf. Arist. *Met.* i. 5. 985ᵇ 17 (adopting Diels' H, ⊏ for Z, N). 'The letter A differs from H in *figure* (σχήματι) ; AH differ from HA in *order* (τάξει); while ⊏ differs from H in position' (θέσει) the ⊏ being but H lying on its side. Probably διαθιγή is dialectic=δια- θήκη, i.e. διάθεσις, and not = 'contact' (√ θϊγ-), as Gomperz after Mullach renders. The primary qualities of each atom *per se* for Democritus were (*a*) *physical*, viz. weight and solidity ; (*b*) *geometrical*, viz. figure and magnitude. Not only colour, but all other secondary qualities of body, depend on these primary qualities, as well as on the τάξις, σχῆμα, and θέσις, of the atoms which compose the body. Gomp. *G. T.* i, 568.

Vision due to pupillar image.

'Seeing,' according to Anaxagoras[1], 'takes place by reflexion of an image in the pupil of the eye, but this image is not reflected in a part of the pupil of like colour with the object, but in one of a different colour[2]. In the majority of eyes, the requisite difference of colour between organ and object exists in the daytime, but in some it exists by night; whence it follows that the latter see keenly by night. In general, the night is more in keeping than the daylight with the actual colour of the eyes. In the daytime objects are reflected in the eye, because light is a condition of such reflexion. But (whether by night or day) the colour which predominates in the object seen is, when reflected, made to fall on the part of the eye which is of the opposite colour[3].' According to the general rule the colours of the eye are dark, i. e. of the hue of night; hence more fit for reflecting images, and therefore for seeing, by day than by night; although to this rule there are exceptions. Anaxagoras held with Empedocles that persons with gleaming eyes (γλαυκοί) see better at night than those with dark eyes. Empedocles, however, based this view, not on the ground that like is perceived by unlike, but on the principles that fire is a visual agency[4], and that the conditions are, in some cases, more favourable for its action at night than by day.

Theophrastus' criticism of Anaxagoras' theory of vision.

§ 21. Theophrastus[5], in criticizing the visual theory of Anaxagoras, says: 'As regards the reflexion in the eye, his opinion is not different from that of most other thinkers; for the majority hold that seeing results from the formation

[1] Theophr. *de Sens.* § 27.

[2] For this difference of colour see Democritus, § 13, p. 26, n. 4 *supra*, and Theophrastus' criticism of Democritus, § 14, p. 29.

[3] τὴν δὲ χρόαν τὴν κρατοῦσαν μᾶλλον εἰς τὴν ἑτέραν ἐμφαίνεσθαι. Here we are reminded by τὴν κρατοῦσαν that, according to the doctrine πᾶν ἐν παντί, all colours as well as all other sensible qualities are in every object, but in different degrees of prominence; and that each object is perceived and named according to that sensible quality which is *predominant* in it. Thus the seeds of *all* colours are in the object, yet red for example may predominate; whence we perceive it as red and call it so.

[4] See Empedocles, *supra* § 9.

[5] *De Sens.* §§ 36–7 (Diels, *Dox.*, p. 509).

of an image in the eye by reflexion.   They do not, however, provide in their theory for these facts, viz. that (*a*) the real magnitudes seen are not symmetrical with the reflected magnitudes ; (*b*) it is impossible for a plurality of reflexions to take place in the eye simultaneously with their contraries ; (*c*) though *movement, distance,* and *magnitude* are visible none of these reflects an image ; (*d*) some animals, e. g. those which have scales on the eyes, and those which live in water, have no image reflected in the eye and yet they see.   Besides these points, if such reflexion were the sufficient reason of seeing, many inanimate things would see ; for reflexion takes place in water, bronze, and many other things.   Anaxagoras also teaches that colours are all reflected in one another, but a strong colour in a weak rather than conversely ; so that while either the strong or the weak ought to see, yet a black eye should see better than one of any other colour: and, in general, an eye of weaker, better than one of stronger colour[1].   Wherefore he describes the organ of seeing as being of the same hue as night, and light as the cause of the reflexion of an image in the eye.   But, in the first place, we see light itself without the need of such reflexion ; and, in the next, we see black colours just as well as white, though the former do not contain light (which according to Anaxagoras is needful to produce the reflected image)[2].   Again, in the case of other things (apart from optical reflexion), we see that reflexion of images takes place in that which is brighter and purer (than the object reflected) ; and, accordingly, Anaxagoras himself declares that the membranes covering the eyes are delicately fine and bright.'

§ 22. The *object* of vision: colour.   'As regards colours[3] Anaxagoras : no

---

[1] 'The 'weakest' colour, as would appear from this, is black according to Anaxagoras and Theophrastus.   This, therefore, represents all other colours by reflexion.

[2] Some such word as ἀλλά or καίτοι seems to have been lost before οὐκ ἔχει in the sentence ἔπειτα οὐδὲν ἧττον τὰ μέλανα τῶν λευκῶν οὐκ ἔχει φῶς.   This, as it stands in Wimmer's and Diels' texts = *non minus nigra quam alba lucem non habent,* makes no sense.   I have translated according to what I conceive the true reading.

[3] Theophr. *de Sens.* § 59 (Diels, *Dox.,* p. 516).

express theory of Colour: indirect information regarding it.

Empedocles held that *white* consists of fire, *black* of water. The others confined themselves to asserting that white and black are the elementary colours, the remaining colours being generated by mixtures of these two. For Anaxagoras has expressed himself quite generally respecting them[1]. He held[2] that the elements of all things were originally confused in one mass infinite in number and severally infinitesimal in bulk. This being so, we must conceive that (for him) many and multifarious seeds of things exist in all bodies—seeds with all sorts of shapes, and *colours*, and savours. . . . Before they were separated from the mass, and while all were still together, no single determinate colour was yet discernible.' ' Colours, according to Anaxagoras, are not self-subsistent or separable from coloured *things*. Each colour requires a substrate. It is not possible that all things whatever should be separated from one another; the process of discrimination[3] is no absolute separation[4]; wherefore it is impossible that walking[5], *colour*, and, in general, the qualities and states of things, should be really separated from their substrates (τῶν ὑποκειμένων)[6].' It is plain that, owing to his theory of πᾶν ἐν παντί, Anaxagoras could not hold that there is in nature any pure or simple colour[7].

---

[1] ἁπλῶς εἴρηκε. Prantl, pressing the γάρ before Ἀναξαγόρας here, infers from the sentence that Anaxagoras with the others held white and black to be primary colours.

[2] Simpl. *ad Arist. Phys.* 184ᵇ 15–188ᵃ 5, pp. 34–5, 156, 175–6 (Diels); Prantl, Περὶ Χρωμ. p. 58.

[3] i. e. that effected by νοῦς.

[4] οὐ γὰρ παντελὴς διασπασμός ἐστιν ἡ διάκρισις.

[5] βάδισις here seems to mean 'movement' in general, which is impossible, according to Anaxagoras, without something that moves.

[6] Simpl. l. c. Prantl, *Arist.* Περὶ Χρωμάτων, p. 59, remarks that it was probably this conviction of the inseparableness of qualities from substance that led Anaxagoras to make his famous assertion that snow is black. To the sensible impression that snow is white, he opposed the rational view that snow is water frozen, and that water—the Homeric μέλαν ὕδωρ—is black; hence snow is really black. The meaning and object of this paradoxical assertion were quite misunderstood by many ancient writers; e. g. Cic. *Acad. Quaest.* iv. 23. 31.

[7] Cf. Arist. 187ᵇ 2 seqq. διό φασι πᾶν ἐν παντὶ μεμεῖχθαι . . . εἰλικρινῶς

Diogenes of Apollonia.

§ 23. Diogenes held that the ultimate agency in Nature Diogenes' (which included for him Mind in all its manifestations) is view of Air as the Air. Thus thought and sensation are activities of the foundation intra-organic air (especially that in or around the brain) of mental and in relation with the outer, or extra-organic air, which physical activity. operates in nature generally. The air in the particular The intra- organs conducted the sensory impressions to that near the organic air the cause of brain, as their central organ; which, again, seems, in certain perception. cases at least, to have co-operated with the air in the breast, Pupillar image the or near the heart. Perception is more perfect the finer chief factor is the intra-organic air, and the more freely the structure of of vision. Points of the vessels promotes its passage to and fro between the agreement between brain, the thorax, and the various parts of the bodily Diogenes, system. Anaxa-goras, and

'Seeing takes place, according to Diogenes[1], by the re- Empe- flexion of objects in the pupil of the eye; for this, by being docles. No theory of mixed (μειγνυμένην) with the internal air[2], produces the sense colour. of vision; a proof of which is that when there is inflammation of the vessels of the eye, the mixture with the air within being interrupted, vision is impaired, although the image is reflected in the pupil as usual.' 'Those animals see most keenly which have the air[3] within them fine and the veins fine likewise (such fineness of the air and the air-vessels being the general conditions of perfect sense), and those which also have the eye itself as bright as possible[4]. The colour which is contrary to that of the eye is best reflected in it[5]: wherefore those whose eyes are black see best by day,

---

μὲν γὰρ ὅλον λευκὸν ἢ μέλαν ἢ κτέ. . . . οὐκ εἶναι· ὅτου δὲ πλεῖστον ἔχει ἕκαστον, τοῦτο δοκεῖν εἶναι τὴν φύσιν τοῦ πράγματος.

[1] Theophr. de Sens. § 40 (Diels, Vors., p. 344).

[2] More especially τῷ περὶ τὸν ἐγκέφαλον ἀέρι.

[3] Theophr. l. c. § 42.

[4] ὅσα τε τὸν ἀέρα (sc. λεπτόν) καὶ τὰς φλέβας ἔχει λεπτάς, ὥσπερ ἐπὶ τῶν ἄλλων (sc. αἰσθήσεων), καὶ ὅσα τὸν ὀφθαλμὸν (sc. ἔχει) λαμπρότατον. Diels should have placed a comma after ἄλλων, as ὥσπερ ἐπὶ τῶν ἄλλων is parenthetical.

[5] For this doctrine see Democritus, supra § 13; Anaxagoras, § 20.

and see bright better than dark objects; while their opposites see better by night. That the internal air, which is a small part of the god [1], is what perceives, is shown by the fact that often, when we have our minds directed to other things (than the object), we neither see nor hear [2].' Diogenes thus agrees with Empedocles and Anaxagoras in making those see best by day whose eyes are black, and those whose eyes are bright, or gleaming grey, see best at night. The reasons for which Empedocles and Anaxagoras held this view have been stated; why Diogenes shared it we are not informed.

Diogenes has left us no theory of *Colour*. It is manifest that he laid great stress on the phenomenon of ἔμφασις— the reflexion of an image in the eye—as a factor of vision. Theophrastus [3] asserts that Diogenes' theory that we see by virtue of the internal air is futile. ' While Diogenes ' (he goes on) 'confutes, after a fashion (ἐλέγχει πως), those who take the mere reflexion in the pupil for a complete explanation of vision, he fails himself to render a satisfactory account of the latter.' For him, it is evident, the conditions of vision were summed up in the reflexion of the image, and the communication between this and the air within the brain and organism in general. Air as first principle, both of nature and of mind, was endowed by him with intelligence.

Plato.

The general attitude of Plato un-
§ 24. For empirical psychology Plato had only the regard of a stepmother. He was averse to physical studies, and Democritus, whose whole life-work was given to these,

---

[1] ὁ ἐντὸς ἀὴρ αἰσθάνεται, μικρὸν ὧν μόριον τοῦ θεοῦ.

[2] The meaning of this is not, at first, clear. But Diogenes believed that Νοῦς in each man is Air—ὁ ἐν ἡμῖν θεός—and a part of the universal Νοῦς, ὁ θεός, which, of course, is also Air. When the individual νοῦς is engaged on its own thoughts, if we then have neither ears nor eyes for external objects, it follows that the operation of these senses is included in that of νοῦς : as it is νοῦς (ὁ ἐντὸς ἀὴρ) that thinks, so it is the same that perceives. He does not here *argue*—he assumes—that νοῦς in each person is ὁ ἐντὸς ἀήρ.

[3] *De Sens.* § 47 (Diels, *Dox.*, p. 512).

he seems to have disliked. At all events he never names him. Accordingly we find comparatively little in Plato's dialogues bearing on this subject, and that little not always up to the standard of what was to be expected from a writer of his transcendent genius. A few scattered references and observations; an interesting disquisition in the *Theaetetus* (which, however, aims not at psychological but rather at epistemological results); and a discussion in the *Timaeus*, for which the author practically apologizes[1], form the chief contributions of Plato to the subject of empirical psychology. Plato's physics were submerged in metaphysics. We cannot, therefore, so clearly distinguish the ruling physical ideas which governed his psychology as we could do and have done in the cases of Empedocles, Democritus, and Anaxagoras. When he proceeds to treat of psychology he descends from first to second causes, and finds himself on uncongenial ground. It is not easy to discover a principle of union between his psychology and his idealism, any more than between his psychology and any ruling physical principles. His physics is virtually contained in his account of the nature and construction of matter, in its four forms, given by him in the *Timaeus*. He accepts the four Empedoclean forms, earth, air, fire, water ; but does not regard them as primitive. These were constituted by the Demiurgos out of fundamental triangles, by a geometrical process doubtless borrowed from the Pythagoreans. The primitive triangles are the right-angled isosceles, and the right-angled scalene. From these are first constructed the pyramid, the cube, the octahedron, and the eikosahedron. The cube, then, is made to form the foundation of earth, as it is the most solid element; the pyramid forms that of fire ; the octahedron that of air ; the eikosahedron that of water. These four 'elements' stand to one another in continuous proportion : as fire is to air, air is to water ; and as air is to water, so is water to earth[2]. Plato's psychology

Marginal note: favourable to empirical psychology : his physics immersed in metaphysics. Account of the soul given in the *Timaeus*.

---

[1] The theory of colour in the *Timaeus* comes in only as a part of the φρόνιμος παιδιά in which the author indulges. Cf. *Tim.* 59 D.

[2] *Tim.* 32 A–B.

also is set forth in the *Timaeus*, in his attempted deduction
of the individual from the cosmic soul. This deduction is
on the face of it metaphysical, and indeed fanciful in the last
degree. When the Demiurgos makes over to the newly
created gods the task of fashioning mortal bodies to be
joined with immortal souls, we see Plato at a loss how to
connect his metaphysics with his physics by any satisfactory
rational or scientific tie. The inferior gods borrowed from
the Cosmos portions of the four elements [1], and of these
they compacted the organic body. Into this body they
introduced the immortal soul with its double circular rotations—the circles of the Same and of the Different. This
soul they located in the cranium, which is spherical, like
the Kosmos, in its external form, and admits no motion
but the rotatory. The body had all the varieties of motion,
backward, forward; upward, downward; right, left. In it
were set up the movements of nutrition and sensation,
which, however, interfere with, and disturb, the movements
of the rational soul in the cranium. Thus its rotations
in the circles of the Same and the Different are caused to
convey false information. In the course of time, and by
the process of education, this state of things is made to
improve. Philosophy attempts to restore the mathematical
exactitude of the intellectual movements. To all this
Plato subjoins a particular account of the senses—their
organs, functions, and *objects.* This will be now given as
far as it concerns the sense of seeing.

*Function
and organ
of vision.
Plato, like
Empedo-
cles, neg-
lects the
pupillar
image.*

§ 25. Neglecting the pupillar image 'Plato held that
seeing takes place in virtue of a coalescence between (*a*) the
rays of the intra-ocular light emanating from the eyes
to some distance into the kindred (i. e. illuminated) air;
(*b*) that which, reflected from external bodies, moves to meet
it; and (*c*) that which is in the intervening air, and which,

---

[1] It is noticeable how great a hold this doctrine of the four elements
(which Empedocles first propounded) took upon the Greek mind. It
pervades the whole period from Empedocles to Aristotle, for though
not of course accepted in its original form by all writers, it was something with which all had to reckon; and which influenced even those
who rejected it.

owing to the diffusibility and nimbleness of the latter,
extends itself in lines parallel with the fiery current of
vision [1].' 'Of the organs first they wrought light-bearing
eyes, and bound them fast in the causal scheme as follows.
That part of fire which has the property of not burning,
but yielding an innocuous light, they contrived to fashion
into a substance homogeneous with the light of day [2]. For
the fire within us, being twin with this, they caused to
flow through the eyes in its pure form, smooth and
dense, having constructed the whole, and especially the
central part, of the eyes in such wise as to confine all
the remainder, i. e. the denser portion, of the fire within,
and to filter forth only such fire as that above described,
by itself, in its purity. Whenever, accordingly, there is
daylight around the visual current (= the light which flows
out from the eyes), this current, issuing from the eyes and
meeting with its like, becoming compacted into union with
the latter (i. e. with the homogeneous external daylight),
coalesces with it into one homogeneous whole [3] in the line
of vision, i. e. in the direction in which the current issuing
from within meets front to front with, and presses against,
any of the external objects with which it comes into
collision. The whole then, owing to the essential homo-
geneity of its constituents, becomes sympathetic, so that
whenever it takes hold of anything, or when anything takes
hold of it, it transmits the movements of such thing into
the whole body as far as the soul [4], and so produces
a sensation, viz. the experience on having which we say

[1] τοῦ περὶ τὸν μεταξὺ ἀέρα εὐδιάχυτον ὄντα καὶ εὔτρεπτον συνεκτεινομένου
τῷ πυρώδει τῆς ὄψεως, Stob. Ecl. i. 52 ; Plut. Epit. iv. 13 (Diels, Dox.,
p. 404). Prantl (Arist. Περὶ Χρωμάτων, p. 75) remarks that συναύγεια, the
term above translated 'coalescence of rays,' seems to have come into
vogue in the later Academy or among the Neo-Platonists. This passage
of the Placita sums up fairly enough the doctrine set forth in the
following passage of the Timaeus (45 B–46 A) itself.

[2] There is a play on the terms ἡμέρα and φῶς ἥμερον.

[3] ἐκπῖπτον ὅμοιον πρὸς ὅμοιον ξυμπαγὲς γενόμενον.

[4] μέχρι τῆς ψυχῆς: up to the 'seat of consciousness,' an expression
of which great use is made by most Greek psychologists, and which
covers the greatest mystery of psychology.

commonly that we *see*. But when the kindred fire without has departed into night, the visual current from within is cut off; since, on issuing from the eye and meeting what is unlike it, it becomes itself changed in quality and extinguished : it becomes no longer homogeneous with the neighbouring air, as the latter now contains no fire.'

Sleep and dreaming.

§ 26. 'Therefore it ceases from seeing and tends to bring on *sleep*. For when the eyelids, whose structure the gods devised as a protection for the sight, are closed, they imprison the force of the fire within ; and this force weakens by diffusion, and so calms, the internal movements ; and when they have become calm, quietude succeeds. If this quietude is profound, the sleep which descends upon us yields but scanty *dreams* ; but if certain of the greater movements have been suffered to remain, these, according to their quality, and that of the regions of the body in which they remain, produce " phantasms " of corresponding quality and number, fashioned within us like unto objects seen, and referred outwards to them by us in memory when we awake [1].' ' Does not dreaming (asks Plato in the *Republic*) consist just in this, that one, whether asleep or awake, regards that which is like something not as merely being *like* it, but as being the *very* thing itself which it resembles [2] ?'

Plato's theory of visual fire compared with that of Empedocles.

§ 27. As Mr. Archer-Hind, ad loc., observes, there are three fires concerned in the above account of vision : (1) that which streams from the eye (τὸ τῆς ὄψεως ῥεῦμα) ; (2) the fire of daylight in the air; and (3) the fire which is the colour of the object seen. The visible object is immersed in the μεθημερινὸν φῶς, which, with χρῶμα, streams from it to the eye. This stream meets τὸ τῆς ὄψεως ῥεῦμα, and both united in one whole (often spoken of as simply ὄψις) convey the impression of the object to the soul. But the fire of daylight, which intervenes between eye and object as a sort of medium, conforms itself somehow to these conjoint currents, supporting and substantiating them, as is stated in the extract given above (§ 25) from

[1] ἀφομοιωθέντα ἐντὸς ἔξω τε ἐγερθεῖσιν ἀπομνημονευόμενα.
[2] *Rep.* 476 C.

the *Placita*. In all this, as well as in Plato's disregard of the pupillar image, there is much that reminds one of Empedocles (see § 29 *infra*). He, too, speaks of a fire issuing from the eye. He, too, says that colour comes as an ἀπόρροια from the object, and Plato, in the *Menon* (cf. § 10 *supra*), seems to accept this account of it while ascribing it to Gorgias and his master. But Empedocles has not left anything to show the part which he would attribute to the daylight in connexion with vision. Nor is it easy to single out in Plato's account of the matter the separate parts played by the fire from the object and the fire of daylight. The one is not to be absolutely separated from the other. The fire from the object ceases if the fire of daylight departs. The colour and the light in which it is seen are intimately connected for Empedocles, as for Plato. Although, therefore, it may be that Plato distinguished his visual theory from that of Empedocles by the part which he makes the daylight play in fusion with the visual light, yet, in the absence of information as to Empedocles' view on this matter, we cannot be quite sure. There seems nothing in the theory of the latter *inconsistent* with the Platonic view. Finally the Empedoclean doctrine was that by each element within us we perceive the same element without, 'fire by fire, earth by earth, &c.'; and Plato was an adherent of the same theory. Aristotle tells us [1] that Plato, in the same way as Empedocles, regards the soul as formed of the elements, on the principle that 'like is known by like.' Plato's 'elements,' however, in the formation of ψυχή, were not material, and were far other than those of Empedocles [2].

§ 28. Light, the *medium* of vision, is a subject of interest to Plato, not however from a physical or psychological standpoint so much as from that of metaphysics. 'We see,' he says [3], 'with the organ of seeing, and hear with the organ of hearing, and with the senses generally perceive their respective objects; but the great Artist who fabricated the senses and their organs has, with regard to seeing, gone more expensively to work than in any of the other

The medium of vision. (Plato seems to speak as if there were no medium of hearing.)

[1] 404[b] 16.　　[2] Cf. *Tim.* 35 A seqq.　　[3] *Rep.* 507 C–508 B.

senses. The organs of hearing and sound need no third [1] thing in order that the former may hear and the latter be heard; nothing, the absence of which would prevent the one from hearing and the other from being heard. The other senses also are exempt from any such need. But the faculty of seeing and the object of this have need of such third thing. For the power of seeing may be in the eye, and the man who possesses it may strive to exercise it, also colour may be present in the object ; but if a third thing called light be not present, the eye can see nothing ; the colour must remain invisible. Light is the precious medium by the intervention of which the object and the organ of vision are brought into conjunction for the exercise of this faculty. The visual organ is not the sun, though the most sunlike (ἡλιωδέστατον . . . ὀργάνων) of the sensory organs [2]; but it receives from the sun, when the latter illuminates the sphere of vision, all the visual power which it possesses. Light wells forth from the sun as from a fountain.'

The object of vision : Colour. § 29. The object of vision is *colour*. If the eye sees, what it primarily sees is this [3]. The visual agency according to Plato [4] consists of fire. Its visible object too is of the same nature. 'The body of the created world is tangible and visible: that it should be tangible it must consist, in part, of earth : that it should be visible it must have an ingredient of fire [5].' '*Colour*, therefore, he regards as a sort of flame from bodies, having its parts symmetrical [6] with

---

[1] It is strange that Plato should here reason as if only this one faculty of sense required a medium—light—between object and organ : as if no medium were required for hearing or smelling.

[2] Cf. Goethe, *Farbenlehre*, Introduction :

> 'Wär' nicht das Auge sonnenhaft,
> Wie könnten wir das Licht erblicken ?
> Lebt' nicht in uns des Gottes eigne Kraft,
> Wie könnt' uns Göttliches entzücken ? '

[3] In *Charmid.* 167 C χρῶμα μὲν ὁρᾷ οὐδὲν ὄψις οὖσα is given as an absurdity. [4] Theophr. *de Sens.* § 5.

[5] χωρισθὲν δὲ πυρὸς οὐδὲν ἄν ποτε ὁρατὸν γένοιτο, *Tim.* 31 B.

[6] Theophr. l. c. We are here (as Th. remarks) reminded of Empedocles, who required συμμετρία between the ἀπόρροιαι and the pores of the organs.

those of the visual current [1] ; so that (since an emanation [2] takes place from the objects seen, and this emanation and the visual fire must harmonize with one another) the visual agency, going forth to a certain point, forms a union with the emanation from the body, and thus we *see*. Hence Plato's visual theory would stand midway between that of those who merely say that the visual current impinges upon the objects [3], and that of those who teach merely that something is conveyed to the eye [4] from the objects seen.' 'Plato's theory of colour approximates to that of Empedocles, since the symmetry which Plato requires between the parts of the colour and the visual current is like the harmonious fitting (ἐναρμόττειν) of the ἀπορροαί into the pores required by Empedocles. . . . It is strange that Plato should simply define colour as a flame ; for, though the particular colour white may be like this, yet black would seem to be the very reverse [5].' We have seen that Plato seems to approve [6] of the definition quoted in the *Menon* from Empedocles [7]. Black and white are recognized by Plato as opposite colours [8]. Hence, too, colours admit of gradation, not *quantitative*, in the sphere of μέγα or πολύ, but *qualitative*, i. e. in point of καθαρότης [9].

---

[1] τῇ ὄψει=τῷ τῆς ὄψεως ῥεύματι.

[2] ὡς ἀπορροῆς τε γιγνομένης κτέ. This, if Theophrastus expresses Plato's doctrine correctly, brings the latter into closer relationship with Empedocles than Mr. Archer-Hind (Plato, *Tim.* p. 156) is inclined to admit. Theophr. *de Sens.* § 91 περὶ δὲ χρωμάτων σχεδὸν ὁμοίως Ἐμπεδοκλεῖ λέγει. τὸ γὰρ σύμμετρα ἔχειν μόρια τῇ ὄψει τῷ τοῖς πόροις ἐναρμόττειν ἐστὶν [ἴσον?].

[3] Who are meant ? Probably Alcmaeon and the Pythagoreans.

[4] Probably those who held with Democritus the theory of visual δείκελα, or εἴδωλα.

[5] Theophr. *de Sens.* § 91.

[6] *Menon* 76 D ἔστι γὰρ χρόα ἀπορροὴ σχημάτων ὄψει σύμμετρος καὶ αἰσθητός.

[7] Prantl (who, objecting to Theophrastus' comparison of Plato's colour theory with that of Empedocles, says that *das Ganze bei Platon mehr dynamisch betrachtet wird*) would have us believe that the Empedoclean definition of colour is only accepted in a spirit of Socratic irony. *Vide* his *Arist. Farbenlehre*, p. 57.

[8] *Phileb.* 12 E, *Protag.* 331 D.   [9] *Phileb.* 53 B.

Genesis of particular colours.

§ 30. 'A fourth[1] department of sensibles yet remains whose many varieties we have to distinguish. These as a class [2] we call *colours*, being a flame [3] streaming off from bodies each and all, having parts symmetrical with those of the visual current, so as to be capable of being perceived [4]. We have already, in what precedes, set forth the causes which explain the origin of vision. Here, then, it is most natural and fitting to discuss the probable theory of *colours*, showing how the particles which are borne from external things, and impinge upon the visual organ, are some smaller, some larger than and some equal to the parts of this visual organ itself [5]; that, moreover, those of equal size are unperceived, and are accordingly called *transparent*, whereas the larger and smaller, the former contracting the visual current and the latter dilating it [6], are analogous respectively to things *cold* and *hot* in application to the flesh [7], and to things which, in their effects on the tongue (sc. the organ of taste), are *astringent*, or from their heating effect on it are called *pungent* [8]. These are the colours *black* and *white*: affections of the parts of the visual current which are, as has been said, identical in principle with those of temperature and taste but in a different sense-modality [9],

---

[1] Reading αἰσθητόν. The three preceding departments were those of *Taste, Odour, Sound*.   [2] Plato, *Tim.* 67 C–68 E.

[3] Prantl (Περὶ Χρωμ., p. 75) blames Theophr. § 86 for inaccuracy in giving, as Plato's definition of χρῶμα, φλόγα ἀπὸ τῶν σωμάτων σύμμετρα μόρια ἔχουσαν τῇ ὄψει, and says that Plato would not have used φλόξ thus. But in fact Theophrastus is merely repeating the words of *Tim.* 67 C.

[4] 'Lit. with a view to perception,' πρὸς αἴσθησιν.

[5] By 'organ' for Plato here has to be understood not the eye, but the ὄψεως ῥεῦμα.

[6] The 'diacritic' effect of white, and the 'syncritic' effect of black on the visual current would seem to have their psychological meaning in the power of visual *discrimination* which light gives, and the confusion, or loss of discrimination, between colour διαφοραί which results from darkness.

[7] i.e. in reference to the organ of touch which for Plato was the σάρξ.

[8] He does not pursue the parallelism of *white* to *hot* and *black* to *cold* into the modality of taste, so that e.g. *white* should be to *sweet* as *black* to *bitter*, nor could he do so consistently with his own account of sweet and bitter, *Tim.* 65 D, 66 E.   [9] ἐν ἄλλῳ γένει.

and presenting themselves to the mind as specifically different on account of the above-mentioned causes[1]. Thus, then, we must characterize them. That which dilates the visual current is *white*; the opposite is *black*[2]. When a more rapid motion (than that of white), belonging to a different kind of fire, impinging on and dilating the visual current right up to the eyes[3], forcibly distends and dissolves the very pores of the eyes, causing a combined mass of fire and water—that which we call a tear—to flow from them, and being itself fire meeting the other fire right opposite: then, while the one fire leaps forth as from a lightning-flash[4], and the other enters in and becomes extinguished in the moisture, colours of all varieties are generated in the encounter between them, and we feel what we call a *dazzling* sensation[5], to the external stimulus of which we apply the terms *bright* and *glittering*.

---

[1] I cannot refer ἐκείνων (E, l. 3) to anything but τοῖς τῆς ὄψεως μέρεσιν above. Stallbaum takes it of θερμὰ καὶ ψυχρά; Mr. Archer-Hind of τὰ συγκρίνοντα καὶ διακρίνοντα. The μόρια of the φλόξ from objects stand in a relation of size to the parts of the ὄψεως ῥεῦμα: if they are equal to the latter, they, or rather the objects, are transparent, and have no χρῶμα; if they are greater, they cause it to contract, and the colour seen is black; if they are smaller, they expand or dilate it, and the colour white is seen. These conditions of sensation are fulfilled at the moment of coalescence, we must suppose, between the ῥεῦμα ὄψεως and the μόρια from objects. But how are we to conceive this coalescence in accordance with the description? If the μόρια when equal to the parts of the ῥεῦμα ὄψεως cause no appreciable disturbance, how is it that they do so when smaller? There seems to be here a confused repetition of the 'pore' theory of Empedocles, who taught that ἀπόρροιαι must actually fit the pores to cause sensation; that if too small they pass through without any appreciable effect: if too large they do not pass in at all. This is fairly intelligible as regards actual 'pores' in the organ; but when applied to the ῥεῦμα in a free medium is not so easy to envisage to the imagination.

[2] Cf. Arist. 119ᵃ 30, 1057ᵇ 8-11. See also *Phileb.* 12 E, *Protag.* 331 D. That which is *merely* διακριτικὸν τῆς ὄψεως is, as we are here told, *white*: but we learn further on that if it διακρίνει τὴν ὄψιν μέχρι τῶν ὀμμάτων it is sparkling *bright*—λαμπρόν.

[3] διακρίνουσαν τὴν ὄψιν μέχρι τῶν ὀμμάτων. The meaning is plain from *Tim.* 45, where ὄψις is shown to consist of the amalgamated fires from the eye and from the object, what Prantl (*Arist.* Περὶ Χρωμ.) calls 'die Doppelbewegung der ἀπορροαί zwischen Object und Subject.'

[4] οἷον ἀπ' ἀστραπῆς. μαρμαρυγὰς τὸ πάθος προσείπομεν.

A kind of fire, again, midway between these two (viz. that producing λευκόν and that producing στίλβον), when it reaches the humour of the eyes, and is blended with it, but does not glitter, produces a sanguine colour [1], when its fire mingles with [2] the brightness in the moisture of the eyes, and to this colour we give the name *red* (ἐρυθρόν) [3].' The remaining colours are compounded of these four—*white, black, bright*, and *red*. ' Bright, when mixed with red and white, becomes *golden-yellow* (ξανθόν). What the proportion of parts in the several possible mixtures is, one should not say even if one knew; since there is no necessary law —no plausible account—which one could set forth with even moderate probability respecting them. Red, blended with black and white, gives *violet* (ἀλουργόν). If these (sc. the red, black, and white which form violet) are mixed and burnt, and black has been thus added in greater amount, the result is a *dark-violet* (ὄρφνινον). *Auburn* (πυρρόν) is produced by the mixture of golden-yellow and grey [4]. *Grey*, again, is formed by the mixture of white and black. *Yellow* (ὠχρόν) by that of white with golden-yellow. When white meets bright and is immersed in intense black, a *deep-blue* (κυανοῦν χρῶμα) is produced. When this deep-blue is mixed with white, the *glaucous* tint—greyish blue—(γλαυκός) results. When auburn is mixed with black the product is *leek-green*. It is clear, from what precedes, to what combinations the remaining colours are to be reduced, so as to preserve the verisimilitude of our fanciful account (μῦθον). If, however, one should endeavour to investigate and test our theories by practical experiment, he would show himself ignorant of the difference between the human and

---

[1] χρῶμα ἔναιμον. In 80 E red is named τῆς τοῦ πυρὸς τομῆς τε καὶ ἐξομόρξεως ἐν ὑγρῷ φύσις, the colour of blood being due, as Archer-Hind says, to the commingling of fire and moisture.

[2] i. e. is not *quenched* in it, as in the preceding case.

[3] In this attempt to discover the origin of *red*, the first of the properly so-called colours, Plato becomes more in earnest with this subject than Aristotle anywhere does.

[4] It is not easy to find English names exactly suitable for these terms. Thus φαιός here is rendered 'grey.' So Mr. Archer-Hind renders it. ὠχρός he translates ' pale-buff.'

the divine nature; for God has knowledge and power[1] to blend the many into one and resolve the one into many, but no man is able, or ever will be able, to accomplish either of these things.'

§ 31. Plato's account of the production of leek-green (πράσινον or πράσιον) by the mixture of auburn and black receives no support from Aristotle at all events. In the *Meteorologica* the latter tells us[2] that there are three colours—*crimson* (φοινικοῦν), *leek-green* (πράσινον), and *violet* (ἀλουργόν), which painters cannot produce artificially by any process of blending. These are the three principal colours of the rainbow[3]. According to Democritus (§ 17 *supra*), however, leek-green can be produced from purple (πορφυροῦν) and woad-blue, or else from pale-green and purplish (πορφυροειδές).

When Plato above calls colour a 'flame,' and speaks of fire as proceeding from the visible object to the eye, we must bear in mind how many apparently different things he understood under the name *fire*—particularly these three: *flame, light,* and *glow*. He says[4]: 'We must understand that there are many genera of fire, such as (1) *flame* (φλόξ), and (2) that which proceeds from flame, which does not burn but gives *light* to the eyes; and (3) that which, when the flame has died down, is left of the fire in the glowing embers.' He treats σέλας and φῶς as identical[5]. For him, just as nothing would without earth be tangible, so nothing would be visible without having fire in it[6]. Plato held[7] the smooth (λεῖον) like the white (λευκόν) to be capable of dilating, or distending, the parts of the visual current (διακριτικὸν τῆς ὄψεως); but

Plato differs from Aristotle and agrees with Democritus as to the compospositeness of leek-green: what Plato means by πῦρ. Plato in general agrees with Aristotle as to the optical effects of τὸ λεῖον or the lustrous. Colour not a merely subjective quality for Plato (in Timaeus), as it was for Democritus.

---

[1] Cf. *supra* Democr. § 17.

[2] 372[a] 7.

[3] Xenophanes, first of the writers whom we know, singled out these rainbow colours:

ἤν τ᾽ Ἴριν καλέουσι, νέφος καὶ τοῦτο πέφυκε,
πορφύρεον καὶ φοινίκεον καὶ χλωρὸν ἰδέσθαι.

Xenoph. *Frag.* 32 (Diels, *Vors.*, p. 56).

[4] *Tim.* 58 C.　　　[5] *Cratyl.* 409 B.

[6] *Tim.* 31 B.　　　[7] *Tim.* 60 A.

as it has a bright and glistening appearance this must be taken (in accordance with *Tim.* 67 E) to mean that it so affects the visual current up to and into the eyes themselves (μέχρι τῶν ὀμμάτων). This account of the smooth was accepted by Aristotle also, who says that 'smooth things have the natural property of shining in the dark, without, however, actually giving light [1].' Prantl [2] says that the account of colour given in the *Timaeus* would appear at first to be founded on atomism. Yet, as he points out, the dynamic import of the two factors—the σύγκρισις and διάκρισις—must be borne in mind ; and it has further to be remembered that Plato does not really explain the structure of the elements atomistically but geometrically. His employment, however, of the term ἀπόρροιαι (common to him with Democritus and Empedocles) indicates on his part a line of explanation which really throws his dynamic account of colour into the background. He treats certain colours as natural to certain things : e. g. red is the colour of blood [3]. So certain colours are naturally connected with certain other sensible qualities, e. g. with *bitterness* [4]. In the *Timaeus* and *Republic* Plato, unlike Democritus [5], regards colours as actually existing in things, not as having a merely subjective existence dependent on φαντασία [6]. The qualitative change (ἀλλοίωσις) which is so important in the colour theory of Aristotle plays but a small part in that of Plato. We find, however [7], the change of whiteness into another colour (μεταβολὴ τῆς λευκότητος εἰς ἄλλην χρόαν) given as an example of ἀλλοίωσις, one of the kinds of μεταβολή into which κίνησις is divisible for Plato as well as for Aristotle.

From the standpoint of sensationalism, colour and

§ 32. Plato [8] finds in the consideration of colour from the Protagoreo-Heraclitean standpoint a suitable illustration of the absence of objectivity in our merely sensible

---

[1] 437ᵃ 31.   [2] Arist. Περὶ Χρωμ., p. 69.
[3] *Tim.* 80 E.   [4] *Tim.* 83 B.
[5] It is another question how far he could really have held any such view consistently with the doctrine of sensible perception set forth, after Protagoras and Heraclitus, in the *Theaetetus* : see next paragraph.
[6] Cf. *Rep.* 508 C.   [7] *Theaetet.* 182 D.
[8] *Theaetet.* 153-7.

experience; and from this standpoint he develops provision- all other
ally a fierce attack upon the fact, or even the conception, sensible qualities
of science or objective knowledge of any kind. In the are (as well
course of this discussion a good deal of interesting informa- as the so-called
tion is given us as to the degree to which the colour 'things')
conception had been analysed by psychologists, and the merely subjective.
character of colour, as a 'secondary quality,' impressed upon
the popular science of the time. The ἀπόρροιαι of colour
and the εἴδωλα of things are (it would appear from this
discussion) of such a kind that they consist and exist *only
in the interaction* between object and subject. The object
is only the ξυναπτοτίκτον. White (λευκόν) and whiteness
(λευκότης), e. g., are but the product of this interaction, and
last only while it lasts. 'If the doctrine of Heraclitus is
applied to perception, and especially to vision, it will be
found that what we call white colour neither exists in our
eyes nor in any distinct thing existing outside them.
It has not even place or position. To see what colour
really is, if we proceed on the principle of Heraclitus that
"all is becoming," we shall find that white, black, and all
other colour arises from the eye meeting some appropriate
motion ; and that what we call a colour is in each case
neither that which impinges upon, nor that which is
impinged upon, but something which *passes*—some relation
—between them, and is peculiar to each percipient. For
the several colours can scarcely appear to a dog or to any
animal as they appear to a human being ; nor, indeed, do
they appear to one man as they do to another ; or even
to the same man at one time as they do at another. What
happens in the generation of colour is this. The eye and
the appropriate object meet together and give birth to
whiteness on the one side, and, on the other, the *sensation*
connatural with it, both of which could not have been
produced by either eye or object coming into relation with
aught else ; then, when the sight is flowing from the eye,
whiteness proceeds from the object which combines with it
in producing the colour, so that the eye is fulfilled with
sight and sees, and becomes (not sight but) a *seeing eye* ;

and the object which lent its aid to form the colour, is fulfilled with whiteness, and becomes (not whiteness but) *a white thing*, whether wood or stone or whatever the object may be which happens to be coloured white. And the like is true of all sensible objects, hard, warm, and so on ; which are similarly to be regarded, not as having any absolute existence, but as being all of them, of whatever kind, generated by motion in their intercourse with one another ; for of the agent and patient, as existing in separation, no trustworthy conception can be formed. The agent has no existence till united with the patient, and the patient none until united with the agent ; and, moreover, that which by uniting with something becomes an agent, by meeting with some other thing is converted into a patient. From all these considerations arises the conclusion that there is no one self-existent thing, but everything is becoming and relative. Being must be altogether cast out of our thoughts, though from habit and ignorance we are compelled—even in this discussion—to keep the term. Great philosophers, however, assure us that we should not allow even the term "something," or "belonging to something," or "to me," or "this," or "that," or any other term which implies the stationariness of things, to be employed in the language of nature and truth ; since all things are being created and destroyed, coming into being, and passing into new forms ; nor can any name fix or detain them ; he who attempts to fix them is easily refuted ; and all these things are true not only of particulars but of classes and aggregates such as are expressed in the general terms made use of in language[1].'

Aristotle.

The *object* of vision ; in general = *colour*, i. e. that which is

§ 33. Aristotle commences his account of the special senses with the sense of *sight*. According to his custom, he examines first the object of seeing. This, stated most generally, is the *visible* (τὸ ὁρατόν)[2], or, as he defines it more

---

[1] Jowett's phraseology has for the most part been adopted.

[2] 418ª 26 seqq. οὗ μὲν οὖν ἐστιν ἡ ὄψις τοῦτ᾽ ἐστιν ὁρατόν. Seeing, by a power common to it and the other senses, perceives contraries : therefore it perceives also the *invisible* (ἀόρατον). By this 'invisible,' however,

closely, 'that which is seen in the light.' So defined, the *object* of sight is *colour* [1]. This is the most general name for the immediate and proper object seen in the light. Colour, unlike certain other things [2] (fire and phosphorescent substances), cannot be seen in darkness. Hence in order to understand colour—the object of vision—we must obtain a true view of the medium of vision—light. Colour overspreads the surface of all that is visible. Now every colour *sets up a motion in the diaphanous medium between each coloured thing and the eye which sees it* [3], *when the said medium exists actually, not merely in potency*. This is the essence of colour. By the motion thus set up in the actualized, i. e. illuminated, diaphanous medium, vision is normally stimulated ; not, as was held by Empedocles, Democritus, and Plato, by ἀπορροαί, or εἴδωλα, from the objects of vision.

seen in the light. The sense of seeing perceives the invisible: how? To understand colour, we must understand light.

§ 34. In order to understand light, therefore, we must consider the nature of the diaphanous, its medium [4]. This is a thing which is, indeed, visible, but not always or directly ; owing its visibility, when it has it, to colour produced in it from without [5]. Instances of the diaphanous are found in *air*, *water*, and many solids [6] ; which *are* diaphanous or trans-

The diaphanous medium; light and darkness. Light does not travel through space, as

is here meant not the *absolutely* invisible, but only σκότος (cf. 421ᵇ 3, 422ᵃ 20-2) ; and even τὸ σκοτεινόν is only μόλις ὁρώμενον (418ᵇ 29) ; as is also τὸ λίαν λαμπρόν, which is ἀόρατον in a different way from σκότος. Cf. *Met.* 1022ᵇ 34 ἀόρατον λέγεται καὶ τῷ ὅλως μὴ ἔχειν χρῶμα καὶ τῷ φαύλως.

[1] Not that the object of sight, thus restricted, and colour are absolutely identical. Cf. *Phys.* 201ᵇ 4, *Met.* 1065ᵇ 32 ὥσπερ οὐδὲ χρῶμα ταὐτὸν καὶ ὁρατόν. Their λόγοι, as Simplicius says ad loc., are διάφοροι.

[2] As will appear there are three kinds of ὁρατά : (1) colour (seen only in light) ; (2) fire (seen both in light and darkness) ; (3) phosphorescent things (seen only in the dark).

[3] πᾶν χρῶμα κινητικόν ἐστι τοῦ κατ᾽ ἐνέργειαν διαφανοῦς καὶ τοῦτ᾽ ἐστιν αὐτοῦ ἡ φύσις, 418ᵃ 31.

[4] This is at the basis objectively of *light* and *colour*, and subjectively of *vision*.

[5] Either by fire or by τὸ ἄνω σῶμα (see note 1, p. 58) : ὁρατὸν ... δι᾽ ἀλλότριον χρῶμα.

[6] As we shall see (p. 60), the diaphanous in bodies is the *vehicle* of the colour regarded as *in* these bodies ; not, like the free diaphanous, the *medium* which propagates the colour movement to the eye.

parent, not *qua* water or air, but because they have inherent in them the same natural substance which exists in the eternal body of the celestial sphere[1]. The actualization of this diaphanous *qua* diaphanous is *light*, just as its mere potentiality is *darkness*. Thus darkness is potentially wherever light is actually, and conversely. Light is thus, too, a colour, belonging incidentally to the diaphanous medium when the latter is actualized by the agency either of *fire*, or of a substance of the same nature as the celestial fire which has in it a principle or element of identity with the terrestrial. As colour can stimulate only the *actually* transparent or diaphanous, it is only in the actuality of this, i. e. in the light, that it can be seen. Fire, however, and certain other things mentioned below, can be seen in darkness. Such, then, is the diaphanous: and accordingly light is not fire, nor a body, nor an emanation from body[2], but the *presence* of fire or some such thing in the diaphanous[3]. Colour is a phenomenon in light, as light is a phenomenon in the diaphanous. Darkness, on the other hand, is the privation (στέρησις) of light—the absence from the diaphanous of that state which when present in it is light. Light is a *presence*, and therefore those are wrong who like Empedocles suppose it to move locally, and come by a process unperceived by us through successive places from the sun to the earth. Reason and observation are both opposed to this view. If, indeed, the interval said to be thus traversed were a short one, light, if it moved, might traverse it without our perceiving the lapse of time it took; but not so when the intervening distance is so

---

[1] ὅτι ἐστί τις φύσις ἐνυπάρχουσα ἡ αὐτὴ ἐν τούτοις ἀμφοτέροις καὶ ἐν τῷ ἀϊδίῳ τῷ ἄνω σώματι. This σῶμα belongs to the region extending from the ἀήρ to the moon and thence upwards to the empyrean in ever increasing brightness and purity. Cf. *Meteor.* i. 3. 340[b] 6 τὸ μὲν γὰρ ἄνω μέχρι σελήνης (the 'upper region' viewed *downwards* as far as the moon) ἕτερον εἶναι σῶμά φαμεν πυρός τε καὶ ἀέρος (Ideler, i. p. 344), *de Cael.* 286[a] 11, and the notes of Trendelenburg and Wallace on *de An.* ad loc.

[2] οὔτε πῦρ οὔθ᾽ ὅλως σῶμα οὐδ᾽ ἀπορροὴ σώματος οὐδενός, directed against Plato, *Tim.* 67 D.

[3] πυρὸς ἢ τοιούτου τινὸς παρουσία ἐν τῷ διαφανεῖ.

great as that of East from West [1]. Hence vision is perfect
at any instant and involves no temporal process [2].

§ 35. Light has been defined as the colour of the diaphanous, incidentally [3] belonging to it, and depending on
the presence in it of something of the nature of fire. The
presence of this in the diaphanous *is* light ; the privation of
it, darkness. This diaphanous is something not peculiar to
air or water or any of the bodies called diaphanous or 'transparent,' but is a kind of universally diffused natural power [4]
not capable of existence apart from body [5] but subsisting in
the things mentioned, and in all other bodies, in varying
degrees. As the bodies in which it subsists have an external
limit or superficies, so has this also its external bounding surface. Light subsists in the diaphanous generally, when the
latter is actualized, and is as it were, indirectly, its colour [6] ;
and so too the exterior boundary of the actualized diaphanous in determinate bodies is their colour, as observation
shows. It is the diaphanous in bodies, then, that causes them
to have this quality of colour. In all bodies colour either *is*
the limiting surface, or *is at* this surface. The Pythagoreans [7]
chose the former alternative, and defined the *surface* of a body
—its external manifestation [8]—as its *colour* (χροιά). But they
were wrong. The colour, though *at* the superficial boundary [9]
of a body, is not *identical with* the boundary of the body
as such, but rather with the exterior limit or boundary

The diaphanous in bodies determinately bounded explains their colour. Pythagorean geometrical view of colour as = superficies. Aristotle's two definitions of colour.

---

[1] For this polemic against Empedocles (in which, says R. Bacon,
A. only contends that light is not a body, not that it does not travel)
see further 446ᵃ 26. Galen, *de Plac. Hipp. et Plat.* § 638, agrees with
Arist. here, ὀρθότατα καὶ πρὸς Ἀριστοτέλους εἴρηται περί τε τῆς παραχρῆμα
μεταβολῆς τῶν οὕτως ἀλλοιουμένων, ὡς κινδυνεύειν ἄχρονον εἶναι.

[2] *Eth. Nic.* 1174ᵃ 14, ᵇ 12.

[3] For what follows see Arist. 439ᵃ 18 seqq.

[4] κοινή τις φύσις καὶ δύναμις. One thinks of the 'luminiferous ether.'

[5] χωριστὴ μὲν οὐκ ἔστι.

[6] τὸ φῶς ἐστι χρῶμα τοῦ διαφανοῦς κατὰ συμβεβηκός 439ᵃ 18 : cf. 418ᵇ 11.

[7] Cf. 131ᵇ 32 ἔσται γὰρ κατὰ τοῦτο καλῶς κείμενον τὸ ἴδιον· οἷον ἐπεὶ ὁ
θέμενος ἐπιφανείας ἴδιον ὃ πρῶτον κέχρωσται κτέ. The colour is therefore the property, or essential mark, of the surface of a body. But as
every surface has colour and every determinate body has surface, every
such body has colour. Void space has no colour, *Phys.* 214ᵃ 9.

[8] ἐπιφάνεια.   [9] ἐν τῷ τοῦ σώματος πέρατι.

of the diaphanous, which permeates the whole body from surface to centre, and which, *at* the surface, takes the aspect of colour. Even the indeterminate diaphanous of air and water has colour, viz. the lustre (αὐγή) or brightness which they exhibit. In them indeed, owing to their indeterminateness [1], the colour varies according to the variation in the beholder's standpoint or distance. Thus we explain the ever changing hues of sea or sky. But determinately bounded body has a fixed colour and the impression of colour (ἡ φαντασία τῆς χρόας) which it conveys is fixed, viewed from whatever standpoint; unless, indeed, something in the environment of the object, i. e. in the air or water through which it is seen, causes it to change its apparent colour. In both cases, in bodies with determinately bounded surfaces, and in the others, such as sea and sky, whose surfaces are not so bounded, the vehicle of colour is the same [2], viz. the *diaphanous*. Accordingly, we may define colour as *the surface limit of the diaphanous in determinately bounded body* [3]. This second definition of colour is quite consistent with that already quoted (p. 57), as *that which stimulates the actualized diaphanous between the object and the eye*. The latter, however, defines colour in relation to vision and to the medium of vision; the former defines it conceived as it exists in objects prior to vision. The diaphanous is for the one definition regarded as the *medium* whereby colour-stimulation is conveyed to the eye; for the other, it is the *vehicle* which in bodies at once constitutes and contains colour.

Colour a genus; its

§ 36. Colour is a *genus* of which the different colours are

[1] Prantl, Περὶ Χρωμ. p. 96, refers the words ἐν ἀορίστῳ τῷ διαφανεῖ (439ᵃ 26) to the *qualitative* indeterminateness of air or water. The reference is rather to the indeterminateness of their boundaries. The boundary of water is not fixed, but liable to constant fluctuation: that of air is still more indefinite. The relation of χροιά and ἐπιφάνεια is one of the cardinal facts in the colour-theory of Aristotle. Hence, though it is true that the διαφανές, to be a faithful medium for all colours, must itself have none (unless the ἀλλότριον χρῶμα called φῶς), this is not to the point here.

[2] τὸ αὐτὸ κἀκεῖ κἀνθάδε δεκτικὸν τῆς χρόας.

[3] χρῶμα ἂν εἴη τὸ τοῦ διαφανοῦς ἐν σώματι ὡρισμένῳ πέρας.

species[1]. It is a quality, and hence has no existence apart from a *substratum* of which it may be called an affection (πάθος). As a rule, Aristotle would apply the general term ποιότης to the permanent colour, while to the transitory (as redness in blushing) he would give the name πάθος or παθητικὴ ποιότης[2]. Yet he can speak of *all* sensible qualities, including colour, as τὰ παθήματα τὰ αἰσθητά in reference to their substrates[3]. There are *seven* distinct species of colour[4], viz. *white, black, golden-yellow* (ξανθόν), *crimson* (φοινικοῦν), *violet* (ἀλουργόν), *leek-green* (πράσινον), *deep-blue* (κυανοῦν). If *grey* (φαιόν) be regarded as a species of black and *golden-yellow* as a species of white, the species are reduced to *six*. If, on the other hand, grey and golden-yellow be counted separately, the species are increased to *eight*. The limitation of colour to a certain number of species (εἴδη) arises from a cause affecting all sensibles (αἰσθητά). Every αἰσθητόν is a genus with species lying between extremes which are contraries[5]. *Outside* these contrary extremes there are no colours. *Inside* them the species are limited by them as boundaries. Nor can we by dividing and subdividing the scale between these fixed extremes get an infinite number of colours. Their proper division is specific, since an αἰσθητόν is a discrete, not a continuous quantity, what continuity it has being merely that of its substrate. A line or other continuous μέγεθος is properly divisible into an infinite number of unequal parts : a genus, being discrete quantity, is divisible only into species which are finite in number. But if we try, by *improper* division (i. e. by the division of *the substrate* in which the αἰσθητόν inheres),

Margin note: species limited. This limitation due to (a) the fact that all αἰσθητά are discrete, not continuous, quantities; and (b) that each αἰσθητόν lies between ἐναντία which limit it. Those who represent species of colour as infinite (? Democritus and Plato) are wrong. Colour inheres in a substratum, which is permanent throughout the succession of alternating colours. Yet only the substratum, properly speaking, changes.

[1] 109ᵃ 36, 227ᵇ 6.

[2] Cf. 8ᵇ 25–10ᵃ 24 : ποιότης is fourfold (1) ἕξις or διάθεσις (the former being the more, the latter the less permanent state), (2) ὅσα κατὰ δύναμιν (καθ᾽ ὃ πυκτικοὺς ἢ ὑγιεινοὺς λέγομεν), (3) παθητικαὶ ποιότητες καὶ πάθη, (4) σχῆμά τε καὶ ἡ περὶ ἕκαστον μορφή.

[3] 445ᵇ 4 seqq.

[4] 442ᵃ 20. The view of Alexander is that we should read either ἕξ (so Susemihl) or ὀκτώ. Cf., however, Theophr. *de Causs.* Pl. VI, iv. 1.

[5] To the class of τὰ ἀντικείμενα belong (1) relatives (τὰ πρός τι), (2) contraries (τὰ ἐναντία), (3) στέρησις and ἕξις, (4) assertion and negation (κατάφασις and ἀπόφασις), *Cat.* 11ᵇ 17–19.

to get an infinity of such αἰσθητά, we fail, for the following reason. One does not by halving a white object get a half-white: each half is as white as the whole. If, however, we go on subdividing, we do reach a point where the colour is no longer perceptible *actually*; a point at which it is only potentially perceptible. This, however, does not alter the colour. For if the potentially perceptible magnitudes thus produced by subdivision be re-aggregated, they again form actual white. We have reached no new colour. Therefore by no process of subdivision of this kind can we increase the number of colours. It is not by the division of their substrates, but by the discrimination due to the eye, that the parts of colour are distinguished. Democritus and Plato (to whom Aristotle seems here to refer) were, therefore, wrong in teaching that the *kinds* of colour are infinitely variable. They are a limited number of species—limited by the bounding extremes between which they fall; their quality is not changed by their being reduced to mere potentiality by subdivision of their substratum [1]. There can be no species outside the limits of the black and white; and within these limits the species that the eye distinguishes are limited: nor can any one species be divided into subspecies by mere division of the substratum in which it inheres [2]. If one of the contraries, white or black, is actual in the *substrate*, the other cannot be present at the same time, but may be so at a different time; i.e. one of the two is *potentially* present when the other is *actually* so. The possibility of change (μεταβολή) in a substance from one contrary quality to another is axiomatic for Aristotle. This change in the case of colour

---

[1] As Prantl (Περὶ Χρωμ., p. 113) puts it: 'Die Mischung nun ist bei Aristoteles Ursache einer endlichen Zahl von Farben, und zwar einer endlichen darum, weil das zwischen den Gegensätzen Eingeschlossene nicht an sich ein continuirliches ist, und nicht bloss potenziell sondern auch actuell Gefühlsobject sein muss.'

[2] But κίνησις is infinitely divisible, and the process of μεταβολή from black to white or from white to black would seem infinite in *gradations* according to the amounts of ingredients used; which is what Plato and Democritus had in mind.

is ἀλλοίωσις[1]. The transition from mere potentiality of blackness (i. e. from white) to actuality of blackness is effected through successive degrees which run through the species of colour. The substrate wherein these degrees of colour and their extremes inhere is *one*[2]. Properly speaking this substrate is what is changed (ἀλλοιοῦται) in respect of its colour. In this the colours *alternate*, i. e. give place one to another. Thus the psychology of colour takes us into the domain of physics. As there can be no colour without body, so there can be no body without colour.

§ 37. Colour is not for Aristotle, as for Democritus, something purely subjective[3]. If it depends upon the eye, it depends also upon the object. Actual colour consists in the concurrent realization of the potentialities of these two. Aristotle finds no word corresponding to ὅρασις (actual seeing) which would express ' coloration ' or the ' actualization of colour.' The αἰσθητικόν, or potentiality of perceiving, realizes itself in αἴσθησις: the αἰσθητόν, or potentiality of being *perceived*, realizes itself in ποίησις αἰσθήσεως, for which as regards colour there is no one word[4]. The coloured thing, as object in nature, prior to its being seen, is *qua* visible, only a potentiality of coloration : in the act of vision it is the ἐνέργεια of this. But *as* potentiality it exists and has its place in nature apart from any visual act. Colour, as apprehended by the seeing eye, stands to the object while yet unseen as ἐντελέχεια (or ἐνέργεια) to δύναμις. The *perception* of colour is the realization of the faculty: the χρῶμα as perceived is the realization of the δυνάμει ὁρατόν. But χρῶμα in the object, even when not yet perceived, exists δυνάμει. What effects the transition from potentiality

Marginal note: Colour not for Aristotle, as for Democritus, something merely subjective.

---

[1] There are four kinds ot μεταβολή : (1) αὔξη, φθίσις (κατὰ τὸ ποσόν), (2) φορά (κατὰ τόπον), (3) ἀλλοίωσις (κατὰ τὸ ποιόν), (4) γένεσις, φθορά : *vide* 319[b] 31 seqq.

[2] 217[a] 22–5 ὕλη μία τῶν ἐναντίων . . . καὶ οὐ χωριστὴ μὲν ἡ ὕλη.

[3] 426[a] 17 οἱ πρότεροι φυσιολόγοι τοῦτο οὐ καλῶς ἔλεγον οὐθὲν οἰόμενοι οὔτε λευκὸν οὔτε μέλαν εἶναι ἄνευ τῆς ὄψεως . . . τῇ μὲν γὰρ ἔλεγον ὀρθῶς, τῇ δ' οὐκ ὀρθῶς.

[4] That is, Aristotle misses a word corresponding to ὅρασις as ψόφησις corresponds to ἄκουσις: cf. *de An.* iii. 425[b] 31 seqq.

to actuality (both between ὁρατόν and χρῶμα, as seen, and between τὸ ὁρατικόν and ὅρασις) is the κίνησις through the diaphanous medium starting from the ὁρατόν and affecting τὸ ὁρατικόν, or ἡ ὄψις. It is light that at once transforms the potential colour to actuality, and the potentially seeing to an actually seeing eye [1].

Phosphorescent things: only seen in darkness. Reason of this. Explanation of the intra-ocular light. Fire as object of vision.

§ 38. Certain objects of vision [2] different from colour, and not seen in the light, have been already (§ 33 *supra*) mentioned [3]. These are perceived only in darkness; they are not grouped under one class-name, but consist of such things as the sepia of the cuttle-fish, fungus, pieces of horn, heads, scales, and eyes of fishes, and so on. In none of these, when seen in the dark, is a colour, properly so-called, visible. All these things possess in common the quality of smoothness (λειότης) and have the natural property, therefore, of *shining* in the dark, yet without *giving light*. Among such phenomena Aristotle (knowing nothing of the properties of the optic nerve or retina) includes the flash seen within the eye when moved rapidly, or struck, when it is closed or in darkness. This flash is, he says, due to the 'smoothness' of the pupil and its consequent power of shining in the dark. A quick movement, he thinks, makes the eye to duplicate itself, so to speak, and thus to become both observed and observer, when the latter, the percipient, sees the shining of the former, the object perceived [4]. Fire, also, is an object of vision and visible even in darkness [5]. The fiery element which ordinarily stimulates the potential *diaphanous* to actuality (i.e. produces daylight), described shortly by Aristotle as of the same nature with the celestial bodies, is not identical with our ordinary fire [6]. It is probably (see p. 58, n. 1) identical with the

---

[1] 430ᵃ 17 τρόπον γάρ τινα καὶ τὸ φῶς ποιεῖ τὰ δυνάμει ὄντα χρώματα ἐνεργείᾳ χρώματα: where νοῦς is, in the manner of Plato (*Rep.* 507 E seqq.), illustrated by φῶς.

[2] Known to *us* as *phosphorescent*. They are 'fiery' in their nature: ἐν τῷ σκότει ποιεῖ αἴσθησιν, οἷον τὰ πυρώδη φαινόμενα καὶ λάμποντα.

[3] 419ᵃ 2, 437ᵇ 6.   [4] 437ᵃ 31.

[5] 419ᵃ 23–5.

[6] τὸ ἄνω σῶμα ἕτερον πυρός τε καὶ ἀέρος 340ᵇ 6.

αἰθήρ, the (afterwards so-called) πέμπτον στοιχεῖον, or πέμπτη οὐσία. This fiery element, in its effect upon the diaphanous medium, is the originative cause of *colour*.

§ 39. As regards the four ordinary elements:

(*a*) Fire—the hot and dry—is distinctively (i. e. in its *finest* form) *white*[1].

The colours of the four elements.

(*b*) Air—the hot and moist—is also white, a quality which it probably owes to its affinity with fire[2].

(*c*) Water—the moist and cold—is *black*, since it is without the fiery element which actualizes the potential diaphanous. From its smoothness, however, it has the power of 'shining,' and also of *reflecting* and *refracting* light-rays (both of which processes come for Aristotle under the head of ἀνάκλασις).

(*d*) Earth—the cold and dry—has neither the λειότης of water, nor the heat of fire and air. It is, therefore, the utter negative of white colour[3]. Throughout these elements in their relations to colour the opposition of ἕξις and στέρησις prevails, as it does in the colour scale itself. In the latter the positive, or ἕξις, is the white; the στέρησις, the black. In the elements relatively to colour the ἕξις is τὸ πῦρ, or, strictly, τὸ οἷον τὸ ἄνω σῶμα; the privation, or στέρησις, is γῆ. In thus holding that black is the colour of water and white of fire Aristotle is quite orthodox: the same view was held by Anaxagoras and Empedocles.

§ 40. Reflexion (ἀνάκλασις) is an important mode of the production of colours, requiring separate treatment. The presupposition of reflexion is the straightness of the light-ray. Aristotle predicates straightness of the ray proceeding to or

Reflexion of light : visual ray proceeds in a straight line : so all

---

[1] We must, however, for Aristotle (134[b] 28) as also for Plato distinguish under 'fire' three things: ἄνθραξ (*glow*) καὶ φλὸξ (*flame*) καὶ φῶς (*light*). This last is τὸ λεπτομερέστατον τοῦ πυρός. Ἀὴρ διαφαινόμενος λευκότητα ποιεῖ, 786[a] 6. But μάλιστα ... πῦρ ἢ φλόξ, αὕτη δ' ἐστὶ καπνὸς καιόμενος, 331[b] 25. The colour called πυρώδης is opposed to white : λευκὸς ἀλλ' οὐ πυρώδης, 'white, not fire-coloured,' is said of ἥλιος, 341[a] 36.

[2] ὁ ἀὴρ πρὸς τἆλλα πῦρ, 466[a] 24.

[3] In the un-Aristotelean tract Περὶ Χρωμάτων fire is spoken of as light yellow, while all the other elements are named white.

other rays rectilinear, unless reflected. Why the seawater shines at night when struck by an oar. The rainbow explained as phenomenon of reflexion of light.

from the eye [1], and *assumes* it of all other rays [2]. All phenomena of illumination, by fire or light, are explained by the reflexion of light—a matter of which the ancients were very ignorant [3]. Reflexion is always and everywhere taking place. If it were not so we should not, as at present, have universal illumination : we should have only a bright spot where the sun's rays fell unimpeded, while, in the rest of the space before us, there would be total darkness [4]. The *smooth* is the cause of reflexion (as it is also an essential cause or condition of whiteness), which therefore regularly occurs in *water* and in *air* (if the latter has any consistency) [5]. If the water of the sea be struck, e. g. with an oar, at night, it appears to shine and sparkle. We cannot see this in the daytime, when the stronger light of the sun effaces it. This is a phenomenon of reflexion. The visual ray is reflected from the water upon some (smooth, and hence) bright surface [6] which returns it to the eye.

In such a *smooth* element a continuous mirror can be formed whose elementary parts (particles of air, or water drops) are so small that only *colour*, or the *gleam of light*, but not the *form of things*, can be reflected in them. Thus the visual ray is reflected from the cloud to the sun. So the rainbow is seen [7]. That in all this Aristotle by ὄψις

---

[1] He was compelled, in spite of his own theory of vision, to employ the term ὄψις (which he found in vogue for visual-ray) in such a manner as to seem to commit himself to the view that the eye sees by rays issuing from a native fire within it. For his optical mathematics, 373ᵃ 5–18, this does not matter: he corrects what he thinks wrong in it, when he deals with the subject of vision and with ὄψις in its psychological sense.

[2] Prantl, p. 118, 656ᵇ 29 ἡ δ' ὄψις εἰς τὸ ἔμπροσθεν· ὁρᾷ γὰρ κατ' εὐθυωρίαν.

[3] 370ᵃ 16, 438ᵃ 9.

[4] 419ᵇ 29 τὸ φῶς ἀεὶ ἀνακλᾶται, οὐδὲ γὰρ ἂν ἐγίγνετο πάντῃ φῶς, ἀλλὰ σκότος ἔξω τοῦ ἡλιωμένου.

[5] 372ᵃ 29 ἡ ὄψις ἀνακλᾶται ὥσπερ καὶ ἀφ' ὕδατος οὕτω καὶ ἀπὸ ἀέρος καὶ πάντων τῶν ἐχόντων τὴν ἐπιφάνειαν λείαν: 372ᵇ 15 γίνεται ἡ ἀνάκλασις τῆς ὄψεως συνισταμένου τοῦ ἀέρος.

[6] 370ᵃ 17 φαίνεται γὰρ τὸ ὕδωρ στίλβειν τυπτόμενον ἀνακλωμένης ἀπ' αὐτοῦ τῆς ὄψεως πρός τι τῶν λαμπρῶν.

[7] 373ᵃ 18 seqq. τὸ νέφος ἀφ' οὗ ἀνακλᾶται ἡ ὄψις πρὸς τὸν ἥλιον· δεῖ δὲ

means the ray of light *per se*, not as something belonging
either to the object or to the eye exclusively, appears
when he tells us that it makes no difference whether
it is the object seen, or the visual agency that changes [1].
Every case of reflexion is conceived as a *weakening*, and
to that extent a *negation*, of the action of the light-ray;
and hence it is reflexion that produces the black, which
then, mingled with the light, produces colours [2].

To this weakening of the ray is ascribed the curious
phenomenon of the *Doppelgänger* [3], as when a person sees
his own image reflected from the air in his vicinity. By this,
too, is explained the halo that forms around lamp-burners
alight, the darkened appearance of clouds when seen
reflected in pools of water, &c. The mixture of the light
with the darkness of the mirroring surface, as well as the
weakening of the ray by or in reflexion, is a cause of
the various gradations of colour. Colour effects in the
atmosphere, and especially halos and rainbows, are explained
by Aristotle in accordance with these observations [4]. In
the three grades of weakening of the rays of light (or of
their mixture with the darker element of the mirror) consist the three colours of the rainbow, crimson (φοινικοῦν),
leek-green (πράσινον), and violet (ἀλουργόν). The iris that
forms round lamps is to be explained on similar principles;
also the rainbow colours seen in a cloud of spray thrown
up, e. g. by an oar [5]. It would not be relevant here to
follow Aristotle into all the bearings in which he discusses
this subject; but he pursues it in its connexion with various
kinds of matter organic and inorganic: the various classes

*The phe-
nomenon
of the
Doppel-
gänger, a
case of
reflexion.
Reflexion a
source of
colour, as
distinct
from
brightness;
halos, &c.,
rainbow
colours—
red, green,
violet.*

νοεῖν συνεχῆ τὰ ἔνοπτρα, ἀλλὰ διὰ μικρότητα κτλ.: 372ᵃ 33 seqq. τῶν
ἐνόπτρων ἐν ἐνίοις μὲν καὶ τὰ σχήματα ἐμφαίνεται, ἐν ἐνίοις δὲ τὰ χρώματα
μόνον: 373ᵇ 15 seqq.

[1] 374ᵇ 22 διαφέρει δ' οὐθὲν τὸ ὁρώμενον μεταβάλλειν ἢ τὴν ὄψιν,
ἀμφοτέρως γὰρ ἔσται ταὐτόν: and 377ᵇ 11 διαφέρει γὰρ οὐθὲν διὰ τοιούτων
ὁρᾶν ἢ ἀπὸ τοιούτων ἀνακλωμένην.

[2] 373ᵇ 1 γίνεται δὲ (ἡ ἀνάκλασις) ἀπὸ μὲν ἀέρος ὅταν τύχη συνισταμένος·
διὰ δὲ τὴν τῆς ὄψεως ἀσθένειαν πολλάκις καὶ ἄνευ συστάσεως ποιεῖ ἀνά-
κλασιν.   [3] 373ᵇ 4 seqq.

[4] 342ᵃ 34 seqq., 377ᵃ 34 seqq.   [5] 374ᵃ 29 seqq.

of plants and animals, their colours at succeeding stages of existence or development: the colour of hair, feathers, saps of plants, &c.[1]

Particular colours: white is the actualization of the diaphanous in the surface of a determinate body: black is the στέρησις of this. White and black in such body are what light and darkness are in the diaphanous generally. § 41. Such is Aristotle's account of colour *in general*, and of the diaphanous as its vehicle in determinate bodies. He also gives an account of *particular* colours, and sets forth and compares the possible, or conceivable, modes of their generation in nature. It has been already stated[2] that the presence of a certain fire-like element, identical in principle with the celestial body, is the cause of light in the diaphanous, e. g. in the atmosphere, by day. The total or partial absence of this is darkness, as in the same diaphanous by night. Now in determinate bodies, in all of which the diaphanous inheres or resides in varying degrees[3], and whose colour (as already explained) is the limit of this diaphanous coinciding with their geometrical surface, we may assume something corresponding to the presence and absence of the fiery element, with consequent variations in the aspect of the bodies. Its total absence means darkness in the atmosphere, *blackness* in a determinate body. In the atmosphere its full presence is daylight, in a determinate body, it means *whiteness*. Thus in determinate bodies blackness is privation of whiteness. Again, what its geometrical superficies is to the solid body, its colour is to the whole diaphanous element inherent in and conterminous with such body[4]. The degree in which this diaphanous is actualized in a determinate body constitutes in this body such colour as it possesses[5].

---

[1] In what precedes Prantl's exhaustive account of Aristotle's *Farbenlehre* has been used. Those who wish to see set forth in detail *all* that Aristotle has said on the subject of colour may read Prantl's *Prolegomena* to the Περὶ Χρωμάτων.

[2] For what follows cf. Arist. 439[b] 18 seqq.

[3] ὑπάρχει δὲ μᾶλλον καὶ ἧττον ἐν πᾶσι.

[4] So Alex. Aphr. Ἀπορ. κ. Λυσ. i. 2, p. 5 (Bruns).

[5] Aristotle (like Plato) speaks of white as χρῶμα διακριτικὸν τῆς ὄψεως, black as χρ. συγκριτικὸν τῆς ὄψεως: *Met.* 1057[b] 8 ... οἷον εἰ τὸ λευκὸν καὶ μέλαν ἐναντία, ἔστι δὲ τὸ μὲν διακριτικὸν χρῶμα, τὸ δὲ συγκριτικὸν χρῶμα. Cf. also *Top.* 119[a] 30.

§ 42. Thus black and white are contraries within the one genus or sensory province of colour. All sensory modalities involve contraries in this way[1]. From these two contraries the other colours are to be explained[2]. The transition from white to black is possible through continuous degrees of privation: that from white to black is likewise possible by an ascending scale in the positive direction. The various colours are species which fall between the two contraries, and are generated of certain combinations of these[3]. It is an axiom with Aristotle that nothing acts on or is acted upon by any *casual* thing, nor is anything generated by any other thing *casually* (τὸ τυχὸν ὑπὸ τοῦ τυχόντος). White is generated from what is not white, yet not from *every* not-white, but only from either black, or the intermediate colours. Everything that is generated, and everything that is destroyed, passes *from its contrary* or *to its contrary*, or *to the intervening states*. These intervening states again are generated from the contraries, as colours from the white and the black. In the province of colour, if we are to pass from white to black, we must come first to crimson (φοινικοῦν) and grey (φαιόν). The successive stages, too, in either direction mark grades of contrariety. The intervening parts of the scale serve for relative extremes, hence change can start from any intermediate stage. An intermediate can serve as

Marginal notes: Black and white, the ἐναντία in the genus colour. Continuous transition between these extremes. The various colours are species generated by combinations of black and white. Three different conceptions of the origin of intermediate colours.

---

[1] Cf. Bonitz, *Met. Arist.*, pp. 430-4 ; Arist. *de Sens.* 442ᵇ 17.

[2] Cf. *De Sens.* iii. (Aristotle's official *Farbenlehre*), also *Phys.* i. 5. 188ᵃ 3-188ᵇ 21 ; *Met.* 1057ᵃ 23 ; Prantl, *Arist.* Περὶ Χρωμ., p. 109 seqq.

[3] The placing of black and white in the colour scale, and assuming that the colours of the spectrum lie between these as extremes, with the implicit confusion between *luminosity* and *colour*, strikes one immediately on reading this. We need not criticize it here, however, but we may observe that Goethe held fast to Aristotle's view. A further criticism (or aspect of the same criticism) is that Aristotle sometimes (not always: cf. 374ᵇ 13 τὸ μέλαν οἷον ἀπόφασίς ἐστιν) treats black like white as a *positive*. It is not, however, necessary for him to assume this. His theories of mixture can be understood well enough on the assumption of the negativity of the black: the addition of a black ingredient need be regarded as no more than the subtraction of a certain amount of whiteness. The term 'mixture,' indeed, is awkward, but that is all. See p. 74 *infra*, n. 5.

a contrary to either extreme. Thus grey is white as compared with black, black as compared with white [1].

The origin of the intermediate colours may be sought for along three different lines.

(a) *Juxtaposition of whites and blacks atomically small.* (a) The *Atomic* theory of colour, or the theory of atomic *juxtaposition* (ἡ παρ' ἄλληλα θέσις). It is conceivable, e. g. that two particles, one of white and one of black, so small as to be separately invisible, should when placed side by side become visible in combination, as a composite whole ; and that it is by juxtaposition (on the same plane relatively to the eye) that the existing varieties of intermediate colours are really produced in nature. For if a white and a black are so juxtaposed, and are visible, *some* colour must result ; and as this colour cannot be either white or black, it must form some third species of colour. The colours thus produced may vary in ways as numerous as the possible proportions of whites and blacks in such combinations. For instance, three particles of white might be juxtaposed with two, or four, of black ; and so on. Or the combinations might be formed not in numerically expressible ratios of this sort, but according to some scale of excess or defect by which the component amounts would stand in no calculable ratio to one another, i. e. in none which could be represented in integral numbers, but could only be expressed by a surd. In fact, it is conceivable that the composition of colours may be to some extent analogous to that of tones in chords [2]. The particular colours formed of components brought together in ratios capable of expression by integral numbers, like tones

---

[1] 224ᵇ 30 ἐκ δὲ τοῦ μεταξὺ μεταβάλλει· χρῆται γὰρ αὐτῷ ὡς ἐναντίῳ ὄντι πρὸς ἑκάτερον, and 229ᵇ 14 ὡς ἐναντίῳ γὰρ χρῆται τῷ μεταξὺ ἡ κίνησις ... τὸ γὰρ μέσον πρὸς ἑκάτερον λέγεταί πως τῶν ἄκρων. The middle grades *properly* have, owing to their relativity, no contraries : cf. 10ᵇ 16 τῷ γὰρ πυρρῷ ἢ ὠχρῷ ἢ ταῖς τοιαύταις χροιαῖς οὐδὲν ἐναντίον ποιοῖς οὖσι. One may ask : if κίνησις be infinitely divisible (see 240ᵇ 8 seqq.), and the process from one contrary in colour to the other be as above described, a κίνησις, why there is not an infinite number of colours. For Aristotle's answer, cf. 445ᵇ 3-446ᵃ 20. But he only denied an infinity of colour *species*.

[2] For 440ᵃ 3 cf. von Jan, *Mus. Scr. Gr.*, pp. 47 n. and 132.

similarly combined in chords, may be those colours which
are generally felt as pleasing to the eye, such as purple
and crimson ; and if such are comparatively few amid the
whole multitude of existing colours, this may be so for just
the same reason for which harmonious sounds also are few
among the possible combinations of sounds. Non-pleasing
colours may be those not founded on numerical ratios.
Or, if one supposes that all composition of colours has
a numerical basis, only that while some colours are arranged
in a certain order, others are in no certain order, it is conceivable that the compounds themselves, whenever they are
not 'pure' (μὴ καθαραί), owe this to the fact that the numbers
on which they rest are not 'pure'[1]. This, then, is one conceivable mode of the production of the intermediate colours.

[1] 440ᵃ 3–5 ἢ καὶ πάσας τὰς χρόας ἐν ἀριθμοῖς ... διὰ τὸ μὴ ἐν ἀριθμοῖς
εἶναι τοιαύτας γίγνεσθαι. If τοιαύτας here goes with γίγνεσθαι, to avoid
contradiction, ἐν ἀριθμοῖς at the close of the sentence must mean something different from what it means in the first part. Biehl suggests
inserting τοῖς αὐτοῖς before it in its second occurrence ; C. Bitterauf,
Dissertatio Inauguralis (Monachii 1900), p. 21, thought of reading
εὐλογίστοις after it. This of course is the direction in which one
would look for the general sense. The second hypothesis is one such
as a Pythagorean, who held that all things *are*, or *are modelled on*,
numbers, would adopt. Even for him, however, there should, according
to Aristotle, be a distinction between numbers which are expressible
in integral units and those not expressible otherwise than as surds.
Arithmetic was based on geometry ; the original unit was a line of
a certain length, e. g. a foot long : or else a power of this, e. g. a square
foot, or a cubic foot. The idea of an abstract unit, the foundation
of the science of monadic number, or arithmetic proper, came later.
Both views of number presented themselves to the popular mind, even
as late as Aristotle. Thus all composition of blacks and whites might
be based on ἀριθμοί, but in two ways. The ἀριθμοί might be such as
are expressed in monadic units ; as if we were to have e.g. three times
as many blacks as whites in the mixture ; or the ἀριθμοί might be
incapable of representation monadically, as if e. g. blacks were to be
represented by the square root of 2 and whites by the square root of 3.
In this latter case, √2 and √3 being unattainable, we could not reach
the monadic ratio of the blacks to the whites. Such may be the
difference between ἐν ἀριθμοῖς in the two places here. We may, to
make the text more lucid, adopt either of the above suggested readings,
or before τοιαύτας insert τοιούτοις, taking it, in reference to ἀριθμοί, to
mean numbers and ratios expressible in monadic units, and assuming

(b) We have called the first mode that of juxtaposition of the separately invisible blacks and whites; the second mode may be called that of the *superposition* (ἡ ἐπιπόλασις) of black and white. Painters sometimes lay one surface of colour over another for the purpose of producing a particular colour effect. For instance, when they wish to represent an object as submerged in water, or as seen through a hazy atmosphere, they paint a duller colour over the brighter, in order to obtain the required effect. Thus too, in nature, the sun, which *per se* is white, shows crimson when shining through a misty or smoky atmosphere. By such superposition, then, nature's colours may have been produced. If this be so, their varieties can be explained in the same way as in the case of atomic juxtaposition, according, that is, to the various ratios, or irrationality, of the proportions in which the surface colours are combined with those beneath. This second

it to have been lost before τοιαύτας as it might easily have been. See Plato, *Theaetet.* 147 D–E (L. Campbell); also Arist. *Met.* xii. 6. 1080ᵇ 16–20 (Bonitz). But what does τεταγμένας ... ἀτάκτους mean ? Alexander (p. 54, Wendland) says that the ἄτακτοι χρόαι arise (according to the reasoning here) not by *incommensurableness* in the excess of blacks above whites or vice versa (οὐκ ἐν τῇ τῆς ὑπεροχῆς ἀσυμμετρίᾳ), but by *disorder* (cf. *Probl.* xix. 38; von Jan, *op. cit.*, p. 47 n.) in the way in which they are juxtaposed (ἐν τῇ τῆς παραθέσεως ἀταξίᾳ). We may juxtapose 10 blacks beside 5 whites in many ways; and though the ratio of 10 : 5 held good for all, yet the colours would be different according to the mode of παράθεσις. 'By μὴ καθαραί Aristotle (says Alexander) must mean juxtapositions of [i.e. colours based on juxtapositions of] unlike parts. The juxtaposition would be καθαρά, if e. g. beside every two whites one black were to come throughout; it would be μὴ καθαρά if we had one black sometimes with two, sometimes with three, whites, and sometimes with one white.' This imports a different idea, by which from a partly Pythagorean we pass to a merely atomistic explanation of the 'impurity' of colours. For Democritus, sensible qualities all rest on διαθιγή, ῥυσμός, τροπή, i. e. τάξις, σχῆμα, θέσις. The *ratio* of the total numbers of blacks to whites may remain, but the order in which the units are brought into juxtaposition may nevertheless vary, with consequent variation in the aesthetic character—the 'purity'—of the χρόα. Thus, even when the χρόαι were ἐν ἀριθμοῖς εὐλογίστοις they might still be 'impure,' if they were ἄτακτοι. This sense can be obtained without changing the text, if we are content to take τοιαύτας (=τεταγμένας) with εἶναι, and render γίγνεσθαι as simply = 'are produced.'

theory is preferable to the first, says Aristotle, for it does
not require us to assume the invisible magnitudes and
imperceptible intervals of time which the first requires,
in order that the successive and diverse stimulations coming
to the eye from the blacks and whites severally should
reach us without our recognizing their diversity or
succession, and should, from their presenting themselves,
or seeming to present themselves, simultaneously, create
in our minds the impression of their being one single colour
only. In the second case we have not to do with invisibly
small units: we have a surface of actually visible colour,
with another below showing through it ; and the κινήσεις
of both are from the first combined in their effect on
the medium. The surface colour would not, of course,
affect the medium, and so stimulate the sense of sight,
in the same way when acting *per se* as it would when
modified by the other colour underlying it [1]. Hence, with
a white surface, for example, showing through a black,
the colour seen will be different from either white or black.

§ 43. (c) Neither of these two theories is, however, in
Aristotle's opinion satisfactory. Both assume a mere
combination of the κινήσεις of blacks and whites, not
a κρᾶσις of the ὑποκείμενον, or matter, of which the black
and white are qualities. He states a third which he
himself adopts. This is the theory of the *complete blending*[2]
of the coloured bodies with consequent blending of their
qualities. For bodies are not mixed in nature as some[3]
think, by a juxtaposition of their least parts, whose
infinitesimal size renders them separately imperceptible to
an observer; but in such a way that they undergo, both in
matter and form, a process of complete and absolute mutual
interpenetration. When the things said to be mixed are still
preserved in small quantities having their former qualities,

(c) Aristotle's own theory: the matter of which black and white are qualities is blended, and so its qualities are blended.

---

[1] 440ᵃ 24 τὸ ἐπιπολῆς χρῶμα ἀκίνητον ὂν καὶ κινούμενον ὑπὸ τοῦ ὑποκειμένου οὐχ ὁμοίαν ποιήσει τὴν κίνησιν.

[2] 440ᵇ 3 ἡ πάντῃ πάντως μεῖξις. Cf. ᵇ 11 τῷ πάντῃ μεμεῖχθαι.

[3] The difficulty of referring this, as Alexander (p. 56, Wendland) does, to the atomists, is that according to them the atoms have *no* colour.

we ought not to call such a process mixture. It may be a composition (σύνθεσις), but neither a mixing (μεῖξις) nor a blending (κρᾶσις). When things are mixed, then all the parts in the new whole are homogeneous[1]. In a true mixture, as of colours, the contraries tend to efface one another's identity[2]. If the former (i. e. σύνθεσις) were nature's mode of mixing, it is always conceivable that an eye of Lyncean keenness[3], if properly placed, would still detect the elements in the mixture, whose constituents would be really blended in no other way than horses and men are blended when a crowd of both come together: for this crowd might, to a person at a distance, seem but one mass, if too far off for the individuals composing it to be discerned[4]. But such mixture is not absolute. The horses and men are, indeed, juxtaposed, but no individual is mixed with any other individual: each horse and each man retains its or his separate entity. The mode of mixture which in nature gives rise to the variety of colours is not this, but one in which no individual part of the compound retains its former qualities unmodified. When things are *materially* mixed in this way, their colours too are blended. Only such blending—not mere juxtaposition or superposition—can produce colours which cannot be even conceived as varying in appearance according as the observer is far or near, but will retain a constant character at all distances alike. In this case, moreover, as in the two former, we may suppose the elements in the compounds of black and white to be combined in any of the various ways there described; that is to say, some in numerically definable ratios, others in degrees which are not expressible in integral numbers[5].

[1] 328ᵃ 5 seqq. φαμὲν δ', εἴπερ δεῖ μεμεῖχθαί τι, τὸ μειχθὲν ὁμοιομερὲς εἶναι.

[2] 447ᵃ 20 ἀφανίζειν ἄλληλα.

[3] Aristotle's hypothetical equivalent for our microscope.

[4] Cf. Lucretius, ii. 312–32.

[5] The tract Περὶ Χρωμάτων, ch. 3, gives a different account of the origin of the various colours. Mixture of primary colours is indeed a leading mode of their production, and their variety is made to depend on the varied proportions in which the ingredients are combined. But the primary colours are in this tract not the *white* and *black* only: to

§ 44. The colour called *grey* (φαιόν) is sometimes spoken of by Aristotle as if it stood mid-scale between black and white: but[1] it is also referred to as relatively a kind of black. *Golden-yellow* also is represented as falling under white[2], to which it is allied as the succulent (τὸ λιπαρόν) is to the sweet (τὸ γλυκύ) in the sphere of taste. *Red* is the colour produced by light streaming through black, as when the sun shines through smoke or through a fog[3]. *Purple* (πορφυροῦν) is distinguished from *crimson* (φοινικοῦν) by its having more of the dark ingredient. Sometimes the light of a lamp shows not white but purple, the ray that is sent from it being feeble, and being reflected from a dark colour. This increasing weakness of the ray brings

Remarks on the particular colours: grey, golden-yellow, red, purple, green, violet. Different account of colour production given in Περὶ Χρω-μάτων. Here colours are generated

them is added *golden-yellow* (ξανθόν). The white and the golden-yellow are colours of the elementary kinds of matter. Fire is golden-yellow: air and (contrary to Aristotle's view) earth and water are white; black is partly bare negation, and partly a positive colour produced in the process by which (e. g. by burning) the elements are transformed into one another. An account is given of the methods of mixture, whether of these primary colours or of those which are derived from them, to explain the multitude of existing colours. These are said to be the effects of: (1) the quantitative preponderance of light or shade in the ingredients, (2) the strength of the ingredients, (3) the proportionality of the ingredients, (4) the brilliancy of the mixed colours, (5) the friction and mechanical force employed, (6) burning, dissolving, melting processes, (7) smoothness and shadows (?: the text is doubtful), (8) combination with external light or reflexion of other colours, and especially in connexion with the influence of the medium in which it takes place. The colours of plants, hair, feathers, &c., are discussed. The two modes of producing colour rejected in *de Sens*. iii. ἡ παρὰ ἄλληλα θέσις and ἡ ἐπιπόλασις, are accepted here and made to play an important part. Light is seemingly conceived as corporeal, in direct contravention of Aristotle's teaching in the *de Anima*. The tract assumes a mixture of the colours with the rays of light: so the distinctive colours of feathers are produced. Colours are said to change their appearance according as they are 'mixed with the sun's radiance or only with shadows.' Prantl finds an incongruity between the two views of black colour, in one (791[b] 3) of which it is regarded as (σκότος) mere στέρησις of light, while in the other (791[b] 17) it is (μέλαν χρῶμα) a positive colour, produced, for example, by burning. Zeller, however, thinks the inconsistency only apparent. *Vide* Zeller, *Arist*. ii. 490, E. Tr.; Prantl, Περὶ Χρωμ., pp. 167 seqq. and pp. 107-9.

[1] 442ᵃ 22. [2] Ibid.
[3] 342ᵇ 4 seqq., 374ᵃ 3, ᵇ 10, 440ᵃ 10.

from
primary
black and
white:
there, from
the colours
of the
elements.
The phe-
nomena of
positive
after-
images;
complemen-
tary
colours;
contrast.
Effects of
this latter
illustrated.
us from purple to *leek-green* and *violet*, successively. The
stronger ray yields *crimson* against the dark ground
(or when mixed with dark); the next in strength gives
leek-green; the weakest, *violet*. In the tract Περὶ Χρωμάτων,
ὄρφνιον is mentioned as containing even a greater proportion
of black than violet has. From the seven colours described
above all the others (according to the doctrine of Aristotle)
are generated by mixing [1]. In the Περὶ Χρωμάτων, however,
though these colours play their part, they are secondary to
the colours of the elements [2]. Visual impressions, primary
positive after-images, continue in the eye after it has ceased
from looking at the object. If we gaze long and steadily
at a bright object, that to which we transfer our gaze at
first appears of the colour of the former object. If when
we have looked steadily at the sun, or some other bright
object, we close the eyes and look as it were straightforward
(with the eyes closed) in the same line of vision, at first we
see the object of the same colour as before: this alters
soon to crimson; the latter changes to purple; till at last the
colour becomes black, and vanishes [3]. In this place Aristotle notices what are called *complementary* colour effects,
though his account of them is not exact. The golden-
yellow of the rainbow is explained by him as a subjective
effect of *contrast* [4]. The space between the φοινικοῦν and
the πράσινον in the rainbow often shows ξανθόν. This is
due to their being next to one another. For φοινικοῦν
beside πράσινον appears white. As a proof of this we may
observe that the rainbow which appears in the blackest
cloud has the purest colour tints (μάλιστα ἄκρατος), and there
too it happens that the φοινικοῦν shows most clearly the tint
of the ξανθόν—the colour between the φοινικοῦν and the
πράσινον. The φοινικοῦν in such a cloud appears white as
contrasted with the surrounding black; and also when (as
the rainbow is fading) the φοινικοῦν is being dissolved it shows
white. A further confirmation of this effect of contrast is

[1] 442ᵃ 25 τὰ δ' ἄλλα μεικτὰ ἐκ τούτων.
[2] Cf. 792ᵃ 4 seqq.   [3] 459ᵇ 5 seqq.
[4] 375ᵃ 7 seqq. Not, as Prantl (Περὶ Χρωμ., p. 156) says, as a *com-*
plementary colour.

that the iris around the moon appears very white; which is
owing to the twofold fact that the colours are *in a cloud*
(which is dark) and seen besides *at night*[1]. Further effects of
contrast are seen by placing white wool side by side with
black : and also in the way in which (as embroiderers say)
lamplight causes illusions as to colour, owing to the peculiar
nature of the illumination shed by it upon the objects[2].

§ 45. Aristotle decisively rejects[3] the definition of *Aristotle rejects the emanation theory of colour:*
colour given by Empedocles[4] and followed by Gorgias, as
apparently by Plato also in the *Menon* (and, with modifica-
tions, in the *Timaeus*), viz. that colour is an 'emanation from *curious resemblance between this emanation theory and the Newtonian emission theory of light. He cannot have held an undulation theory, for he asserts, against Empedocles, that light does not travel.*
the object of vision symmetrical with, and therefore
perceptible by, the organ of vision.' Since those philosophers, who hold this theory of visual perception by ἀπορροαί,
in any case reduce the perception of colour to a mode
of *contact* between the organ and the object (of which
a particle thus comes to, and touches, the eye), it would
have been better if they had assumed such contact to
take place through a medium, rather than by ἀπορροαί
travelling from object to organ. For all the sensory
functions indirectly are, or involve, a mode of contact[5],
but all except the organ of touch itself[6] operate through
a medium[7]. In rejecting this view of colour, and the
theory of ἀπορροαί on which it was based[8], Aristotle
rejected as if by anticipation the Newtonian emission
theory of light. There seems at first sight to have been
before his mind a glimmering of the now accepted
undulation theory; but this impression cannot be sustained
when we find him, against Empedocles, vigorously denying
that light *travels*[9] (cf. p. 59, n. 1 *supra*).

[1] 375ᵃ 19.   [2] 375ᵃ 22 seqq.; Prantl, Περὶ Χρωμ., 157–8.

[3] 440ᵃ 15–20.   [4] Cf. Karsten, *Emped.*, p. 488.

[5] 435ᵃ 18 καίτοι τὰ ἄλλα αἰσθητήρια ἁφῇ αἰσθάνεται, ἀλλὰ δι' ἑτέρου.

[6] For the questionableness even of this exception cf. *de An.* ii. 11.
422ᵇ 22 seqq.

[7] For the emanation theory of colours cf. further Lucretius, iv. 72–86
with Giussani's notes.

[8] So Bäumker, *Des Aristoteles Lehre von den äussern und innern
Sinnesvermögen*, p. 40.

[9] In 418ᵇ 16 he maintains that light is a παρουσία, or that,

§ 46. The diaphanous (described §§ 34–5 *supra*) is the objective medium of vision. As in the cases of smelling and hearing, so in that of seeing, there is an extraorganic medium, intervening between the organ and the object [1]. Without such medium the object could not produce its characteristic effect upon the organ, or the latter be excited from its potentiality to its realization as an organ. Thus if the coloured object be placed directly and immediately on the surface of the eye it cannot be seen [2]. In order, therefore, to be affected at all by the colour, the eye requires a medium. This medium is *light*, or the *actualized diaphanous*. The object must excite a movement (not, however, a *local movement*) in the diaphanous medium, whether air or water (for either of these may be media of vision), and this movement must communicate itself somehow to the eye. This medium being absolutely required if we are to see at all, it was a mistake for Democritus to think that if there were a vacuum (neither *air* nor *water*) between the eye and its object one would see with the maximum of accuracy: 'that we could see even an ant in the sky [3].' The contrary is the fact: without the medium one could see nothing [4]. Air and water are both *media* of colour. Through them we see because—in virtue of the diaphanousness common to both—

Side note: Necessity of a *medium* of vision: this is the *actualized diaphanous*. Democritus wrong in thinking that we could see best in a *vacuum*. Air and water, as varieties of the diaphanous, both mediate colour vision. Need of internal medium—diaphanous within the eye itself. Hence eye 'consists of water.' The medium of all colours is itself colourless.

though it were a κίνησις, it is still not the particular form of κίνησις called φορά, which involves local movement, but an ἀλλοίωσις or qualitative change, which he thinks can take place simultaneously in all parts of the diaphanous medium.

[1] 438ᵇ 3 ἀλλ' εἴτε φῶς εἴτε ἀήρ ἐστι τὸ μεταξὺ τοῦ ὁρωμένου καὶ τοῦ ὄμματος, ἡ διὰ τούτου κίνησίς ἐστιν ἡ ποιοῦσα τὸ ὁρᾶν.

[2] 419ᵃ 12 ἐὰν γάρ τις θῇ τὸ ἔχον χρῶμα ἐπ' αὐτὴν τὴν ὄψιν οὐκ ὄψεται.

[3] 419ᵃ 15 ὁρᾶσθαι ἂν ἀκριβῶς καὶ εἰ μύρμηξ ἐν τῷ οὐρανῷ εἴη.

[4] Only for the medium of vision has Aristotle a distinctive name— τὸ διαφανές. He does not name the media of sound and odour, though media are equally necessary for those senses. By later writers they were called (on the analogy of τὸ διαφανές) τὸ διηχές and τὸ δίοσμον respectively. It is remarkable that Aristotle (*de Sens.* vi. 446ᵃ 20–ᵇ 27) is quite ready to admit respecting these media, what he denies so stoutly of τὸ διαφανές, that in them the stimulus of sense travels locally and takes time to come from object to organ.

the stimulation (κίνησις) produced by colour is conveyed
through them to the organ of vision, which is thus on
its part stimulated to activity. The medium of colour is
the same as that of light, sc. the διαφανές. This belongs
to both water and air, not *qua* water or air, but *qua*
partaking in common of the nature of the celestial element,
or αἰθήρ[1]. Fire and this αἰθήρ, or τὸ ἄνω σῶμα, stimulate
the potential diaphanous and render it actual[2]; colour
stimulates the actual diaphanous and so becomes visible.
But this diaphanous is also a *subjective* medium of vision.
It exists not only outside, but also inside the eye[3]. It
remains to be noticed that that which is to be a fitting
medium of all possible colours must itself be colourless.
This rule has its analogue in the cases of all the other
senses. The medium of sound—air—must be actually
soundless; that of odour, inodorous; that of taste, tasteless.
So water is tasteless *per se*.

§ 47. The *organ* and *function* of vision. Like all other
organs, the eye is defined by its function. All organs are true
to their definition only while capable of discharging their
functions; e. g. the eye, only as long as it can *see*. A dead
person's eye is no longer an eye in the true sense, but only
in an ambiguous sense, of the word[4]. The eye is the
particular organ affected by the stimulation (κίνησις) set
up by colour in, and propagated through, the diaphanous
medium: affected, i. e. in such a way as to have the
sensation of colour. But the κινήσεις thus set up in the
eye must be in some way conveyed to ' the soul '[5].

The diaphanous medium, therefore, which operates

The organ of sight: its nature and meaning: its structure, and various parts. The function of the 'pupil,' the essential part of the eye. Covering of the pupil: 'Hard-eyed' animals.

[1] οὐ γὰρ ᾗ ὕδωρ οὐδ' ᾗ ἀήρ, διαφανές, ἀλλ' ὅτι ἐστί τις φύσις ὑπάρχουσα
ἡ αὐτὴ ἐν τούτοις ἀμφοτέροις καὶ ἐν τῷ ἀιδίῳ τῷ ἄνω σώματι, 418b 7.

[2] And also visible so far as *light* is its colour.

[3] So, as we shall see (p. 114), the ear has within it a cell of air which
is a means of continuing inwards the external medium of sound.

[4] *Meteor.* iv. 12. 390a 10 seqq.; *de An.* ii. 1. 412b 20 ἡ ὄψις· αὕτη γὰρ
οὐσία ὀφθαλμοῦ ἡ κατὰ τὸν λόγον . . . ἧς ἀπολειπούσης οὐκ ἔστιν ὀφθαλμὸς
πλὴν ὁμωνύμως, καθάπερ ὁ λίθινος.

[5] For the question whether or how far the sensations realize themselves in the separate organs without stimulating the faculty of central
sense, see the chapter on the *Sensus Communis*, § 48.

objectively or externally, is also employed on the *subjective* side within the eye itself, for the purpose of transmitting inwards the κινήσεις received by this organ from without. The eye as a living functioning whole [1] is named ὀφθαλμός and sometimes ὄμμα. It is an organ, consisting of heterogeneous parts [2]. But the part of this whole which is properly concerned in vision—that ᾧ βλέπει—is the part generally named ἡ κόρη, which we usually render the *pupil* (*vide supra* § 2, p. 9 n.), but by which, at least from the time of Empedocles forward, the Greek psychologists meant the 'crystalline lens.' Round this internal moist part called ἡ κόρη comes what Aristotle calls τὸ μέλαν, probably the *iris*; and outside of this again is the *white* [3]. The pupil and vision are to the eye what body and soul respectively are in the economy of the ζῷον as a whole [4]. The κόρη is the material part most intimately concerned in seeing. Therefore, for its protection, it is covered with a membrane so thin and clear as not to obstruct vision, and has in higher animals a further protection afforded by the eyelids. The need of this precautionary protection arises from the humid constitution of this visual part [5]. There are creatures whose eyes are even better protected, viz. by scales [6], but these suffer for it in having less acute vision [7]. The primary organ of touching, in relation to the flesh as medium, is compared with the pupil (as the primary organ of vision) in relation to the whole diaphanous [8]. If the external medium of vision were organically attached to the pupil, both would form one whole, comparable to that formed of the organ of touch proper and the organically connected environment of flesh which is its medium.

---

[1] 413ᵃ 2 seqq. ἡ κόρη καὶ ἡ ὄψις.

[2] μόριον ἀνομοιομερές. Cf. 647ᵃ 4 seqq. For its anatomical structure according to Aristotle, see Philippson, ὕλη ἀνθρωπίνη, pp. 230 seqq.

[3] 491ᵇ 20 τὸ δ' ἐντὸς τοῦ ὀφθαλμοῦ τὸ μὲν ὑγρὸν ᾧ βλέπει, κόρη, τὸ δὲ περὶ τοῦτο, μέλαν, τὸ δ' ἐκτὸς τούτου, λευκόν.

[4] Cf. 413ᵃ 2: add 108ᵃ 11 ὡς ὄψις ἐν ὀφθαλμῷ, νοῦς ἐν ψυχῇ.

[5] *De Part. An.* ii. 13, 657ᵃ 30 seqq.

[6] 657ᵇ 34 τὰ σκληρόφθαλμα.

[7] 421ᵃ 13, 657ᵇ 36.   [8] *De Part. An.* ii. 8. 653ᵇ 23 seqq.

§ 48. For perfect vision (i. e. both *far-sight* and *clear-* Structural
sight) there must be a due proportion of moisture in conditions
the eye. Those that have too little are the creatures of perfect
vision.
with gleaming (γλαυκά) eyes : those that have too
much are the black-eyed (μελανόμματα). The former
see well by night but badly by day, owing to the eye,
from its defective amount of ὑγρόν, being over-stimulated
in daylight. The latter see well by day but badly by
night, because of the small proportion of the fire to the
water in the eye, and the weakness of the light in the air
at night[1]. Besides this the membrane which covers the
pupil should be transparent, white, and of even superficies.
It must be *thin*, in order that the stimulating process from
without may pass straight through it. It must be *even*, that
it may not cast shadows, as it would if wrinkled. One
reason why old persons do not see keenly is that the
membrane covering the pupil of their eyes, like the whole
epidermis, becomes wrinkled and thick with age. This
membrane again must be *white* ; for if black it would not
be diaphanous. The very essence of black is non-diaphanous-
ness : lanterns would not show light if their sides were
black. The moisture in the eye, moreover, must be pure
(καθαρόν) and 'symmetrical' with the movement of stimulation. If this is not so, and if the δέρμα or membrane
be not as described above, the eye will not be clear-sighted,
i. e. distinguish accurately between visible objects, but may
be long-sighted[2]. Creatures with protruding eyes are short-
sighted ; those with deep-set eyes are long-sighted, the
sockets serving as a tube to combine and direct the movement of the visual ray. This explanation holds good whether
the ray proceeds outwards, from the eye, or inwards, from the
object.

§ 49. The physical constitution of the visual organ *Physical*
proper interested Aristotle as well as his predecessors. constitu-
Empedocles and Plato had followed Alcmaeon (§ 4 *supra*) visual

---

[1] Cf. 779[b] 34 seqq., 780[a] 25 seqq.
[2] 780[b] 22. In this requirement of συμμετρία between the κίνησις and
τὸ ὑγρόν we are reminded of Empedocles.

organ proper. Democritus' attitude. The 'image' reflected in the pupil not the essential factor of vision, as Democritus and others thought. It is a merely external thing: a phenomenon of *reflexion.* The eye does not consist of fire. True explanation of the 'intraocular flash': a phenomenon of *reflexion.*

in holding that it consists essentially of *fire.*   Aristotle[1] preferred to hold with Democritus that it consists of water[2].   Democritus, indeed, came to this conclusion on false grounds.   He thought that the eye consists of water because he supposed vision to be merely the mirroring (ἡ ἔμφασις) of external objects in the eye, which consisting of water acts as a mirror.   The mirroring which does take place is, however, merely due to the smoothness (λειότης) of the surface of the eye; and, as a fact, does not find its full explanation merely in the reflecting surface of the eye in which the image is seen, but requires account to be also taken of the spectator's eye which alone sees this image. In short this is only a case of the reflexion of light[3], a subject but imperfectly understood by Democritus and his contemporaries[4].   Democritus, too, should have asked himself why[5], if vision were merely reflexion, the other surfaces which reflect images do not see as well as the eye. The visive part of the eye is, therefore, of water, but vision takes place not by *mirroring* in this water, but by the diaphanousness of the latter—a property which it possesses in common with the air and water of the external world.

As for the theory that the eye consists of fire, Aristotle not only regards it as false, but considers himself to have traced the error to its source.   This error is due, he says, to the well-known but misunderstood fact that if the eyeball be suddenly moved or pressed when the eye is closed, or when there is darkness, a flash ('phosphene') as it were of fire or light is seen within the eye.   If this (from which some conclude that the eye consists of fire) gave a real ground for the popular conclusion, and if vision were due

[1] *De An.* iii. 1, 425ᵃ 4 ; *de Sens.* ii, 438ᵃ 5 seqq.

[2] Among the many signs of spuriousness in the *Problems* we find that in 960ᵃ 32 the visual part of the eye is said to be of fire, ἡ μὲν ὄψις πυρός.

[3] ἀνάκλασις, which sometimes means *refraction*, e.g. 373ᵇ 10 seqq.

[4] 438ᵃ 9, 370ᵃ 16 οὗτοι μὲν οὖν οὔπω συνήθεις ἦσαν ταῖς περὶ τῆς ἀνακλάσεως δόξαις.   For Aristotle's account of it and its relationship to vision and colour see § 40 *supra.*

[5] Democritus (as we have said) would have replied that the soul which sees belongs to the whole organism, not to the eye alone.

to the eye's being of fire, the question at once arises why one sees this fire only when the eye is suddenly and rapidly moved. Again, why does not the eye *always* see itself, as it does in such a case? It is impossible to reply that it does so, indeed, but that we are not aware of it ; for we could not be unaware of it if it were true. If a person in full consciousness sees, he must be aware that he sees. To put this phenomenon of the fire-flash in its true aspect, we need only observe that the surface of the pupil, like many other smooth objects, naturally *shines* in darkness, without, however, *giving light*. The phenomenon is one of *reflexion* (ἀνάκλασις) of light[1]. Hence it is only when the eyeball is rapidly moved that this shining becomes visible, because only then could it as it were duplicate itself, from one becoming two, so that the eye seeing becomes as it were different from the eye seen, and the latter becomes object to the former as percipient. Besides, if the visual part of the eye were really fire, and vision were to be thus fully explained, as Empedocles and Plato held, the eye should see in darkness, not merely in light : their notion being that light issues from the eye, which Empedocles, at least, compared to a lantern.

§ 50. To say with Plato, in answer to this, that the visual current, when it issues by night from the eye, is extinguished in the darkness, is sheer folly. For *fire* may be extinguished but not light—such fire, that is, as is made of coals, and its flame may be thus extinguished by the cold or moist (ψυχρῷ ἢ ὑγρῷ)[2]; but neither one nor the other of these (sc. πῦρ ἀνθρακῶδες and φλόξ) exists as an element in *light*. Should it be said that they do exist in it, but in quantities so small as to be imperceptible, the answer is : if this were true, light should on the above grounds be sometimes extinguished by day, e. g. in wet weather, or in water, and in very cold weather there should regularly be darkness by day, as under such circumstances ignited bodies and flame

(marginal note:) Polemic against Plato and Empedocles. Light not *extinguished* by night, as Plato held : only flame and 'glow' *can* be 'extinguished' at all, and these are not elements of light. Vision not

---

[1] ἐκείνως αὐτὸς αὑτὸν ὁρᾷ ὁ ὀφθαλμὸς ὥσπερ καὶ ἐν τῇ ἀνακλάσει.

[2] 437[b] 12 seqq. Fire had three great varieties : φλόξ, ἄνθραξ, and φῶς. *Vide supra* pp. 53, 65 n. 1. Only the two first could be ' quenched.'

due to a
light sally-
ing forth
from the
eye towards
or to the
object.
There is no
σύμφυσις,
such as
Plato held,
of light
with light.

are extinguished. No such thing happens to light, however, under these circumstances. Further, to say with Plato that the eye sees by means of light issuing forth from it [1]; that this light either extends and prolongs itself as far as the stars, as Empedocles would seem to say [2]; or that (as Plato held) when it has reached a certain point outside it organically coalesces with (συμφύεσθαι) the light coming from the objects seen—this is all idle talk. If there were to have been such coalescence of internal with external light, it were better that it should take place, to begin with, inside the eye itself. Yet even this is but a vain notion. For what is, or could be, meant by the 'organic coalescence [3]' of light with light? Such 'organic coalescence' does not take place between any random things, but according to fixed laws. And how could it happen when, as in the case before us, a membrane, covering the pupil, intervenes between the outer and the inner light? Hence this popular notion that the visual part of the eye is of fire must be abandoned. False in itself, it has been adopted on mistaken grounds, and can be maintained only by fallacious reasoning.

Why the
eye consists
of *water* in

§ 51. To resume: the pupil consists of water, because water as diaphanous [4] is homogeneous with the external

---

[1] Aristotle himself uses ὄψις in the *Meteorologica* in such a way as to make one think at first sight that he held the theory here condemned. See Bonitz, *Index Arist.* 553[b] 30; Ideler, *Arist. Meteor.* i. 6. 3, p. 384 'Hoc igitur loco Aristoteles videtur lumen ex ipso oculo emittere ut hac ratione singulae res visibiles fiant, quod etiam magis patet ex iis quae sequuntur: οὐ δύνασθαι τὴν ὄψιν τῶν ἀνθρώπων φέρεσθαι κλωμένην πρὸς τὸν ἥλιον. Sententiam hanc ab Empedocle et Platone propositam ipse Aristoteles improbavit, *de Sens. et sensili* c. 2. 437[b] (cf. Theophr. *de Sens.* § 7 seqq.) longeque aliam proposuit (*de An.* ii. 7. 418[b]).' Ideler rightly (cf. 374[b] 22, 781[a] 3), however, holds that Aristotle is there, for his special purpose (i.e. elucidation of certain 'optical' facts), adopting the current view of ὄψις, which served his turn quite as well as his own view would, while avoiding unnecessary or irrelevant matter of dispute.

[2] See, however, § 7 *supra*, p. 18.

[3] συμφύεσθαι: the Greek word involves associations which are not contained in the English 'coalescence,' but which are vital for Aristotle's argument.

[4] εἴπερ μὴ πυρὸς τὴν ὄψιν θετέον, ἀλλ' ὕδατος πᾶσιν, 779[b] 19; 780[a] 4 ἡ τούτου τοῦ μορίου κίνησις ὅρασις, ᾗ διαφανὲς ἀλλ' οὐχ ᾗ ὑγρόν, 438[b] 5 seqq.

medium of vision. Air, which is likewise diaphanous, might
conceivably have served for the purpose of an internal
medium of vision [1]; but air is not so easily or conveniently
as water packed into a small space and confined within a
capsule. At all events, facts show that the water is in the
eye. When eyes are decomposed or mutilated, that which
flows from them is seen to be water. In embryonic eyes,
too, this water is particularly cold and bright. In sanguineous animals the white of the eye is adipose, simply in order
to keep this water from becoming congealed. This same
object is effected by the hard scale on the eyes of bloodless
animals [2]. The function of this water in the visual organ
is as follows. The cause of sight is a stimulus from the
object propagated through the medium to the organ of
vision. This is impossible without light. But light is
required not only in the atmosphere without us but also
within the eye itself. Hence the external medium of
vision, normally air, has its function taken up internally by
another medium, water. The internal and external media
are homogeneous in this respect that both are diaphanous,
i. e. possess the one quality essential to the conveyance of
the visual stimulus. The external light, which is the
condition of seeing externally, is continued in this way
into the organ. This must be done if the stimulus is
to reach 'the soul'; for the soul, or its visual organ, is not,
to be sure, situated at the outermost extremity of the eye,
but somewhere within [3], rendering it needful that light

particular, and not of air, which is also diaphanous. Facts which prove the eye to be essentially of water, and also indicate the light-bearing function of this water. The sudden flash caused by cutting the optic πόροι. The water in the eye a secretion from the brain.

[1] In pronouncing here against *air*, Aristotle would seem to reject
the theory of Diogenes of Apollonia, who made air constitute the
essential organ of seeing, as of all other senses.

[2] 779[b] 15-28. 'Empedocles is not right in ascribing the γλαυκότης
(gleam) of some eyes to the fire they contain : the blackness of others
to the greater amount of water. Such colours depend altogether on
the greater or less quantity of water in the pupil. That eye is best
which has the due proportion of water in it.'

[3] What 'within' here means is sufficiently seen from 491[b] 20 τὸ
δ' ἐντὸς τοῦ ὀφθαλμοῦ τὸ μὲν ὑγρόν, ᾧ βλέπει, κόρη. It does not refer
to the organ of *sensus communis* or imply that each organ—here the
eye—is not *per se* capable of having the sensations which belong to it,
or even that each special organ involves in its action the immediate or
concurrent co-operation of the central organ.

should be conveyed to it through some medium. That
light is really conveyed inwards in this way is proved by
the accidental experience of those who have received
a slash with a sword across the temple, severing 'the
passages of the eye[1].' Such persons have experienced
a brilliant illumination, immediately followed by total
darkness, as if a lamp had suddenly flared up within them,
and then, all at once, gone out. What really takes place
in such cases is, that the diaphanous medium, the 'pupil,'
which *is* a sort of lamp, is suddenly cut away. The water
on which depends the continuation inwards of the outer
diaphanous medium is, for Aristotle, secreted to the eye
from the brain. The eye, like the organ of smelling, is
formed by an off-growth from the brain[2]. For the brain
is the moistest and coldest of all parts in the organism.
From this some of the purest of its moisture is conducted
through the 'pores' which connect the eye with the
membrane surrounding the brain[3]. Hence it is fitting that
the organ of sight, being like the brain moist and cold,
should have its seat near the brain. The eye in its
embryonic stage is, like the brain, *over*-moist and *over*-large;
and again in its later development it, like the brain, gains
in consistency, while it is reduced in size.

Vision—
the result
of a process § 52. Vision is effected, according to Aristotle, by
from object a process from object to eye, not conversely[4]. Seeing is
to eye not the result of a mathematical or other abstract relation
through between object and eye, such as the relation of equal to

---

[1] 438[b] 14 ὥστε ἐκτμηθῆναι τοὺς πόρους τοῦ ὄμματος. Aristotle here
speaks of πόροι : what were they? Some think of the optic nerves,
which are said to have been first known to Alcmaeon by dissections.
Even if Aristotle did mean these by what he here calls πόροι, we still
must not imagine that he understood their function *as* nerves. Such
knowledge did not come till after his time. Cf. Dr. Ogle's note to his
translation of Arist. *de Part. An.* ii. 10, pp. 176–7 : 'On the whole
I think it is most probable that by πόροι in this place (sc. *de Part. An.*)
Aristotle means no more than openings or *foramina*'; but he goes on
to add that, in our passage *de Sens.* ii and in *de Gen. An.* ii. 6, by πόροι
are meant the optic nerves as anatomical phenomena.

[2] 438[b] 28.   [3] 744[a] 9 seqq.

[4] ὁρῶμεν εἰσδεχόμενοί τι, οὐκ ἐκπέμποντες, 105[b] 6.

equal. If it were so, the distance, for example, of the object should make no difference to vision, any more than it does to the equality of one equal to another[1]. The process from without is not, however, a conveyance of ἀπορροαί, but a κίνησις—more precisely an ἀλλοίωσις—in the diaphanous medium between the object and the eye. As to the nature of the κίνησις, as a fact of physics, modern science has far outrun the simple and vague notions of Aristotle. It is now known how light travels and is reflected: how rays from an object, directed through the refractive apparatus of the eye, produce an image on the retina, which, since Descartes'[2] time, has been recognized as the cardinal objective fact for the explanation of vision. Thus the physics and the physiology of vision have been really harmonized, to some extent, as Aristotle tried but failed to harmonize them. But as to the nature of the further κίνησις which connects the retinal image with the sensorium, or the magic change by which the retinal image in B's eye (as it appears to A) becomes a field of vision (as it is for B); how that which, externally regarded, is but a tiny picture is translated into a fact of consciousness, no more is known now than was known in Aristotle's days.

Margin note: a medium. The relation of object to eye is a physical, not merely abstract, e.g. mathematical, relation. But the physical process is not one of emanation, but of a kind of κίνησις.

§ 53. *Biologically*, the sense of *touch* is more important than that of sight: it is the most fundamental of all the senses. It is the essential criterion of animal existence. It sentinels and defends the seat of life, and without it animals would perish. Next to touch stands *taste* in point of vital importance: indeed it is according to Aristotle a *mode* of touch. The other senses—*smelling, hearing,* and *seeing*—are not only biologically useful, and conduce to the preservation of the animal's existence; but they also contribute to its *well-being* on an implied higher level of development[3]. Creatures which, besides life, have sense-

Margin note: Comparative values of the senses. Touch and taste biologically most necessary to animals; the other senses necessary for their well-being. Connexion between locomotive

---

[1] *De Sens.* vi. 446ᵇ 10 seqq.

[2] See the Fifth Discourse of his *Dioptrique*.

[3] *De An.* iii. 12. 434ᵇ 11 seqq. ; *de Part. An.* ii. 10. 656ª 6 seqq.
ὅσων μὴ μόνον τοῦ ζῆν ἀλλὰ καὶ τοῦ εὖ ζῆν ἡ φύσις μετείληφε· τοιοῦτο δ' ἐστὶ τὸ τῶν ἀνθρώπων γένος· ἢ γὰρ μόνον μετέχει τοῦ θείου τῶν ἡμῖν γνωρίμων ζῴων, ἢ μάλιστα πάντων. Cf. also *Top.* iii. 2. 118ª 7 seqq.

power and mediated sense-perception in animals. Both developed *pari passu* in the animal kingdom. Hence the primary organ of sense-perception and the primary organ of locomotion are identical in animals. Of externally mediated senses, sight has highest *biological* value. It is in its direct consequences also of highest value *psychologically*. In indirect consequences, however, hearing is more valuable psychologically, for on hearing depend learning by oral instruction and the use of language.

perception possess a form of existence which is richer in variety and more highly endowed in different degrees. On the possession of locomotive power seems to rest the need or chief usefulness of the externally [1] mediated senses—hearing, seeing, and smelling. Accordingly the internal principle or seat of locomotion and that of sense in general are for Aristotle the same—the heart, in sanguineous animals, and in non-sanguineous the 'part analogous.' As the locomotive faculty is developed and its powers differentiated, corresponding development seems to occur in the faculty of sensation. It is to animals which possess locomotive power that seeing, hearing, and smelling are particularly important, enabling them to take timely precautions against danger, and to perceive their prey in advance.

But of all the senses which perceive through external media, seeing is of highest biological as well as psychological importance. In the latter aspect, i. e. in its bearing upon the development of knowledge and experience, the superiority of this sense is most striking. Even apart from its practical uses the exercise of the senses is desired by us for its own sake, that of the sense of seeing, however, more than all the rest. For this most of all leads to knowledge, disclosing to us multitudinous qualities of things, and showing us their natures [2]. Its superiority to hearing is intrinsic and indisputable, as a vehicle of first-hand intelligence. Yet hearing may incidentally have more effect in education. Hearing is that which makes learning possible [3]; and it is through learning that general truths are chiefly reached, while seeing gives us the particulars whence they are derived. Thanks to the fact that all bodies are coloured, all are visible ; and it is chiefly by the sense of seeing that we perceive the *common* sensibles figure, magnitude, motion, number. Animals that can remember distinct visible qualities of things store up the knowledge thus derived, and from the storehouse of memory

[1] All are mediated, not all externally mediated.
[2] *Met.* i. 980^a 21–^{b} 26.
[3] τὸ μανθάνειν : the Greek pupil was an ἀκροατής.

experience is elaborated; from this and by this again comes scientific knowledge, which arises as the details of experience become organized under general conceptions[1]. The matchless clearness and distinctness of visual impressions, to which all perceptions of form are primarily due[2], renders these peculiarly suitable not only for being remembered but also for being arranged, i. e. grouped and classified, under such conceptions. Nevertheless, owing to the part played in mental development by teaching and learning, hearing, on which the use of language depends, has in some ways the advantage over seeing. Thus it is found that persons who are congenitally blind are intellectually better developed than those who are congenitally deaf (436[a] 15).

§ 54. The evidential value of sight[3] is in certain cases *The objective evidential values of seeing and touching.* superior to that of touch, and corrects the illusions of the latter sense. For example, if two fingers of the hand are crossed, and a small object placed between them so as to be in contact with both, it will to the sense of touch *The tactual illusion of the crossed fingers exposed by the sense of sight.* appear as if two objects. The sense of sight proves that it is only one. The sense of sight is also superior to touch in purity; hence the pleasures of seeing are morally higher than those of touching[4]. Possession of sight is 'more *Ethical superiority of sight to touch.* choiceworthy' than that of the olfactory sense[5]. Sight being our most 'evidential' sense (ἐναργεστάτη) its results *Sight guides our movements in space, and determines our notions of direction.* as affecting our feelings—exciting passions and emotions— are proportionately vivid[6]. Passions or emotions artificially stimulated through this sense approach nearest to the impressiveness of reality. The ideas of danger which it conveys inspire fear with an immediacy and force not to be equalled by those of the other senses[7]. Sight, too, is of *Illusions of*

[1] 981[a] 5 ὅταν ἐκ πολλῶν τῆς ἐμπειρίας ἐννοημάτων μία καθόλου γένηται περὶ τῶν ὁμοίων ὑπόληψις.   [2] *Top.* ii. 7. 113[a] 31.

[3] Cf. 460[b] 20, 956[a] 36, 1011[a] 33. Heraclitus (*apud* Polyb. xii. 27, *Fr.* xv, Bywater) says ὀφθαλμοὶ τῶν ὤτων ἀκριβέστεροι μάρτυρες, an opinion founded on the theory that the eyes contain more fire.

[4] *N. E.* x. 5. 1176[a] 1.

[5] *Rhet.* i. 7. 1364[a] 38.   [6] *Probl.* 886[b] 10–37.

[7] Cf. Horace, *Ars Poet.* 180–1 :
      Segnius irritant animos demissa per aurem,
      Quam quae sunt oculis subiecta fidelibus.

sight, not as to its proper αἰσθητά, but as to objective matters, e. g. the *distances* of objects, and their *magnitudes*. We all see the sun as only a foot in diameter. Sight and touch err regarding the 'Common Sensibles' in general. Such are rather errors of inference than of sense-perception. Aristotle knew nothing of colour-blindness.

primary importance as *directing* our movements in space [1]. It is by this sense that the notions of 'before' and 'behind' are determined. Moving 'forward' means moving in the direction in which the eyes naturally look. 'Even crabs which move sidewards may be said in a way to move forward, since they move in the direction in which their eyes naturally look.'

Yet this sense, too, is subject to illusions, as is every individual sense taken by itself when it refers its immediate *datum* to an object [2]. Thus regarding the fact that the colour seen is white, the sense of seeing is almost incapable of error: but as regards the distance at which the white, referred to an object, is from us, or as regards the object to which it is referred, error is frequent. So, too, with regard to the magnitude of objects. Thus the sun's disk appears almost invincibly as if it were but a foot wide. This impression is not due to any pathological state, nor is it the result of scientific ignorance on our part [3]. In the best of health and with sound knowledge of the facts, this is the momentary impression given us by sight as we look at the sun [4]; and thus it is that we are liable to err as regards each and all of the '*common* sensibles.' Such errors, however, as well as those committed in attributing the immediate data of sight to wrong objects, are not really errors of vision: they are errors of judgment. Surreptitious judgments tend to become inextricably mixed up with the immediate impressions of seeing as of other senses. Of errors arising from *colour-blindness*, or of this phenomenon itself, Aristotle seems to have had no notion.

Visual illusion (or

§ 55. A remarkable case of illusion is referred to in the

[1] *De Incess. An.* 712[b] 18.

[2] ὥσπερ τὸ ὁρᾶν ⟨ἐπὶ⟩ τοῦ ἰδίου ἀληθές, εἰ δ' ἄνθρωπος τὸ λευκὸν ἢ μή, οὐκ ἀληθὲς ἀεί, 430[b] 29 (we must either read so, inserting ἐπί or περί before τοῦ ἰδίου, or at least make the gen. one of 'respect.' It goes with the predicate. 'The seeing of the particular quality' is an ungrammatical translation): cf. 428[b] 18, 442[b] 8.

[3] Galen observes the omission on Aristotle's part to determine anywhere the manner by which we perceive the position, magnitude, and distance of objects. Cf. Galen. *de Placit. Hipp. et Plat.* § 638.

[4] 458[b] 28.

Meteorologica[1]. 'Owing to the feebleness of the visual ray (ὄψις) it is often refracted by the air even when not condensed in the way described. Such was the case in the strange experience of a certain person whose sight was weak at the time, and to whom, as he walked, it appeared as if his own image always preceded him, and kept looking back towards him[2]. This illusion was due to the visual ray being bent back from the air around him which (just as distant, or thick, air often does) became like a mirror, so that the ray could not displace or penetrate it, and hence was compelled to return to the eye[3]. So capes at sea sometimes seem raised above the water, and heavenly bodies loom larger when near the horizon.' In the *Problems*[4]—an un-Aristotelean work—many curious but trifling remarks occur on this and similar subjects. The most important concern (*a*) the difficulty, or impossibility, of moving one eye voluntarily without at the same time moving the other in the same way; (*b*) the fact that one object appears as two to a person who by inserting the finger beneath the eyeball displaces it[5]; (*c*) that myopic persons write in very small characters; (*d*) that objects appear multiplied to persons in a state of intoxication or mental distraction[6]; (*e*) that straightness in a line is better discerned with one eye than with two, which is explained by reference to the necessary convergence of rays from both eyes when both are used; (*f*) that ὁ μύωψ brings objects near in order to see them, while ὁ πρεσβύτης holds

Marginal notes: hallucination) explained. The two eyes move *together* in the same direction. If one eyeball is displaced by the finger we see objects doubled. Myopic persons write in a very small hand. Intoxicated persons see objects multiplied: explanation of this. One eye discerns straightness in a line better than both eyes. The μύωψ and the πρεσβύτης.

---

[1] iii. 4. 373ᵇ 2–10.

[2] This (as already remarked, p. 67) reminds one of the 'Doppelgänger,' or the 'Brocken-spectre.'

[3] What is very remarkable here is the seemingly frank acceptance by Aristotle of a theory of vision warmly repudiated by him in *de Sens.* ii. We must assume that he in such cases expresses himself from the popular point of view. So *we* have to speak of the sun 'rising' and 'setting.'　　　　　　　　　　　　　　　　[4] 957ᵃ 38 seqq.

[5] Also referred to *de Insom.* 461ᵇ 30; *Met.* x. 6. 1063ᵃ 6–10.

[6] This phenomenon is explained by comparison with the illusion of the crossed fingers representing one object as two. The κίνησις does not come from each eye to the same part of the soul, which accordingly sees twice. The 'different parts of the soul' thus represent what we might think of as non-identical parts of the retina.

them at a distance. In the tract on Dreaming illusions of sight are mentioned which, however, are, it is stated, really errors of judgment for which the sight *per se* is not to blame. Such are hallucinations, and the illusion of those on ship-board to whom the shore, not the ship, seems to be in motion. Aristotle says also [1] that defects of long and short sight are due not to anything wrong with the soul, but to defects in the visual organ itself. If an old man could have a young man's eye he would see as well as the young man. The sensory weakness of old age is caused not by an affection of the soul itself, but by an affection of that wherein the soul resides; as happens in cases of intoxication and illness.

[1] 408^b 21.

THE ANCIENT GREEK PSYCHOLOGY
OF HEARING

Alcmaeon of Crotona.

§ 1. 'WE hear with the ears, says Alcmaeon, because they Function and organ of hearing. have vacuum in them; for this (vacuum) is resonant. The sonant object produces sound in the cavity (of the outer Air within the ear, the factor of hearing: by this external sound reverberated to the brain. ear), and the air (of the intra-tympanic ear) re-echoes (to this sound) [1].' The effect of the external sonant object is first conveyed to the hollow chamber of the outer, i. e. the extra-tympanic, ear, from which the κενόν, or air of the intra-tympanic ear, takes it up and reverberates it to the 'point of sense,' which for Alcmaeon was the brain, or in the brain [2].

§ 2. 'Alcmaeon says that we hear by means of the vacuum The κενόν = the ἀήρ for Alcmaeon. within the ear, for this it is that transmits inwards the sounds (which come from without) at every immission of the soniferous air-waves (into the outer ear). For all vacua are resonant [3].' I have chosen here the text of Pseudo-Plutarchus, which gives κενά, instead of that of Stobaeus,

[1] Cf. Wachtler, *Alcmaeon*, p. 40; Diels, *Dox.* 506. 23; Theophr. *de Sens.* 25 ἀκούειν μὲν οὖν φησι τοῖς ὠσίν, διότι κενὸν ἐν αὐτοῖς ἐνυπάρχει· τοῦτο γὰρ ἠχεῖν. φθέγγεσθαι δὲ τῷ κοίλῳ, τὸν ἀέρα δ' ἀντηχεῖν.

[2] Diels proposes two different corrections—τοῦτο γὰρ ἠχεῖν [φθέγγεσθαι] διὰ τὸ κοῖλον, and τοῦτο γὰρ ἠχοῦν φθέγγεσθαι διὰ τὸ κοῖλον. Neither is necessary. The subject of φθέγγεσθαι should be taken quite generally, as if = τὸ ψοφοῦν. Diels renders our text—'sonum autem edere (sc. τὸ κενόν) cavo, h. e. propter cavernam auris interioris.' But if κενόν here = ὁ ἀήρ, as would seem from Arist. 419ᵇ 33, the form of the sentence forbids us to regard it as subject to φθέγγεσθαι. Nor can τῷ κοίλῳ be the hollow of the intra-tympanic ear; it is rather the external meatus, with the apparatus in general by which the vibrations of the outer air are caught and conducted inwards to the tympanum. Philippson (ὕλη ἀνθρωπίνη, p. 107) saw this when he (unnecessarily however) proposed κόχλῳ for κοίλῳ here.

[3] Diels, *Dox.* 406ᵇ 21, Aët. *Plac.* iv. 16. 2 Ἀλκμαίων ἀκούειν ἡμᾶς τῷ κενῷ τῷ ἐντὸς τοῦ ὠτός· τοῦτο γὰρ εἶναι τὸ διηχοῦν κατὰ τὴν τοῦ πνεύματος εἰσβολήν· πάντα γὰρ τὰ κενὰ ἠχεῖ.

which gives κοῖλα, agreeing in every other respect. As Wachtler says, the κενόν and the ἀήρ are here equivalent terms. He quotes most appositely Arist. *de An.* ii. 8. 419^b 33 τὸ δὲ κενὸν ὀρθῶς λέγεται κύριον τοῦ ἀκούειν· δοκεῖ γὰρ εἶναι κενὸν ὁ ἀήρ. But here the ἀήρ in the κοῖλον or outer part of the ear must be distinguished from the ἀήρ or κενόν of the inner part. The former receives and introduces the sonant stimulus from the atmosphere; the latter catches it up and transfers it to the brain. The transference is referred to in Theophrastus by ἀντηχεῖν, in the passage from Aëtius by διηχοῦν (with the use of which compare τὸ δίοσμον, τὸ διαφανές, and, especially, τὸ διηχές— late terms used to signify the respective media of odour, colour, and sound). The simple ἠχεῖν in both passages denotes the action of the air within the ear—as of confined air generally—in taking up, or ʻechoing,ʼ sound, apart from the notion of transmitting it. No better commentary on these extracts can be found than that contained in Arist. *de An.* ii. 8. 419^b 33–420^a 19. Cf. *infra* § 20.

Alcmaeon represents the formation of the ear as determining sound : the ear not a mere *conduit*.

§ 3. Alcmaeon was, says Wachtler, the first who attempted to explain the phenomenon of sound and our perception of it by reference to the structure of the ear itself, and the manner in which this was affected by air in motion from without. Empedocles to some extent follows or agrees with him. Their successors generally regard the ear as little more than a conductor of air to the sensorium, most of them holding sound, as a perception, to result from a percussion of the brain or other inward organ by the air thus conveyed through the ear [1].

---

[1] In the passage from Aëtius πνεῦμα cannot mean ʻbreath,ʼ yet it is scarcely identical with ἀήρ. It appears to signify the latter *set in motion* by the external sonant object, and entering, with its soundwaves, into the external ear. Cf. Pseudo-Hippoc. *de flat.* 3 (vi. 94 L) πνεῦμα δὲ τὸ μὲν ἐν τοῖσι σώμασι φῦσα καλέεται, τὸ δὲ ἔξω τῶν σωμάτων ἀήρ, from which it appears that πνεῦμα was treated as the general term for air by some writers. Cf. the use of σύμφυτον πνεῦμα in Aristotle. In connexion with the meaning of πνεῦμα here one may perhaps quote a curious observation of Aristotle, *Hist. An.* i. 11. 492^a 13, respecting Alcmaeon: κεφαλῆς μόριον, δι᾽ οὗ ἀκούει, ἄπνουν, τὸ οὖς· Ἀλκμαίων γὰρ οὐκ ἀληθῆ λέγει, φάμενος ἀναπνεῖν τὰς αἶγας κατὰ τὰ ὦτα.

Empedocles.

§ 4. 'Empedocles teaches that hearing is caused by the Function
impact of the air-wave against the cartilage which is and organ
of hearing:
suspended within the ear, oscillating as it is struck, like the *gong*
a gong[1].' For χονδρῶδει ὅπερ (Plut.) Stobaeus has χόνδρῳ (or *trum-pet*) within
ὅνπερ. A variant is κοχλιώδει, for which Pseudo-Galenus, the ear.
Hist. Phil. (referred to by Karsten, p. 483), gives κοχλιώδει Empe-
χόνδρῳ, 'the spiral-shaped cartilage.' Zeller thinks that docles
know
κώδων here means a 'trumpet,' not a gong or bell. But of the
while 'trumpet' might describe the shape of the *outer* structure
of the
ear, or 'concha,' it is not suitable for what seems to have internal
been before the writer's mind in the above passage—some- ear?
thing *inside* the ear which oscillated freely to the impact
of air-waves. The main point, as Karsten remarks, is that
'Empedocles appears to have regarded hearing as conditioned by the external air-wave, or wave of sound,' in
contact with the ear, and by the resonance of a certain
part of the ear itself. In hearing, the ἀπόρροιαι were simply
'air' or particles of air. For the meaning of χόνδρος, cf.
Arist. *Hist. An.* i. 11. 492ᵃ 15 ὠτὸς δὲ μέρος τὸ μὲν (sc. the
intra-tympanic part) ἀνώνυμον, τὸ δὲ (sc. the 'concha')
λοβός· ὅλον δ' ἐκ χόνδρου καὶ σαρκὸς σύγκειται—that is, the
whole of the *external* ear, for he proceeds: εἴσω δὲ τὴν
μὲν φύσιν ἔχει οἷον οἱ στρόμβοι (i. e. spiral shells, κοχλίαι,
ἕλικες) τὸ δ' ἔσχατον ὀστοῦν ὅμοιον τῷ ὠτὶ (i. e. the bony part
farthest in resembles the external ear in form) εἰς ὃ ὥσπερ
ἀγγεῖον ἔσχατον ἀφικνεῖται ὁ ψόφος. It is from στρόμβοι here

From this it might seem as if Alcmaeon actually held that the resonant
medium—the κενόν—received its impulse from the breath—perhaps the
air in the Eustachian tubes—which, therefore, would be the meaning
of πνεῦμα in the passage of Aëtius. Aristotle would hardly—it may
be argued—have insisted as he does against Alcmaeon that the ear
is ἄπνουν, unless the latter had been known to hold this strange view.
Such an idea about αἶγες would have given Alcmaeon the illustration
wanted to confirm his exposition of the above view of hearing.

[1] Diels, *Dox.* 406ᵃ⁻ᵇ 16, Plut. *Epit.* iv. 16, Stob. *Ecl.* 53 ; Karsten,
Emped., p. 483 Ἐμπεδοκλῆς τὴν ἀκοὴν γίνεσθαι κατὰ πρόσπτωσιν πνεύματος τῷ χονδρώδει, ὅπερ φησὶν ἐξηρτῆσθαι ἐντὸς τοῦ ὠτὸς κώδωνος δίκην
αἰωρούμενον καὶ τυπτόμενον.

that the gloss κοχλιώδει would seem to be derived. How far Empedocles attempted (like Aristotle) to distinguish between *inner* and *outer* ear is not plain; yet everything depends on our knowing this if we are to understand him. It is probable, however, that by the χόνδρος he meant some structure which he found by dissecting the internal ear. Neither he nor yet Aristotle seems to have had any accurate knowledge of the 'ossicles'—the *malleus, incus,* and *stapes*—in the tympanic cavity, bridging the way from the tympanic membrane to the *fenestra ovalis,* and transmitting vibrations from the one to the other. This being so, the use of the word αἰωρούμενον here is the more curious.

'Empedocles says that hearing results from the sounds coming from without, whenever the air, being set in motion by the voice, rings within (the ear). For the organ of hearing, which he terms "the fleshy bone," is a sort of gong which rings internally. The air, when it is set moving, beats against the solid parts, and thus causes the ringing sound [1].' The 'solid parts' are the same as the 'gong'. We notice that ἀκοή is used in two senses here; first of the *hearing,* secondly of the *organ* of hearing. ἠχεῖν and ἦχος are used with special frequency of *ringing* sounds, but particularly of those which rever-

---

[1] Cf. Diels, *Dox.* 501-2; Theophr. *de Sens.* § 9; Karsten, *Emped.,* p. 483 τὴν δ᾽ ἀκοὴν ἀπὸ τῶν ἔξωθεν γίνεσθαι ψόφων, ὅταν ὁ ἀὴρ ὑπὸ τῆς φωνῆς κινηθεὶς ἠχῇ ἐντός· ὥσπερ γὰρ εἶναι κώδωνα τῶν ἴσων ἤχων [τιν᾽ ἔσω ἠχοῦντα?] τὴν ἀκοήν, ἣν προσαγορεύει σάρκινον ὄζον [ὀστοῦν]. κινουμένην [κινούμενον?] δὲ παίειν τὸν ἀέρα πρὸς τὰ στερεὰ καὶ ποιεῖν ἦχον. Such is the text as suggested by Diels, *Dox.* l. c. He has not (*Vors.,* pp. 177, 209) adhered to his previous suggestion of ὀστοῦν for ὄζον, but, as the sense requires reference to the inner not the outer ear or 'concha,' we must accept some such correction or force the meaning of ὄζον beyond what it can bear. With regard, however, to Diels' ἔσω ἠχοῦντα for ἴσων ἤχων, is it necessary? He explains (*Vors.,* p. 209) κώδων· σάρκινος ὄζος thus: '*das Gehör ist* gleichsam eine Glocke *der gleichgestimmten* (?) *Töne. Er nennt es* fleischigen Zweig.' Keeping ἴσων, then, we might suppose the meaning to be that the κώδων took up and rang to the ψόφοι with which it was framed by nature to harmonize, or was, as Empedocles would say, ξύμμετρος. There are sounds which we cannot hear, as there are colours which we cannot see, though other creatures may hear or see them.

berate within a cavity. Hence they are here employed
with idiomatic propriety for the ψόφος, or 'external' sound,
reverberated within the aural cavity. What distinguishes
Empedocles' doctrine from that of Alcmaeon is the κώδων
interposed by the former between the outer and inner
stages through which sound-vibrations pass before reaching
consciousness. For both philosophers air is the vehicle of
sound. According to Alcmaeon the air in the outer ear
is set moving by the ψόφος, and in its turn sets in motion
the air in the inner chamber, which transmits the vibration
to the brain. According to Empedocles, as the organ of
vision contains a lantern, so the organ of hearing contains
a bell or gong, which the ψόφος from without causes to ring:
this ringing, as we are vaguely left to suppose, being
conveyed inwards by a subsequent process to the 'point of
sense,' and the feeling or perception of sound being thus
awakened.

§ 5. 'Empedocles explains hearing by stating that it is
due to intra-aural sounds. But it is strange of him to
suppose that he has made it self-evident *how* we hear,
by merely stating this theory of a sound, as of a gong,
within the ear. For suppose that we hear the *outer* sounds
by means of this *gong*; by what do we hear *the gong itself*,
when it rings? For this—the very point of the whole
inquiry—is neglected by him[1].' Karsten too hastily
inferred from ἔσωθεν here that this, not ἔξωθεν, should be
read in the former passage, Theophr. *de Sens.* § 9, ἀπὸ τῶν
ἔξωθεν ψόφων. But probably two different sorts of ψόφοι
are referred to in the two different passages: the ψόφοι
coming from sonant objects in the outer space around us,
and the ψόφοι made within our ears by the 'gong.' The
latter are here referred to, where Theophrastus with the
art of a dialectician pushes the difficulty of such materialistic
psychology home against Empedocles. The 'gong' rings

Side note: Theophrastus criticizes Empedocles' theory of hearing: that hears the internal 'gong'?

---

[1] Theophr. *de Sens.* § 21; Diels, *Dox.*, p. 505 ἀλλὰ περὶ μὲν τὴν ἀκοὴν
ὅταν ἀποδῷ, τοῖς ἔσωθεν γίνεσθαι ψόφοις, ἄτοπον τὸ οἴεσθαι δῆλον εἶναι πῶς
ἀκούουσιν, ἔνδον ποιήσαντα ψόφον ὥσπερ κώδωνος. τῶν μὲν γὰρ ἔξω δι'
ἐκείνου ἀκούομεν, ἐκείνου δὲ ψοφοῦντος διὰ τί; τοῦτο γὰρ αὐτὸ λείπεται
ζητεῖν. Ἔσωθεν rather should be ἔξωθεν. No sound comes *from* within.

to the outer sounds: but to us the sounds of the 'gong'
itself are a fresh *incognitum*: how do we hear *them*? With
another gong?

Object of §6. 'Empedocles treats of all the special senses according
hearing.
Empe- to the same principle, and teaches that we perceive by the
docles' ex- fact of the ἀπόρροιαι fitting duly into the pores of each sense-
planation
of the organ. Whence it happens, according to him, that no one
distinctive sense can discern the objects proper to any other, inasmuch
quality
of each as the pores in the organs of some senses are too wide, in
sensory
object by those of others too narrow, for the alien sensible object which
emana- should enter them, so that in the former case the emanations
tions. How
does the from the object pass right through without touching,
principle while in the latter they are not able to effect an entrance
that 'like
perceives at all[1].' Empedocles and his reporters have given us no
like' bear
on Empe- real clue to the various ways in which his principle that
docles' 'like is perceived by like' was carried out by him in the
doctrine of
hearing? psychology of perception. We can only conjecture how
Theo- he would have applied it in the case of hearing. Probably
phrastus'
criticism. the ἀπόρροιαι of sound, being air, 'fit' the pores of the ear
qua containing air essentially. The principle itself is
a deduction from the metaphysical theory that 'like affects
like,' and seems intended merely to procure for the latter
its psychological application[2]. The smallness of the part
actually given to it in practice, in reference to hearing,
however, is only one among many instances, ancient and
modern, of the difficulty of bringing metaphysical theories
to bear in any real way upon concrete psychical facts.
Theophrastus, whether fairly or not, criticizes its applica-
bility here, as follows: 'It is not by sound (ψόφῳ),' he
says, 'that we perceive sound, nor by odour that we
perceive odour, nor by the homogeneous sensibles in
general that we perceive the homogeneous, but rather
by their contraries, so to speak. For the sense-organ
which is applied must be itself indifferent (ἀπαθῆ) in its
nature. When indeed there are actual sounds within the

---

[1] Theophr. *de Sens.* §7; Diels, *Dox.*, p. 500.
[2] Cf. Theophr. *de Sens.* §2 Ἐμπεδοκλῆς δὲ πειρᾶται καὶ ταύτας (sc. τὰς
αἰσθήσεις) ἀνάγειν εἰς τὴν ὁμοιότητα.

ears, or actual tastes in the organ of taste, or odours in the organ of smell, all these senses become deadened to their office (κωφότεραι), and this the more, in proportion as they contain more of their respective "similars"[1]. From this criticism it would at least seem as if Empedocles had endeavoured to give to his principle of *similia similibus* practical effect. But we have no direct means of judging such attempts or of estimating the fairness of the criticism of Theophrastus. For a similar difficulty as to the application of the principle to the theory of vision, cf. VISION, § 11, p. 22 *supra*.

16 9283

Democritus.

§ 7. In explaining seeing Democritus assumes δείκελα (as εἴδωλα, see p. 29 n. 3) to pass from the object to the eye. In explaining hearing he makes the analogous assumption of 'sounds' (φωναί), as particles thrown off by the sonant body and conveyed by the medium of the air to the ear, and through it 'to the soul.' The sound is a 'stream of atoms[2]' which sets the atoms of the air in motion, and, joining itself with these according to similarity of shapes and sizes, makes its way into the body to the soul. Its chief, but not sole, entrance is through the orifice of the ear. His theory of sound is more reconcilable with his doctrine of primary and secondary qualities than is his theory of seeing.

'He explains hearing somewhat in the same way as other writers do. For he says that the air, when it rushes into the vacuum of the ear, produces a motion there; only that it enters likewise at all parts of the body, but in a special way, and in greatest quantity, through the ears, because there it has the largest vacuum to pass through, and remains least stationary. Wherefore one does not perceive sounds with the rest of the body, but only with the ears. When once it has entered it is dispersed, owing to its rapidity; for vocal sound (physically considered) is due to the air being condensed, and entering with force. Accordingly, as he explains sense by contact externally, so he explains it as due to contact internally.

Function and organ of hearing according to Democritus. Hearing is a mode of contact between the atoms of sound (conveyed through the air into the body by the ear orifice) and the soul atoms in the body.

---

[1] Theophr. *de Sens.* § 19.   [2] See *infra* § 9, p. 102.

Conditions of acute hearing.

One hears most acutely if the external membrane is dense, and the vessels (φλέβια) empty and as free as possible from moisture, and if, moreover, they are well bored, both in the rest of the body and in the head and ears; and if, in addition, the bones are dense and the brain well tempered, and the parts surrounding it as dry as possible. For thus the vocal sound enters in one volume, as it passes in through a vacuum large and without moisture and well bored; and is dispersed swiftly and equably throughout the body, and does not slip out and away[1].' While Democritus agrees with others in the main, his theory has the peculiarity of making the stimulus of hearing affect not merely the organ of hearing proper but the whole bodily organism. On this point Theophrastus afterwards directs his criticism, and to this he here draws attention in the words πλὴν ὅτι κτέ. For Democritus' reduction (in which most φυσιολόγοι agreed) of all senses to modes of one, viz. touching, cf. Arist. de Sens. iv. 442ᵃ 29. It is a question what the 'external membrane,' on the πυκνότης of which hearing so much depends, means. It does not seem to be the tympanum, as, from the tenor of the passage, density of this would appear to be an obstruction to the entrance of the ἀήρ, and therefore to hearing. It is rather the membranous covering of the inner surface of the concha, which has for its office to collect and conduct the ἀήρ inwards. The πυκνότης of this would (from Democritus' standpoint) prevent the ἀήρ from slipping through and being lost (διεκπίπτειν) before it could pass inside and effect its purpose.

The peculiarity of Democritus' theory of hearing criticized by Theo-

§ 8. 'In this Democritus is as indefinite as other philosophers, but the strange and peculiar point in his theory is the *entrance of sound at all parts of the body*, and *its dispersion through the whole body* after it has entered by the organ of hearing; just as if this sense of hearing

---

[1] Theophr. *de Sens.* 55–6; Diels, *Dox.*, p. 515, *Vors.*, p. 391; Mullach, *Democr.*, pp. 212–13, 342–4. The translation is from the text as given by Diels *Vors.*, keeping πυκνουμένου, which suits ἀθρόον a little below, but rejecting Schneider's τῇ ἀκοῇ for καί.

were effected not by its proper organ, but by the body. as phrastus.
a whole. For even if the whole body is sympathetic to the Unfairness of his
operation of the organ of hearing, it does not follow from criticism.
this that the whole body has the sense of hearing. For it
is sympathetic to the operations of all the senses alike, and
not only to those of the senses, but also to those of the
soul. Such then is Democritus' account of seeing and
hearing. The other senses he explains in about the same
fashion as that in which most other philosophers explain
them[1].'

§ 9. In the above extracts from Theophrastus the par- Object of
ticular object of hearing is referred to as φωνή—voice or hearing : sound.
vocal sound. This word is not of course equivalent to Hearing is
sound in general, but it is taken, as often, for the leading a mechani- cal sense.
type of sound[2]. It is chosen simply because speech is one Sound is a stream
of the most interesting and important kinds of sound. of atoms.
Democritus and others regarded sound as affecting the Both the sound
auditory apparatus materially or mechanically, in the form atoms
of an inrush of air. Sound is a stream of atoms emanating themselves and the air

---

[1] Theophr. *de Sens.* 57; Diels, *Dox.*, p. 515, *Vors.*, p. 392; Mullach, *Democr.* 213–14, 345. Theophrastus overlooks the fact that Democritus, according to the previous statement of Theophrastus himself, *denies* that we hear with the rest of the body, and gives the reason why we do not. Mullach renders the words πάσαις γὰρ τοῦτό γε ὁμοίως ποιεῖ, καὶ οὐ μόνον ταῖς αἰσθήσεσιν ἀλλὰ καὶ τῇ ψυχῇ: 'enimvero omnibus (sensibus) hoc similiter ascribit, neque his tantum sed etiam animae,' making the subject of ποιεῖ Democritus instead of σῶμα. The τοῦτό γε ποιεῖ merely = συμπάσχει, which Theophrastus has not wished to repeat. Mullach seems to think that we have here a general reference to the way in which Democritus explained all the senses and the soul materially. What Theophrastus means is that Democritus has just as good or bad reasons for diffusing the operations of the other senses over the whole body, as for doing this with the sense of hearing. In all these operations the whole organism by sympathy has a part, as in psychical operations generally. If, however, as Theophrastus would argue, the whole body cannot on this account be said, for example, to see, neither can the whole body be said to have the sense of hearing. For the possibility of sensory function without sense-organs or even nerves, see Haeckel, *Origin and Development of the Sense-organs,* and G. J. Romanes, *Mental Evolution in Animals,* p. 81.

[2] Cf. Plato, *Charm.* 168 D οἷον ἡ ἀκοή, φαμέν, οὐκ ἄλλου τινὸς ἦν ἀκοὴ ἢ φωνῆς. ἦ γάρ; Ναί.

broken up
by them
into like
forms and
sizes reach
the ear.
Explana-
tion of the
pitch and
purity of
tones.

from the sonant body and causing motion in the air between this and the ear. The sound atoms are not supposed to reach the ear alone, but together with air fragments which resemble them. These fragments, following the law that like consorts with like, come together according to their similarity of shapes and sizes. Probably the purity of sounds depends on the *similarity*, the pitch and volume on the *magnitude*, of their constituents. ' Democritus says that (when sound is produced) the air is broken up into bodies of like form, and, thus broken, is rolled along by and with the fragments of vocal sound [1].' Epicurus says of φωνή that ' It is a stream sent forth from creatures uttering a voice, or from objects which make a ringing sound, or a noise [2].' In terms precisely equivalent to those ascribed to Democritus (from whom no doubt he borrowed his views of the physical nature of sound), he states that *this* stream (not the ' air ') is broken up into ' bodies of like form.' We are left in little doubt what ῥεῦμα—the stream—meant: Gellius, *Noct. Att.* v. 15, speaks of it as ῥεῦμα ἀτόμων (according to the probable conjecture of Burchard, accepted by Mullach and Diels, of ἀτόμων for λόγων). The nature of φωνή, as resulting from a blow (πληγή) struck on a portion of ἀήρ, is dealt with more in detail by Plato [3] and Aristotle. We have no further particulars than those above given to show us what the views of Democritus were on the nature of sound.

[1] i.e. the atoms sent off by the sonant body. Cf. Diels, *Vors.*, p. 389; Plut. *Epit.* iv. 19 § 3 Δημόκριτος καὶ τὸν ἀέρα φησὶν εἰς ὁμοιοσχήμονα θρύπτεσθαι σώματα καὶ συγκαλινδεῖσθαι τοῖς ἐκ τῆς φωνῆς θραύσμασιν. For ὁμοιοσχήμονα cf. Theophr. *de Sens.* § 50 αἱ φλέβες ⟨αἱ⟩ κατὰ τοὺς ὀφθαλμοὺς εὐθεῖαι καὶ ἄνικμοι, ὡς ὁμοιοσχημονεῖν (= 'to conform') τοῖς ἀποτυπουμένοις· τὰ γὰρ ὁμόφυλα μάλιστα ἕκαστον γνωρίζειν. The θραύσματα ἀέρος here are ὁμοιοσχήμονα with those ἐκ φωνῆς, the atoms from the sonant body. If the latter are homogeneous, those into which they mince (θρύπτειν) the air are also homogeneous. Cf. Arist. 419[b] 23 τὴν θρύψιν τοῦ ἀέρος.

[2] Plut. *Epit.* iv. 19; Diels, *Dox.*, p. 408 Ἐπίκουρος τὴν φωνὴν εἶναι ῥεῦμα ἐκπεμπόμενον ἀπὸ τῶν φωνούντων ἢ ἠχούντων ἢ ψοφούντων· τοῦτο δὲ τὸ ῥεῦμα εἰς ὁμοιοσχήμονα θρύπτεσθαι θραύσματα.

[3] For the expression ῥεῦμα applied to φωνή, cf. its application to λόγος by Plato, *Soph.* 263 E τὸ δέ γ' ἀπ' ἐκείνης [τῆς ψυχῆς] ῥεῦμα διὰ τοῦ στόματος ἰὸν μετὰ φθόγγου κέκληται λόγος.

Anaxagoras.

§ 10. 'Anaxagoras held that sense-perception is effected by the action of contraries [1] upon one another, for like is unaffected by its like ... on this same principle he explains smelling and *hearing* [2], the former taking place together with respiration (inhalation), the latter by the fact of sound entering and making its way through the ear to the brain: for the bone which encloses (the brain) forms a cavity into which the sound rushes [3].' Large organs better perceive great and distant objects: small organs the small and near objects. 'The larger animals have more sensory power, and in a word sensory power is proportionate to the magnitude (of the organs of sense). For all animals which have large, clear, bright eyes see large objects and see them at long distances, while those which have small eyes see contrariwise: and it is likewise in the case of hearing. For the large animals hear the great sounds and those coming from afar, while the small sounds escape them, but small animals hear the small sounds and those close by them [4].'

§ 11. 'When Anaxagoras states that the larger animals have greater sensory power, and, in a word, that sensory power is proportionate to the magnitude of the sensory organs, the question arises: if this be true, whether have the small animals or the large animals the more perfect sense? For it would seem to be a mark of more exact

[Right margin notes:] According to Theophrastus Anaxagoras applies the principle 'unlike is perceived by unlike' to explain hearing. Larger animals with larger organs have the advantage over others in perceiving sensory qualities in greater volume.

Theophrastus examines Anaxagoras' statement that animals have αἴσθησις in

---

[1] In this principle Anaxagoras followed Heraclitus, and probably Alcmaeon.

[2] How the principle is applied to hearing Theophrastus does not say.

[3] Theophr. *de Sens.* §§ 27–8; Diels, *Vors.*, p. 323 Ἀναξαγόρας δὲ γίνεσθαι μὲν τοῖς ἐναντίοις· τὸ γὰρ ὅμοιον ἀπαθὲς ὑπὸ τοῦ ὁμοίου ... ὡσαύτως δὲ καὶ ὀσφραίνεσθαι καὶ ἀκούειν τὸ μὲν ἅμα τῇ ἀναπνοῇ, τὸ δὲ τῷ δικνεῖσθαι τὸν ψόφον ἄχρι τοῦ ἐγκεφάλου· τὸ γὰρ περιέχον ὀστοῦν εἶναι κοῖλον, εἰς ὃ ἐμπίπτειν τὸν ψόφον. With Wachtler (*Alcmaeon*, p. 42) I have taken τὸν ἐγκέφαλον as object of περιέχον.

[4] Theophr. l.c. § 29; Diels, *Vors.*, p. 323. The text translated is that given by Diels with Schneider's insertion, accepted by Diels and based upon Theophr. § 34 τὸ μέγεθος τῶν αἰσθητηρίων.

proportion to their magnitude. (Perhaps Anaxagoras did not mean that the larger animals have *finer* sensory discrimination.) sensory power that the small objects should not escape it [1], and it is not unreasonable to suppose that the creature which is able to discern the smaller objects should be able to discern the larger objects as well. Thus it seems that the small animals are better off (on his showing) than the large in respect of some senses, and, so far, the sensory power of the larger animals is inferior to theirs. If, however, on the other hand, it appears that many objects escape the senses of the smaller animals, so far the sensory power of the larger animals is superior [2].' If Anaxagoras for *greater magnitude* had substituted *higher development* his proposition would have been more important. Except so far as size and higher organization accompany one another, there is no fixed relation between the perfectness of sense and the size of the sense-organs or of the animal. It may be, however, that Anaxagoras merely meant that the larger animals have greater, or more voluminous, sensations; not that they have finer sensory discrimination than the smaller animals possess [3].

Object of hearing, sound, is, physically regarded, air set in motion by a shock. § 12. The object of hearing, as already observed, is often referred to under the special name of φωνή—*vocal* sound. 'Anaxagoras held that φωνή is produced by the *breath* (or air in motion) which collides against the *fixed, solid* air and, by a recoil from the shock, is borne onwards to the organs of hearing, just as what is called an " echo " is produced [4].'

---

[1] Cf. Aristotle 442[b] 14.

[2] Theophr. §§ 34–5; Diels, *Dox.*, pp. 508–9. Romanes (*Mental Evolution in Animals*, pp. 80 seqq.) gives 'a general outline of the powers of special sensation probably enjoyed by different classes of animals,' referring to the investigations of Engelmann and Haeckel on the same subject.

[3] For what Aristotle meant by better sensory faculty (ἀκρίβεια αἰσθήσεων) as regards hearing and smelling, cf. *de Gen. An.* v. 2. 781[a] 14–781[b] 29, *infra* § 26.

[4] Diels, *Vors.*, p. 325, *Dox.*, p. 409 Ἀναξαγόρας τὴν φωνὴν γίνεσθαι πνεύματος ἀντιπεσόντος μὲν στερεμνίῳ ἀέρι, τῇ δ' ὑποστροφῇ τῆς πλήξεως μέχρι τῶν ἀκοῶν προσενεχθέντος· καθὸ καὶ τὴν λεγομένην ἠχὼ γίνεσθαι. For this cf. Arist. *de An.* ii. 8. 419[b] 25 seqq., where the production of sound generally is illustrated by reference to the way in which an echo is caused. Aristotle (420[b] 5) distinguishes φωνή

Diogenes of Apollonia.

§ 13. 'When the air within the head is struck and moved by a sound [hearing takes place][1].'

'Hearing takes place when the air within the ears, moved by the external (impression), propagates such motion to the brain[2].' As Diogenes did not regard the brain *per se* as the special organ of intelligence, the last words may be due to Theophrastus. More probably, however, they mean that when the motion set up in the air within the ears has been propagated *to the air-vessels* in the brain, it is thence forwarded to the main air ducts 'in the region of the heart' where conscious perception is awakened. This would be in accordance with the opinions of Diogenes.

'Hearing is most acute in creatures in which the veins are slender, and which have the *meatus* of the ear (analogously to what has been said of the organ of smelling) short, slender, and straight; and which, moreover, have the (external) ear erect and large. For the air within the ears when itself moved moves the air within (the brain)[3]. If the (orifice of the) ear is too wide, when the air within it is moved there follows a ringing in the ear, and the objective sound heard is indistinct, because the body (of air in the ear) on which it (the external impulse) impinges does not remain at rest[4].' 'All creatures live and see and *hear*

Marginal notes: Function and organ of hearing. Motion of air within ear propagated to air around brain. Conditions of acute hearing. The air, the source of all intelligence and all created things, also the source of the faculty of hearing.

---

from ψόφος—ἡ δὲ φωνὴ ψόφος τίς ἐστιν ἐμψύχου· τῶν γὰρ ἀψύχων οὐθὲν φωνεῖ, ἀλλὰ καθ' ὁμοιότητα λέγεται φωνεῖν, οἷον αὐλός κτέ.

[1] Diels, *Vors.*, p. 345, *Dox.*, p. 406 τοῦ ἐν τῇ κεφαλῇ ἀέρος ὑπὸ τῆς φωνῆς τυπτομένου καὶ κινουμένου ⟨τὴν ἀκοὴν γίνεσθαι⟩.

[2] Diels, *Vors.*, p. 344; Theophr. *de Sens.* § 40 τὴν δ' ἀκοὴν ὅταν ὁ ἐν τοῖς ὠσὶν ἀὴρ κινηθεὶς ὑπὸ τοῦ ἔξω διαδῷ πρὸς τὸν ἐγκέφαλον.

[3] In these words we see foreshadowed the doctrine of hearing afterwards elaborated by Arist. *de An.* ii. 8. The air in the ear as a whole is moved by the sound, and this motion is then transferred or propagated to the inner air in the brain. But see p. 259 *infra*.

[4] Diels, *Vors.*, p. 344; Theophr. *de Sens.* § 41 ἀκούειν δ' ὀξύτατα ὧν αἵ τε φλέβες λεπταί, ⟨καὶ ἃ⟩ καθάπερ τῇ ὀσφρήσει κἂν τῇ ἀκοῇ τέτρηται βραχὺ καὶ λεπτὸν καὶ ἰθὺ καὶ πρὸς τούτοις τὸ οὖς ὀρθὸν ἔχει καὶ μέγα· κινούμενον γὰρ τὸν ἐν τοῖς ὠσὶν ἀέρα κινεῖν τὸν ἐντός· ἐὰν δὲ εὐρυτέρα ᾖ, κινουμένου τοῦ ἀέρος ἦχον εἶναι καὶ τὸν ψόφον ἄναρθρον διὰ τὸ μὴ προσπίπτειν πρὸς ἠρεμοῦν.

by the same thing (viz. air), and from this same thing all derive their intelligence as well (τὴν ἄλλην νόησιν)[1].'

Plato.

Function and organ of hearing. The auditory region extends from the head to the liver.

§ 14. 'Plato and his followers think that the air in the head receives a shock, and that this air is then reflected into the intellectual centres[2], and thus the sensation of hearing takes place[3].' This account of Plato's view must be corrected according to the following passages. 'Plato explains hearing through the operation of vocal sound, for vocal sound is a shock, communicated by the air through the ears to the brain and blood, till it reaches the soul; and the motion, caused by this shock, proceeding from the head to the liver, is hearing[4].'

'Hearing, which we have now to examine, is a third mode of sensation within us, and we must set forth the causes to which the affections of this sense are due. Vocal sound in general we must assume to be the shock conveyed by the air, through the ears, to both brain and blood[5], propagated to the soul; and the movement produced by this shock, beginning from the head and terminating in the region of the liver, is hearing[6].'

---

[1] Diels, *Vors.*, p. 350 πάντα τῷ αὐτῷ καὶ ζῇ καὶ ὁρᾷ καὶ ἀκούει, καὶ τὴν ἄλλην νόησιν ἔχει ἀπὸ τοῦ αὐτοῦ πάντα.

[2] The soul, for Plato, perceives *through* the organs of sense (p. 261).

[3] Diels, *Dox.*, p. 406ᵃ 28, ᵇ 28, Plut. *Epit.* iv. 16. Stob. *Ecl.* i. 53 Πλάτων καὶ οἱ ἀπ' αὐτοῦ πλήττεσθαι τὸν ἐν τῇ κεφαλῇ ἀέρα· τοῦτον δὲ ἀνακλᾶσθαι εἰς τὰ ἡγεμονικὰ καὶ γίνεσθαι τῆς ἀκοῆς τὴν αἴσθησιν.

[4] Diels, *Dox.*, p. 500. 14; Theophr. *de Sens.* § 5 ἀκοὴν δὲ διὰ τῆς φωνῆς ὁρίζεται· φωνὴν γὰρ εἶναι πληγὴν ὑπ' ἀέρος ἐγκεφάλου καὶ αἵματος δι' ὤτων μέχρι ψυχῆς, τὴν δ' ὑπὸ ταύτης κίνησιν ἀπὸ κεφαλῆς μέχρι ἥπατος ἀκοήν.

[5] The blood-vessels do duty for sensory nerves.

[6] Plato, *Tim.* 67 B τρίτον δὲ αἰσθητικὸν ἐν ἡμῖν μέρος ἐπισκοποῦσι τὸ περὶ τὴν ἀκοήν, δι' ἃς αἰτίας τὰ περὶ αὐτὸ ξυμβαίνει παθήματα, λεκτέον· ὅλως μὲν οὖν φωνὴν θῶμεν τὴν δι' ὤτων ὑπ' ἀέρος ἐγκεφάλου τε καὶ αἵματος μέχρι ψυχῆς πληγὴν διαδιδομένην, τὴν δὲ ὑπ' αὐτῆς κίνησιν, ἀπὸ τῆς κεφαλῆς μὲν ἀρχομένην, τελευτῶσαν δὲ περὶ τὴν τοῦ ἥπατος ἕδραν, ἀκοήν. Plato's conception of the physiological fact of hearing is thus summarized by Zeller, *Plato* 428 n., E. Tr.: 'The sensations of hearing are caused by the tones moving the air in the inside of the ear, and this motion is transmitted

§ 15. We can hear nothing which does not possess or yield φωνή. 'If the sense of hearing is to hear itself, it must possess φωνή; in no other way could it hear itself[1].' Distinguishing λόγος (rational speech) from διάνοια (thinking), Plato calls the former 'a stream accompanied with sound, proceeding from the soul, through the mouth[2].' 'He defines vocal sound (φωνή) as [on its physical side] air in motion, impelled from the seat of intelligence, through the mouth, and [as physiological stimulus of hearing] a shock caused by the air, through the ears, to the brain and blood, propagated to the soul. Vocal sound, is by an extension of the term, also used in the case of irrational animals and lifeless things, to signify neighings, and mere noises, but properly it is articulate speech, considered as "illuminating" the object of intelligence[3].' 'According to Pythagoras, Plato, and Aristotle, vocal sound is incorporeal. For it is not the air, but the figure bounding the air, or its *surface*, that, in virtue of a certain sort of shock, becomes vocal sound. But every surface is in-

Object of hearing: sound. What vocal sound (φωνή) is: a shock imparted by the air, *through* the ears, *to* the brain and blood, and propagated to the soul; the motion which caused this 'shock' having come from the soul.

through the blood into the brain and to the soul. The soul is thus induced to a motion extending from the head to the region of the liver, to the seat of desire, and this motion proceeding from the soul is ἀκοή.' In this summary two inaccuracies appear. The construction of ἐγκεφάλου τε καὶ αἵματος is not with διά (as Zeller following Stallbaum takes it) but with πληγήν: the conjunctions τε καί were enough to show that these words could not be co-ordinated with ἀέρος after ὑπό or with ὤτων after διά, but must be regarded as objective genitives after πληγήν, thus giving Plato's true meaning, according to the suggestion of Mr. Archer-Hind in his note, which he does not, however, follow in his translation. In the next place Plato does not speak of hearing as 'a motion proceeding *from* the soul.' Like every other form of sensation, it is for him a motion proceeding through the body *to* the soul, involving an *affection* of both conjointly. Cf. *Phileb.* 33 D and *Tim.* 43 C.

[1] *Charm.* 168 D.   [2] *Sophist.* 263 E, *Theaet.* 206 D.

[3] Diels, *Dox.*, p. 407[a] 22, [b] 13, Plut. *Epit.* iv. 19, Stob. *Ecl.* i. 57 Πλάτων τὴν φωνὴν ὁρίζεται πνεῦμα διὰ στόματος ἀπὸ διανοίας ἠγμένον, καὶ πληγὴν ὑπὸ ἀέρος δι᾽ ὤτων καὶ ἐγκεφάλου καὶ αἵματος μέχρι ψυχῆς διαδιδομένην. λέγεται δὲ καὶ καταχρηστικῶς ἐπὶ τῶν ἀλόγων ζῴων φωνὴ καὶ τῶν ἀψύχων ὡς χρεμετισμοὶ καὶ ψόφοι· κυρίως δὲ φωνὴ ἡ ἔναρθρός ἐστιν ὡς φωτίζουσα τὸ νοούμενον. It is noticeable here that καὶ ἐγκεφάλου καὶ αἵματος seems to show that the writer neglected or missed the true construction of the corresponding words of Plato, *Tim.* 67 B.

corporeal. It is moved, indeed, together with bodies, but, in its own nature, it is absolutely bodiless; as, when a stick is bent, it is the material of it that is bent, but its surface is not affected thereby [1].'

§ 16. 'Plato states that vocal sound is a shock communicated by the air through the ears to the brain and blood, propagated to the soul. According as it is swift or slow in its motion, it is shrill or grave in its tone. One vocal sound is in accord with another when the beginning of the slower is similar to the ending of the more rapid [2].' Theophrastus seems to have intended, by the change he introduces into the order of Plato's words, to indicate that which has been above (p. 106, n. 6) given as their true construction. He makes it plain that the shock is imparted to the brain and blood, and that, grammatically, πληγή governs ἐγκεφάλου καὶ αἵματος. The blow—the shock—is, in the case of speech, due to the soul causing the air in the respiratory organs to strike against the sides of the ἀρτηρία, or windpipe (Arist. 420ᵇ 28).

'In the same way we must look for the explanation of sounds, which present themselves to us as shrill or grave according as they are swift or slow, their movements now harmonious, at other times discordant, according to the similarity or dissimilarity of the motion excited in us by them. For when the movements of the preceding and more rapid sounds are ceasing, and have just arrived at a speed similar to that of the movements with which the succeeding sounds, adding their movements to the preceding, stimulate them, then the slower sounds catch them up, and doing so excite no confusion, and introduce no

---

[1] Diels, *Dox.*, p. 409ᵃ 25, Plut. *Epit.* iv. 20 Πυθαγόρας Πλάτων Ἀριστοτέλης ἀσώματον [sc. τὴν φωνήν]. οὐ γὰρ τὸν ἀέρα, ἀλλὰ τὸ σχῆμα τὸ περὶ τὸν ἀέρα καὶ τὴν ἐπιφάνειαν κατὰ ποιὰν πλῆξιν γίνεσθαι φωνήν· πᾶσα δὲ ἐπιφάνεια ἀσώματος· συγκινεῖται μὲν γὰρ τοῖς σώμασιν, αὐτὴ δὲ ἀσώματος πάντως καθέστηκεν· ὥσπερ ἐπὶ τῆς καμπτομένης ῥάβδου ἡ μὲν ἐπιφάνεια οὐδὲν πάσχει, ἡ δὲ ὕλη ἐστὶν ἡ καμπτομένη.

[2] Diels, *Dox.*, p. 525. 17, Theophr. *de Sens.* § 85 φωνὴν δὲ εἶναι πληγὴν ὑπὸ ἀέρος ἐγκεφάλου καὶ αἵματος δι' ὤτων μέχρι ψυχῆς· ὀξεῖαν δὲ καὶ βαρεῖαν τὴν ταχεῖαν καὶ βραδεῖαν· συμφωνεῖν δ' ὅταν ἡ ἀρχὴ τῆς βραδείας ὁμοία ᾖ τῇ τελευτῇ τῆς ταχείας.

alien element; but introducing into them the beginning of
a slower movement, after the pattern of that formerly faster
but now slowing down, they blend and form with them
one single auditory affection of shrill and deep combined;
whence it is that they afford *pleasure* (ἡδονήν) to the foolish,
but *joy* (εὐφροσύνην) to the wise, as the latter contemplate,
in them, the divine harmony, thus showing us its own
copy in mortal movements [1].'

§ 17. In translating this passage, a special difficulty
arises from the want of an English word to distinguish
κίνησις from φορά. To render κινήσεις by 'vibrations [2]'
would be easy, if it did not involve the introduction of
a later scientific conception scarcely comprehended in
Plato's thought. We should not hastily ascribe the scientific theory of the causes of high and low notes to Plato,
Aristotle, or their predecessors. Alexander (Hayduck,
p. 39), commenting on Arist. *Met.* i. 5. 985^b 26, speaking of
the Pythagorean theory of the harmony of the spheres,
represents the high notes in the scale as assigned by the
Pythagoreans to the outer spheres, merely because these
spheres are at the end of longer radii, and therefore move
more rapidly, than those nearer to the centre. Not the
rapidity of vibrations *in* air, but that of the mere onward
movement *of* air or portions of air, seems to have been
for Plato the producing cause of height in tones.

Moreover, Plato, like his predecessors, believed that a
definite portion of air was projected forwards from the
sonant body to the ear; not that a *mere movement* took
place in the medium. Certain physical facts at the basis
of harmonic theory, e.g. that halving the length of a tense
string raises its tone an octave, were no doubt known to
the Pythagoreans and to Plato. That the former had
determined the principal harmonic ratios is plain from
the remains of Philolaus (Boeckh, *Philol.*, pp. 65–86), and
these ratios were known to Aristotle (*de Sens.* iii. 439^b 31).

Plato did hold the modern vibration theory of sound.

---

[1] Plato, *Tim.* 80 A–B.
[2] Wundt does so (*H. and A. Psych.* p. 67, E. Tr.) in alluding to the
psychology of this period.

What is not so certain is how far they had any idea of the physical fact that a sonant object gives rise to a succession of air-vibrations[1], whose frequency and amplitude condition the pitch and loudness of sound.   Mr. Archer-Hind thinks it 'evident from Plato's language that he conceived the acuter sound both to travel more swiftly through the air, *and to have more rapid vibrations,*' thus coming very near the correct explanation of pitch.   But from the way in which Plato connects sounds, cupping-glasses, projectiles, &c., under one formula of explanation, it would seem as if the notion of air-vibration—i.e. vibration in an elastic medium—did not come before his mind at all.   The swiftness or slowness of the sound-movement is for him just like that of the projectile; only that in the former case there is a succession of sound-stimuli, portions of air started off, as it were, one after another from the sonant body at a certain velocity, and at certain greater or smaller, regular or irregular, intervals.   The theory of harmonic ratios in which Pythagoreanism delighted seems to be here unapplied by Plato, though elsewhere he shows himself fully acquainted with it[2].   I have, accordingly, refrained from using 'vibrations' as a rendering of κινήσεις here, because such a rendering would seem to credit Plato with knowing that air is an elastic medium vibrating and transmitting sound by a series of contractions and expansions.   Of this theory, originated by Heraclides or Strato, Plato had no conception.

Ethical value of the sense of hearing.   § 18. From the last extract it becomes apparent that Plato was aware of the ethical and emotional importance of certain classes of sound.   'Harmony and rhythm[3] are

---

[1] The theory of vibration frequencies, as the cause of high or low tones, seems rather to have originated with Heraclides or Strato, according to whom each sound is composed of particular 'beats' (πληγαί) which we cannot distinguish as such, but perceive as one unbroken sound, high tones consisting of more such beats, low tones of fewer.   Plato like Aristotle (contrast, however, Pseudo-Arist. 800ᵃ 1–5) held that high or low in tone depends on the *speed at which the sound travels through the air* towards the ear.   Cf. Zell. *Arist.* ii. 379 n. and 465-6 n., E. Tr.; von Jan, *op. cit.* pp. 135 seqq.

[2] Cf. *Phileb.* 17 C–E.

[3] Cf. Grote, *Plato*, iii. p. 266; Pl. *Tim.* 47 C–E.

presents to us from the Muses, not, as men now employ them, for unreflecting pleasure and recreation, but for the purpose of regulating and attuning the disorderly rotations of the soul, and of correcting the ungraceful and un-measured movements natural to the body.' In the *Republic* and *Laws* also Plato expresses his high appreciation of the educational value of music duly regulated and employed[1]. In this he was in substantial agreement with Aristotle. Indeed he anticipates the dictum of the latter[2] that hearing is more important than seeing for the development of mind and character. 'Of sound and hearing the same account must be given [as has been given of seeing]; to the same ends and with the same intent they have been bestowed on us by the gods. For not only has speech been appointed for this same purpose, whereto it *contributes the largest share*, but all such music as is expressed in sound has been granted for the sake of harmony[3].' The facts that λόγος is (*indirectly*, as Arist. says) an object to the sense of hearing, and that on λόγος higher education chiefly depends, are sufficient of themselves to secure for this sense a paramount place in the development of mind and character.

Its psychological value for the development of intelligence.

Aristotle.

§ 19. Aristotle[4] divides sound under two heads, ψόφος and φωνή. The former is the general name, including noises; the latter is properly used of vocal and articulate sound, but often extended to include musical sounds whether produced by voice or otherwise.

Taking sound first in the more general sense, he distinguishes between its actual and potential aspects. There are certain things which are incapable of producing sound, e.g. wool; others are capable of producing sound, e.g. bronze, and smooth hard substances. As the former are, even potentially, soundless, the latter are potentially sonant,

Object of hearing—sound: divided into ψόφος and φωνή. Cause of sound in former, more general, sense. Three conditions involved: (a) a sonant thing, (b) a

---

[1] Cf. *Rep.* 530 C–531 C, with Adam's Commentary thereon.
[2] *De Sens.* i. 437ᵃ 6–17.
[3] Plato, *Tim.*, 47 C, Archer-Hind's Trans.
[4] For what follows see *de An.* ii. 8. 419ᵇ 5 seqq.

shock communicated to it by a blow from something else, (c) a movement in a medium implied in this. The celestial spheres do not sound: Why? Vibration of hollow bodies. Air and water both media of sound. The blow, not the medium, the chief determinant or factor of sound. Air or water may serve both as medium and as sonant body: how this is.

even when not actually sounding [1]. 'As it is possible for a person possessing the faculty of hearing not to hear actually at some given moment, so a thing may have the property of sounding without always actually doing this. When, however, that which can hear realizes its potentiality, and also when that which can sound does sound, then the realized faculty of hearing and the realized sound both concur; so that the former may properly be named "actual hearing" (ἄκουσις), and the latter "actual sounding" (ψόφησις).' Actualized sound is a local movement of something [2], and involves the relation of some one thing to some other thing, in some third as medium [3]. This third thing is normally air in the case of land animals. That which physically causes sound is a shock or blow. This cannot occur when only one thing is concerned; for that which gives the blow and that which receives it are two different things. That which sounds does so in relation to something else, and in a medium, for the blow implies local movement (φορά). That which moves with a movement of its own may produce sound: that which, as a boat on a river, moves because the thing in which it is fixed moves, produces no sound. Hence the celestial bodies move without a sound, and we need not assume a 'music of the spheres' which none can hear [4]. Sound, then, is not a shock or blow of any casual thing against something else; for wool if struck gives no sound. Bronze on the contrary does produce sound, as do all smooth and hollow things. The bronze sounds because it is smooth; the hollow things sound because after receiving the first blow they produce many, owing to the reverberation (τῇ ἀνακλάσει) taking place when that which has been set in motion within them is unable to find an exit. Sound is heard in air, and in water also. It is not, however, the medium, i.e. the air or the water, that chiefly determines

---

[1] Cf. 425^b 28–426^a 8.

[2] φερομένου τινὸς κίνησις, 446^b 30.

[3] πᾶν ψοφεῖ τύπτοντός τινος καί τι καὶ ἔν τινι, τοῦτο (sc. τὸ ἐν ᾧ) δ' ἐστὶν ἀήρ, 420^b 14, 419^a 32.

[4] 291^a 1–15.

the production of sound. It is the blow or shock ($\pi\lambda\eta\gamma\acute{\eta}$) caused by one body striking against another[1] in the air. The air or water, too, may serve as one of the bodies which by their collision produce sound; but these are less sonant than the solid bodies[2]. They may so serve to produce sound when the air, e. g., holds its ground on being struck, and is not at once dissipated. Hence it sounds only when it is struck quickly and forcibly. The movement of the striker must be too rapid for the dispersion of the mass of air struck. This it may well be; just as one might get in a blow at a moving heap, or whirling vortex-ring, of sand[3] in rapid motion before it could retire from, and so elude, the blow.

§ 20. An echo occurs when the mass of air set in motion by the 'stroke' rebounds like a ball from another portion of air formed into a single mass by some receptacle which confines it within fixed boundaries and prevents it from being suddenly dispersed. It would seem as if echoes must be always occurring, though not always audibly; just in the same way as light is being always reflected, as is proved by its diffusion everywhere.

What is said, and rightly said, to be the chief agent in determining the hearing (as distinct from the production) of sound is *vacuum*[4]. But by this what people generally mean is *air*, not absolute void. The organ of hearing proper consists of air[5]; and the air without us causes us

Echo: how produced. Reflexion of sound compared with reflexion of light. Ancients right in saying 'vacuum' determines the hearing of sound, if by 'vacuum' is meant air. The organ of hearing

---

[1] In what follows Bäumker (*op. cit.* p. 27) seems right in taking Aristotle to mean that sound is producible by means of air or water alone in contact with a solid striking body. Such sound is not so strongly pronounced however. Torstrik is wrong in proposing to strike out ἀλλ' ἧττον. Themistius illustrates by the cracking of a whip, which shows that he took ἐν ἀέρι here to refer to a blow struck by one solid in the mere air or water and yet producing sound. As Torstrik in his clear note on 419ᵇ 20 says, 'iam ei in mentem venit stridor ille vel sibilus quem virga vel flagro efficimus celeriter discusso aere: ibi enim τὸ ἐν ᾧ quodammodo etiam τοῦ πρὸς ὃ vices gerit.'

[2] The terms *fluid* and *solid* are generally opposed *inter se* by Aristotle as well as by moderns.

[3] For ὁρμαθὸν ψάμμου here cf. *Hermathena*, No. xxx, 'Miscellanea,' p. 73.

[4] Cf. 656ᵇ 13–16, together with 420ᵃ 18 seqq.

[5] 656ᵇ 16 τὸ δὲ τῆς ἀκοῆς αἰσθητήριον ἀέρος εἶναί φαμεν, 425ᵃ 4.

proper is formed of an air-chamber built into 'the ear.' This takes up the sound-movements of the outer air, and conveys them to the soul, in its sensorium. Democritus implicitly criticized. Animals do not receive air at all parts of the body: nor do they hear at all parts. Hearing under water; possible conditionally.

to hear when it has been set in motion as one continuous body. Owing to the fact that it is so easily dispersed, this outer air yields no audible sound unless the solid which has been struck is smooth. In this case the air to which the shock is communicated rebounds in a single united mass, owing to the nature of the superficies of the said solid; for the superficies of a smooth body is one. Anything, therefore, which is capable [1] of causing motion in a single mass, of air, which reaches continuously to the organ of hearing, is capable of producing sound [2]. For the organ of hearing proper is physically homogeneous with the air ($\sigma\nu\mu\phi\nu\grave{\eta}s\ \mathring{a}\acute{\epsilon}\rho\iota$) [3]. Since then the air is one [4] it follows that when the outer air is moved, the inner air is moved also [5]. Hence it is not true that an animal hears with all parts of the body [6], nor does the air enter the body at all parts; for the part which should receive the movement, so as to give it effect for consciousness, has not in every part of the body an inner air at its disposal such as it has in the ear [7]. But on this inner air hearing depends. Air in general is soundless owing to its being easily dispersed: when a portion is prevented from being dissipated, and this is affected by the shock of a blow, it yields or transmits sound. Now the air within the ears [8] has been built into its chamber in order that, being undisturbed by the general movement of the atmosphere, it may be sensitive to the different kinds of auditory movements propagated towards

[1] Not all things are so capable : οὐ δὴ πᾶν· οἷον ἐὰν πατάξῃ βελόνη βελόνην.

[2] As Trendelenburg says : the air at the surface of the solid struck is here referred to as being one : that air which propagates the sound to the ear is referred to as one and continuous.

[3] For the above cf. 419^b 5–420^a 4, 656^b 16, 781^a 14 seqq.

[4] 420^a 4: I translate ἕνα ἀέρα, the restoration of Steinhart, cf. 419^b 35.

[5] 420^a 5 : I translate Torstrik's reading ὁ εἴσω κινεῖται.

[6] This implicitly controverts, with the same unfairness as Theophrastus shows, the theory of Democritus. See §§ 7–8 *supra*.

[7] οὐ γὰρ πάντῃ ἔχει ἀέρα τὸ κινησόμενον μέρος καὶ ἔμψυχον.

[8] 420^a 9, 656^b 15, where the expression τὸ γὰρ κενὸν καλούμενον ἀέρος πλῆρές ἐστιν refers to the hollow of the ears in connexion with the whole occiput, or hinder portion of the cranium, which Aristotle strangely regarded as vacant, or containing air only.

it. The external medium which is to receive and transmit all sounds must in itself be free from sound[1]. The outer air is therefore *per se* soundless, a quality which it owes to its being so easily dispersed. But the air within the ear— the portion of air which is the essential element in the organ of hearing—as distinguished from the outer air which is the external medium—has a proper motion of its own. Thus it has a peculiar resonance, like a horn ; and this, while it lasts, is a sign that the auditory faculty is unimpaired. When this ceases, it is a proof of deafness. We can hear to some extent under water ; because the water does not enter the air-chamber of the ear. If it did so, hearing would be at an end. Hearing ceases to be possible, also, if the tympanic membrane is injured, just as blindness ensues if the membrane covering the eye is injured. As the water-holding eye is joined with the watery brain, so the air-holding ear is connected with the air-holding hinder part of the cranium[2]. Perhaps the air in the ear is ultimately connected with that in the lungs—the origin of all the air in the body[3]. At all events the essential part of the organ of hearing is the air-cell which has been thus described as ' built into ' the ear.

§ 21. Is it the striker that sounds, or the thing struck ? The answer is that both do so, each in its own way. Sound is a movement of something mobile ; something that is moved like things which rebound from smooth surfaces. The surface must be smooth, in order that the air may rebound from it in a single mass (ἄθρουν). Sound, unlike light, *travels* in the air from the sonant body to the ear. This is plain from the fact of our seeing a blow struck at a distance, but not hearing the sound of the blow till some time after[4]. Articulate sounds are due to the conforma-

Which sounds— the striker or the thing struck ? Sound, unlike light, travels.

[1] 418ᵇ 26 ἔστι ... δεκτικόν ... ψόφου ... τὸ ἄψοφον.

[2] 491ᵃ 31 τούτου (sc. the whole cranium) δὲ μέρη τὸ μὲν πρόσθιον βρέγμα ... τὸ δ' ὀπίσθιον ἰνίον ... ὑπὸ μὲν οὖν τὸ βρέγμα ὁ ἐγκέφαλός ἐστιν, τὸ δ' ἰνίον κενόν. Cf. 494ᵇ 24, ᵇ 33, 656ᵇ 18 πάλιν δ' ἐκ τῶν ὤτων ὡσαύτως πόρος εἰς τοὔπισθεν συνάπτει.

[3] 781ᵃ 31 διὰ τὸ ἐπὶ τῷ πνευματικῷ μορίῳ τὴν ἀρχὴν τοῦ αἰσθητηρίου εἶναι τοῦ τῆς ἀκοῆς. [4] 446ᵃ 20 seqq.

tion of the moving air. Such sounds are less accurately heard at long distances, because the form of the movement in the air becomes altered on its way to the ear[1].

§ 22. Differences of quality such as sharp and grave are potentially existent in the sounds themselves, but are actualized only in the actual ψόφησις with its correlative ἄκουσις. These two—ψόφησις and ἄκουσις—are two aspects of one fact, only distinguishable by reason. Just as without light colours are not *seen*, though potentially in the coloured objects, so without ψόφησις—the actualization of sound—and its correlative ἄκουσις—the actual perception of sound—the quality of sharp or grave is not heard. These terms, sharp and grave (ὀξὺ καὶ βαρύ), thus applied are really metaphorical, being transferred from objects of touch to those of hearing. The sharp is that which moves the sense much in a little time; the grave that which moves it little in much time. The sharp *as heard* is not literally swift, nor the grave slow; yet the quality of the former as perceived is due to the rapidity of the motion that causes it; while the quality of the latter is owing to the slowness of the corresponding motion[2]. There seems to be an analogy between that which to the touch is sharp or blunt, and that which to the sense of hearing is sharp or grave. The sharp as it were *pierces*, while the blunt *pushes*, because the one effects its movement in a short, the other in a long time, so that incidentally the one sound is swift the other slow[3]. Theophrastus (*apud Porphyr. Frag.* 89) controverts this theory, common to Plato and Aristotle, which accounts for the difference of sharp and grave in sound by more rapid local movement in the stimulus of the former, less rapid in that of the latter. The stimulus of the higher note, he thinks, does not move onward more swiftly than that of the lower[4]. Strato and

Qualitative differences of sound, e.g. pitch, exist potentially in sounds per se: actually, only in sounds qua heard. So with colours. The terms ὀξύ and βαρύ metaphorical in relation to sound. Physical nature of sharp and grave. Origin of theory of vibration-frequencies. The sense of hearing, like all others, is or involves a μεσότης or λόγος: as shown by its perception of the λόγοι of chords. Hence sounds that are too loud impair or destroy this sense.

[1] 446ᵇ 6. Cf. *Probl.* xi. 51, 904ᵇ 27 ἡ φωνὴ ἀήρ τις ἐσχηματισμένος.

[2] Aristotle seems to have in mind here Plato's account of sharp and grave in the *Timaeus.* Cf. 'HEARING,' Plato, §§ 16–17 *supra.*

[3] 786ᵇ 7–788ᵇ 2, where the differences of ὀξύ and βαρύ are explained with reference to male and female voices.

[4] Cf. Zeller, *Aristotle,* ii. p. 379 n. (E. Tr.).

the writer of the tract Περὶ Ἀκουστῶν teach that every sound
stimulus is composed of πληγαί or beating vibrations which
we cannot distinguish as such, but perceive as one unbroken
sound; high tones, whose movement is quicker, consist of
more vibrations, low tones of fewer. But the forward motion
of the stimulus through the air from object to organ is of
the same speed in either case [1].

The sharp and the grave are contraries between which the
object of hearing in general lies. The sense of hearing presides over the province contained within or bounded by these
contraries. Every sense [2] occupies or represents a mean.
Thus hearing stands between any two degrees of pitch, and
on this μεσότης depends its discriminative power. It is a proportion or λόγος of the ἐναντία, and, while indifferently poised
with respect to all, contains in itself the discriminant between
any two different sounds whatever. A concord such as the
octave is a ratio of 1 to 2. But this (as *object* of hearing) and
ἀκοή (as *sense* of hearing) are, at the moment when both are
actualized, *one*; hence the latter, sc. ἀκοή, is also a ratio (λόγος)
(see *infra* § 30). Hence, too, excessively loud sounds are
injurious to the faculty of hearing, as they tend to destroy
the ratio or proportion (the finely balanced, delicately poised
position) which it holds between the ἐναντία, and amongst,
or in relation to, all possible pairs of differences of pitch, and
hence to destroy the μεσότης on which rests its discriminative power. The same is true of each other sense as
regards its object. On the other hand, those composite
objects which in their composition exhibit the qualities
corresponding to the nature of their organ, are pleasure-
giving. Thus concords which themselves involve a ratio, are
pleasing to the sense of hearing; and the same may possibly,
in some unknown way, be true of the relation between each
other special sense (or sense-organ) and its proper object [3],
when the pleasure from the latter is truest and greatest.

§ 23. Thus far we have considered ψόφος or sound φωνή as
distinct

---

[1] Zeller, *op. cit.* ii. 465 n.
[2] Cf. SENSATION IN GENERAL, § 24.
[3] See *de An.* ii. 12. 424ª 27–424ᵇ 1, 426ª 27–ᵇ 12.

from ψόφος. generally. Voice (φωνή) is a special kind of sound pro-

duced by living creatures. Inanimate beings do not utter voice, though by a metaphor a flute is said to do so, as are also other sonant things capable of varieties of tone (ἀπότασις), and hence of producing melody and διάλεκτος, or 'discourse of sound.' 'Απότασις is the genus which includes ἐπίτασις and ἄνεσις, while μέλος is used for notes in the melodic series. It is not so easy to give a direct translation of διάλεκτος as here employed. I have rendered it by a metaphor, as being distinct from μέλος and used to designate the effect of a number of instruments played in harmony or in unison. To 'discourse sweet music' would not unnaturally be expressed by a metaphorical διαλεχθῆναι. Articulation and harmony are terms as suitable for the interplay of ideas in conversation as for that of tones in concert. The voices of animals are possessed of these musical qualities.

Voiceless
animals :
the fish
of the
Acheloüs
have not
real voice :
they only
make a
certain
kind of
noise.
Nature's
twofold
employ-
ment of the
inhaled
air : regu-
lation of
tempera-
ture and
production
of voice.
The organs
of voice
and articu-
lation.

§ 24. There are, however, many animals which have no voice : e. g. those called bloodless, and also fishes. Those fishes which, e. g. in the river Acheloüs, are said to utter voice, merely make a noise with the gills or some such part. It is quite natural that fishes should not have voice ; since, as we have said, sound depends on movement of air, while voice is the sound made by an animal, but not with every given part of its organism, it follows that only those animals which inhale air have voice [1]. Nature employs the air that is inhaled for two objects, just as she employs the tongue for tasting and also for speaking. The two objects for which she employs the breath are (a) the regulation of the internal heat of the body ; and (b) the production of voice. The first of these objects is sub-servient to the purpose of the animal's existence, the second is a condition of its well-being.

The windpipe is an organ of respiration [2]. The organ to

[1] *Hist. An.* iv. 9. 535ª 27–536ᵇ 24.

[2] φάρυγξ is here (535ª 29) used for λάρυγξ (535ª 32). In Aristotle's time these words had not come to be distinguished as they now are. Nor does φάρυγξ here differ substantially from ἀρτηρία (sc. ἡ τραχεῖα) further down (535ᵇ 15), hence I have rendered it by 'windpipe.' 'Αρτηρία of course had not come yet to mean 'artery.'

which this is subservient is the lung, possession of which is due to the fact that land animals have more heat than others. The region of the heart[1] is that which primarily needs respiration and its cooling effects; hence the necessity that the air should enter this region as it does in the process of respiration. One consequence of this arrangement is that a shock can be imparted by the soul, which tenants that region, to the inhaled air; by this shock the latter is struck against the trachea, as it is called[2]; and by the stroke vocal sound is produced.

§ 25. For, as has been said, not every animal sound is vocal sound: not e.g. *clucking* with the tongue, or *coughing*. The production of voice implies that the organ which communicates the shock in the first instance must be animate, and have some mental representation accompanying its action[3]. There must be this representation, because voice is *significant* (σημαντικός) sound[4], and does not merely imply *any* shock imparted to the air inhaled, as for example, in coughing. On the contrary, in uttering voice, one uses the inhaled air in order to make that which is in the trachea strike against the walls of the trachea itself. Hence it is that one cannot utter voice while in the act of inhaling or exhaling, but only while holding the breath. He who thus holds the breath and speaks, excites, in doing this, a movement in the fund of breath held in. Fishes do not inhale; therefore they do not possess a windpipe, and hence they have no voice[5].

Voice is sound produced by animate beings and signifying something. Voice is produced only while one holds the breath. Why fishes are voiceless.

§ 26. 'In accurate hearing as well as in accurate smelling two things are involved: one is the discernment as far as possible of the different qualities of the objects of these senses; the other is the power of hearing or smelling at a long distance. The power of keenly dis-

Meaning and conditions of perfect hearing. The 'connatural spirit.' In

[1] Here the lung is said to be in the 'region of the heart'; cf. 668^b 33 seqq.   [2] πρὸς τὴν καλουμένην ἀρτηρίαν.

[3] δεῖ ἔμψυχον εἶναι τὸ τύπτον καὶ μετὰ φαντασίας τινός. Cf. 786^b 21 τοῦ δὲ λόγου ὕλην εἶναι τὴν φωνήν.

[4] Even the inarticulate sounds of the voice of the lower animals (οἱ ἀγράμματοι ψόφοι οἷον θηρίων) are significant (δηλοῦσί τι). 16^a 28.

[5] For §§ 20–27 cf. de An. ii. 8. 419^b 25–421^a 6.

<table>
<tr><td>

learning
to repeat
from oral
dictation
we act like
a phono-
graphic
record.
Why
persons
yawning,
or violently
exhaling,
hear less
well than
when in-
haling the
breath.
Hearing
affected by
changes of
season,
humidity
of atmo-
sphere, &c.
The sort
of auditory
apparatus
which
favours
perfect
hearing.
Man's
senses
compared
with those
of other
animals.

</td><td>

cerning the qualities of their objects is dependent on the
organs of these senses, just as the corresponding power
depends on the organ of seeing, in which this power
resides if both the organ itself and the membrane enclosing
it be free from alien matter. For the passages of all the
sensory organs, as has been stated in our work On Sensa-
tion, extend towards the heart, or in creatures without
a heart, to the analogous organ. The passage of the
sense of hearing, since the organ of this sense is formed
of air, terminates at the point where the connatural spirit
produces [1], in certain animals, a heaving, pulsating, move-
ment, in others maintains the respiratory process. On
the fact of its terminating here—in the region or seat
of the central or common sense—rests the power we have
of learning from dictation, by which the sounds we make
echo verbatim those which we have heard; which implies
that the movement expressed through our speech is an
exact reflex of a movement which had passed in through
our organ of hearing, as if both were impressions struck from
one and the same die; and thus it is that one utters in
speech exactly that which he has heard.' Thus in repeating
from dictation one acts like a phonographic record.

'Persons yawning or exhaling hear less well than persons
inhaling, because the starting-point (τὴν ἀρχήν) of the organ
of hearing is adjacent to the part concerned in breathing,
and hence, when the organ of breathing sets the breath
in motion, the apparatus of hearing is at the same time [2]

</td></tr>
</table>

[1] For what precedes cf. 456ᵃ 1–29. Τὸ σύμφυτον πνεῦμα: this
pervades the channels of hearing and smelling, and is the medium
by which sounds and smells are conveyed to their respective senses.
Cf. 744ᵃ 3 ἡ δ᾽ ὄσφρησις καὶ ἡ ἀκοὴ πόροι συνάπτοντες πρὸς τὸν ἀέρα τὸν
θύραθεν πλήρεις συμφύτου πνεύματος, περαίνοντες πρὸς τὰ φλέβια τὰ περὶ
τὸν ἐγκέφαλον κτέ.

[2] 781ᵃ 30 seqq. The Didot translation is: 'quoniam principium
sensorii auditus parti spiritali impositum est, et quatitur moveturque
spiritus eodem quo instrumentum movet tempore'—as if τὸ πνεῦμα were
subject to σείεσθαι καὶ κινεῖσθαι. This is a grammatically possible construction, but the sense it gives is irrelevant. It is needless to say that
when the organ of breathing does its office, the breath is moved, and
besides Aristotle's point is that there is a disturbance of *hearing* at

shaken or moved ; for the organ of breathing while exciting movement is itself moved, ⟨and therefore excites movement in the adjacent organ of hearing ¹⟩. The like happens in wet seasons and climates : the ears seem to be filled with breath owing to their proximity to the organ which governs respiration. Accuracy in discriminating the sensible qualities of sounds and odours depends, therefore, on the clearness of the sensory organ and of the membrane which covers it. For, as in the case of vision, so in such cases the movements that take place under these conditions are all plain to immediate intuition.'

As regards the capacity or incapacity of certain animals for hearing or smelling distant objects, the case is likewise analogous to that of vision. 'Animals which have, in front of the sensory organs, as it were, conduits extending to a considerable length through the sensory tracts concerned, are capable of perception at long distances. Hence animals, like Laconian hounds, whose nostrils are long can discern odours keenly at a distance. Likewise animals with ears which are long and projecting, like those of certain quadrupeds, cornice-wise (ἀπογεγεισωμένα) far out from the head, and which have the spiral interior also long, ⟨can hear at great distances⟩; since such ears catch the movement from afar off, and deliver it to the sensory organ. As regards the general perception of distant objects man is inferior to almost all other animals, in proportion to his bodily size ; but on the other hand he is superior to all in the nicety of his discrimination of the sensible distinctions in objects perceived. The cause of the latter is that his sensory organ in each case is purest and least contaminated with earthy or corporeal matter, and he, of all animals, has naturally the most delicately fine skin in proportion to his bodily magnitude ².'

such a time. Hence I take τὴν ἀρχὴν τοῦ αἰσθητηρίου τοῦ τῆς ἀκοῆς again as subject in the second clause, and τὸ πνεῦμα as accus. after κινοῦντος.

¹ The facts referred to by Aristotle are due to the proximity of the Eustachian tubes to the auditory passage : owing to this when we yawn or exhale forcibly we have a feeling of obstruction in the ears, and hearing is for the moment impaired.

² For § 26 cf. *de Gen. An.* v. 2. 781ᵃ 15–ᵇ 29.

Confusing statements. The ears have no passages leading into the brain, but have passages into the hinder part of the cranium called τὸ κενόν; also the ear has a passage leading to the οὐρανός or palate. Yet the organs of hearing and smelling are πόροι, filled with a σύμφυτον πνεῦμα, running into blood-vessels surrounding the brain.

§ 27. Aristotle states, as we have seen, that hearing depends upon *vacuum*, or what is taken for such, i. e. a portion of air enclosed in the inner chamber of the ear. This, however, is somehow connected with the air in the occiput, and the results of the sound-movements in the outer air which affect it are conveyed within; and from this interior air the movements ultimately find their way to the region of the heart, which is the central or common sensorium. Of the passages connecting the external auditory apparatus with the interior of the head, he does not seem to have had a clear conception. 'One [viz. the inner] part of the ear is nameless, the other is called the "lobe." The whole consists of cartilage and flesh. Inwardly its formation is like that of spiral shells, the bone at the inner extremity (into which, as last receiver, sound comes) being in shape like the [outer] ear. This inner ear has no passage (πόρος) into the brain, but it has one to the palate (οὐρανός) and a vein (φλέψ) extends into it from the brain.'

'Certain animals, as was to have been expected, have the organ of hearing situated in the head. For what is called the vacuum in the cranium is really full of air, and the organ of hearing, as we hold, consists of air. Now passages (πόροι) lead from the eyes into the blood-vessels around the brain; and a passage leads back, likewise, from each of the ears and connects it with the hinder part of the head[1].' 'The organs of sight, like all the other organs of sense, are attached to passages (ἐπὶ πόρων), but while the organs of touching and tasting consist either of the body, or of some part of the body, of animals, those of smelling and hearing are themselves passages filled with connatural spirit (πλήρεις συμφύτου πνεύματος) in communication with the external air, and terminating inwardly in the blood-vessels which surround the brain and extend from it to the heart[2].' It is by means of these blood-vessels that the external auditory impulses are finally conveyed to the central sensorium.

---

[1] Cf. 492ᵃ 15–21, and 656ᵇ 13–19.
[2] *De Gen. An.* ii. 6. 743ᵇ 36–744ᵃ 5.

§ 28. Aristotle was even more strongly impressed than Plato with the intellectual, ethical, and aesthetic importance of hearing compared with the other senses. It contributes not only to the preservation of animals, but to their well-being, and, in the case of all those which possess intelligence, assists powerfully in the development of this. 'As regards primary vital needs, the sense of sight is more essential, and more directly contributory, to an animal's security: but, as regards intellectual development, and in its secondary consequences, the sense of hearing takes a higher place. . . . True, the sense of hearing only imparts knowledge of the different sensible qualities of sound, and in the case of a few animals, those of vocal sound ; yet, in its secondary effects and their bearing on intelligence, the part contributed by hearing is greatest of all. For to rational discourse (λόγος) is due the power we have of learning, and such discourse is an object of hearing, not indeed directly, since what we hear is as such merely *sound*, but incidentally, for it is made up of words, and each of these is a significant sound (σύμβολον). Hence if we compare persons congenitally blind with persons congenitally deaf, we find that the former are the better developed intellectually[1].' That learning depends on the sense of hearing, so that those who cannot hear cannot learn, is dwelt upon by Aristotle elsewhere. 'Creatures may be endowed with a certain amount of intelligence without having the power of learning, as is the case with all which are destitute of the faculty of hearing sounds, as, for example, bees[2].' Speaking of the habits and characteristics of the lower animals, after pointing out how these vary in intelligence, he goes on : 'Some of them possess in common with man, to a certain degree, the faculty of teaching and learning, whether from one another, or from mankind ; those, that is, which have

Marginal notes: Biological, intellectual, ethical, and aesthetic values of the sense of hearing. In its secondary effects hearing has higher psychological worth than seeing. Reasons of this. Words are sounds with ideas annexed to them : 'learning' depends on hearing. Animals learn by this sense if they can distinguish significant sounds. Seeing gives us *particulars* in vast numbers; hearing gives us *general notions*.

---

[1] Cf. *de Sens.* i. 437ᵃ 1-17.

[2] *Met.* i. 1. 980ᵇ 22-4 φρόνιμα μὲν ἄνευ τοῦ μανθάνειν, ὅσα μὴ δύναται τῶν ψόφων ἀκούειν. Evidently the connotation of μανθάνειν was less wide than that of our 'learn.'

the auditory sense, and can not merely hear sounds, but also distinguish by this sense (διαισθάνεται) the different qualities of significant sounds[1].' But the importance of hearing as an instrument of education arises chiefly from the fact already mentioned that words (ὀνόματα) are in their nature general (σύμβολα). They are marks of typical mental impressions associated with them by both speaker and hearer. They stand for notions. The impressions of sight, on the other hand, are primarily of the nature of particulars and appeal rather to the individual. Those received from λόγος through the sense of hearing are, almost from the first, of the nature of universals, and therefore almost directly (i. e. so far as we *understand* them) stimulate the faculty of intelligence. But when words are combined in sentences, and form trains of reasoning, their mind-develop-

Written language adds itself to spoken language. ing effect is still more obvious. When to that of spoken words we add the effect of words written, and remember also that language with its symbolic power ranges over the whole tract of ocular as well as other sensible experience, we can easily understand the paramount intellectual effect ascribed by Aristotle to the sense of hearing. He is, how- ever, careful to point out that hearing has not these grand results directly, but only κατὰ συμβεβηκός. Like every other sense its immediate data consist of particulars[2].

Ethical importance of hearing. The modes or kinds of music. Object of hearing alone directly affects the emotions. Musical sound the § 29. In its bearing upon moral character, hearing, which makes us acquainted with music, is in Aristotle's opinion of very great importance. No other sense can compare or compete with it in this respect. 'Why is it' (the writer of the *Problems* asks) 'that the object of hearing alone among the objects of sense possesses character (ἦθος ἔχει), that is, affects the emotional temperament of the hearer? This, he adds, is true of it, even when the music is unaccompanied by words. Neither colour nor odour nor savour has a

---

[1] *Hist. An.* ix. 1. 608ᵃ 15–21.

[2] Hence, in *de Sens.* i. 437ᵃ 13, ἀκουστὸς ὤν belongs to what follows, and the comma should stand not after ὤν, but after μαθήσεως, or else in both places. What the writer wishes to guard against there is the false notion that the full significance of λόγος is matter of immediate perception by the sense of hearing.

similar effect[1].' 'The movements set up in us by music are of the nature of action, and actions are the "notation" of character[2]. We must not merely take our share in the pleasure which all derive from music, but consider whether and how far it has an influence on the mind and character. That it has this influence would be plain if it could be shown that by its means our characters are qualitatively determined (ποιοί τινες τὰ ἤθη γινόμεθα). That this, however, is true is proved not only by many other sorts of music, but particularly by the compositions of Olympus; for these raise the hearers to a high pitch of excitement (ποιεῖ τὰς ψυχὰς ἐνθουσιαστικάς), and such excitement is an affective state of the mind and character (τοῦ περὶ τὴν ψυχὴν ἤθους πάθος). Further, music gives pleasure; and virtue consists in taking pleasure in right objects, as well as in loving and hating rightly.' Our mind and character undergo a change as we listen to the music that we love. Hence the musical modes (αἱ ἁρμονίαι) are naturally distinguishable from one another according as they correspond to different dispositions of character. Some are melancholy, others gay; some produce mental elation, others tend to calm excitement. Hence it is obvious that music has the power of influencing character; from which it follows that it may be a powerful instrument of education[3].

Margin: 'notation' of action: action the 'notation' of character in man. Emotional effect of the compositions of Olympus. Music can give pleasure, and pleasure is intimately connected with morals and character. The modes (ἁρμονίαι) distinguished according as they correspond to distinct moral dispositions. Music a powerful influence in education.

§ 30. An account of Aristotle's views on συμφωνία, or the theory of concords, would lead to a subject with which we are not here concerned—Greek Harmonics. Besides, though we find many allusions to the physical basis of music in the works ascribed to Aristotle, nowhere, except in the unquestionably spurious *Problems*, do we find this subject treated technically. There are, however, in the *de Sensu* a few references which assume on the reader's part familiar

Margin: The Aristotelean account of the pleasure found in συμφωνίαι. Meaning of concords. Nature's analogies

---

[1] *Prob.* xix. 27. 919ᵇ 26-9. Aristotle was not the writer of the *Problems*, yet they were chiefly inspired from his works, and so may serve as evidence for his general doctrine in this and many other matters.

[2] *Prob.* 919ᵇ 35-7 αἱ δὲ κινήσεις αὗται πρακτικαί εἰσιν, αἱ δὲ πράξεις ἤθους σημασία ἐστίν. σημασία is the term for musical notation.

[3] Cf. in general *Pol.* viii. 5. 1339ᵃ 11-1340ᵇ 19, particularly 1340ᵃ 2-ᵇ 12.

for musical concords. Concords among sounds relatively few. The sense of hearing (like every other sense) depends on a ratio or involves one : for on this depends its perception of the concordance of objective sound. A concord perceptible by one ἐνέργεια of hearing. Is this really so? or only apparently? Sound travels. Proof of this. A mathematician with a bad musical ear may be perfect in the theory of harmonics. So those who understand the theory of music may have no real sense

acquaintance with it. We will therefore extract, from the *Problems* and elsewhere, some passages containing certain leading ideas which may at least serve as an adequate commentary on these references.

First of all hearing itself is or involves (§ 22, p. 117 *supra*) a ratio of composition. 'If a concord is a species of vocal sound; and if the sound and the hearing of the sound are (as has been shown) in a certain way one, (though in another way at the same time not one); and if again a concord is a ratio, it follows that hearing (τὴν ἀκοήν) is a ratio of some sort. Hence it is that each excess of either the sharp or the grave spoils the hearing ⟨as it spoils the concord⟩[1].' 'Nature has an eagerness for contraries, and of these, not of similars, composes concord (τὸ σύμφωνον).' 'Art, imitating nature, also brings contraries together. Painting, mixing together *white* and *black*, *yellow* and *red*, renders its representations " consonant " (συμφώνους) with their originals; while music, mixing *sharp* notes (φθόγγους) with *grave*, and short with long-sustained, in sounds of different *timbre* (ἐν διαφόροις φωναῖς), brings to pass one single harmony (ἁρμονίαν)[2].' 'It is the mixture of notes, not the mere sharp or grave, that forms ⟨the pleasing sound we call⟩ concord[3].' 'Concord is a particular kind of mixture of sharp and grave[4].' 'They (concords) are ratios of opposites like the octave and the fifth[5].' 'The concords are few compared with sounds in general ; since they are, of all combinations of sounds, those based on *numerically expressible ratios[6].'

'Mixture is possible among things whose extremes are contraries : it is impossible that there should be—unless in some incidental way—a mixture of *white* and *sharp* : there can be no such mixture of them as of *sharp* and *grave* in a concord[7].' 'The soul perceives the mixture of *sharp*

---

[1] 424ª 27, 426ª 27 seqq.

[2] See *de Mund.* v. 396ᵇ 7-22. This is, however, a non-Aristotelean work.　　　　　[3] *De An.* iii. 2. 426ᵇ 5.

[4] *Met.* viii. 2. 1043ª 10 συμφωνία δὲ ὀξέος καὶ βαρέος μεῖξις τοιαδί.

[5] *De Sens.* vii. 448ª 9.

[6] *De Sens.* iii. 439ᵇ 31-440ª 2.　　　　　[7] *De Sens.* vii. 447ᵇ 1.

and *grave* in a concord with one single act of sense' : it would require two such acts to perceive *sharp* and *white*— data of two different senses[1].

of pleasure from it : no real sense of what it is.

Sound *travels*, however, though light does not. When we see a person at a distance strike a blow which causes a sound, the sound does not reach the ear until after the stroke. So each of a row of listeners, posted at ever greater distances from the person, would hear the blow at successively later times[2]. 'Hence certain theorists say that the sounds (οἱ ψόφοι) which affect the hearing in a concord (συμφωνία) do not all arrive at the point of sense coinstantaneously, but only seem to do so, and that this seeming is due to the fact that the interval separating their different arrivals is too short to be noticeable. . . . This, however, is not the case, for it is impossible that there should be a time-interval too short to be noticeable[3].' Such a theory would involve an instant of blank or vacant consciousness, which we cannot admit.

'The term ἁρμονική is ambiguous, for it may refer either to the mathematical knowledge of music, or to the perception by the ear of musical consonance. Those who have a good ear perceive the facts of such consonance. The mathematicians, on the other hand, know the reasons of these facts. For mathematicians can demonstrate the causes of musical concords, yct it often happens that those who have this power have no perception of the concrete particulars[4].'

§ 31. A writer in the *Problems* asks: Why does the interval between the extremes in the octave (in certain cases) escape the ear, and the composite whole pass for unison? The answer suggested is, that 'this unisonous effect is due to the fact that each sound—the high and the low—seems identical with the other. For in sounds equality arises from proportion, and the Equal is a branch of the One[5].' 'Degrees of consonance (says Chappell) depend upon the

Why does the octave seem unisonous? The reason is that the sounds in it are 'identical in virtue of their ratio to one another.'

---

[1] *De Sens.* vii. 447ᵇ 7.   [2] *De Sens.* vi. 446ᵇ 5–26.

[3] *De Sens.* vii. 448ᵃ 19–26.   [4] *Analyt. Post.* i. 13. 79ᵃ 1–5.

[5] *Prob.* xix. 14. 918ᵇ 7–12 διὰ τί λανθάνει τὸ διὰ πασῶν, καὶ δοκεῖ ὁμόφω-νον εἶναι; . . . ἢ ὅτι ὥσπερ ὁ αὐτὸς εἶναι δοκεῖ φθόγγος; (Didot) διὰ τὸ ἀνάλογον· ἰσότης ἐπὶ φθόγγων, τὸ δ' ἴσον τοῦ ἑνός. (Otherwise von Jan, *op. cit.* p. 85 n.)

Actual basis of this suggestion. Why is the octave the most pleasing of all intervals? Because its ratio is expressible in integral terms, while those of other intervals always involve in one of the terms an improper fraction. The octave can be expressed as the ratio of one to an integral number (sc. two); the other intervals cannot. Fundamental reason of the pleasing nature of συμφωνία. It is a λόγος ἐναντίων, and λόγος involves τάξις, which is φύσει ἡδύ.

proportion that coincident vibrations bear to those which "sound apart" [i.e. are dissonant]. The unison alone is perfect consonance, because therein only do all vibrations coincide [1].' But the degree of consonance in the octave is greater than that in any other interval, because in this, whose total ratio is 1 : 2, the proportion between coincident and non-coincident vibrations is 1 : 1, i.e. greater than in any other. On the proportionality thus maintained of consonant to non-consonant vibrations in the octave appears to rest the 'equality' spoken of above; and on this equality, again, rests the 'approach to oneness' which causes the interval to be unnoticed and the sounds taken for one. Aristotle speaks with less subtlety of this matter. 'It is easier to perceive a thing (in its proper nature) when single than when blended with something else, e. g. wine when unmixed than when diluted, or honey, or a colour, or the note highest in pitch (νήτη) when by itself than when in the octave [2].' 'Also the quarter tone escapes notice: one hears the melodic rise and fall of the voice as a *continuum*, but the interval between the extremes in the quarter tone passes unnoticed [3].' 'Why'—it is asked in the *Problems* [4]—'is the octave the most pleasing of all intervals? Perhaps because its ratios are expressible by integral terms, while those of the other intervals are not so. For since the string of highest pitch, the νήτη, is ⟨in its rate [5] of vibration⟩ double the string lowest in pitch, the ὑπάτη, for every *two* vibrations of the former the latter has *one*, and for every *two* of the latter the former has *four*, and so on. But the rate of vibrations of the νήτη is once and a half that of the μέση. Thus the interval of one to one and a half in which the *fifth* consists is not ultimately expressible in integers; for while the less is one, the greater is so many and a half more. Hence

---

[1] Cf. Chappell, *History of Music*, pp. 221-4; von Jan, *op. cit.* pp. 96, 101 nn.; Wundt, *H. and A. Psych.* p. 69 (E. Tr.).

[2] Arist. *de Sens.* vii. 447ᵃ 17-20.

[3] Arist. *de Sens.* vi. 446ᵃ 1-5.   [4] xix. 35. 920ᵃ 27 seqq.

[5] Only by this parenthesis can the sense be given. The νήτη was but half as *long* as the ὑπάτη. The passage, therefore, implies more accurate knowledge of the vibration of strings than Aristotle possessed.

integers are not compared with integers, but there is a fraction over. The case is similar with the *fourth*: the interval 3 : 4 cannot be expressed as a ratio of *one* to any integral number; it appears $1 : 1\frac{1}{3}$. Or perhaps the octave is most perfect because it is made up of the *fifth* and the *fourth*, and is the measure of the melodic series[1].'

'We are delighted with concordance of sounds because such concordance is a blending of contraries which bear a ratio to one another. But a ratio is a fixed arrangement— a thing which, as has been said, is naturally pleasing[2].' 'If we take two vessels equal and similar to one another, but the one empty, the other half full, and cause them to sound together, they form an octave with one another. Why is this? Because the sound coming from the half full vessel is double the other ⟨in rate of vibration⟩[3].' The *Problems*, from which these extracts are taken, are later than Aristotle, and in some ways represent more highly developed theories of music and of harmonics than those of Plato or Aristotle.

§ 32. It would seem, and has been urged by many, e. g. by Trendelenburg, Arist. *de An.* p. 107 (Belger), that a portion of what Aristotle wrote on the subject of vocal sound must have been somehow lost. In his work *de Gen. An.* v. 7. 786ᵇ 23, we read: 'As to the final cause of voice in animals, and as to what voice and sound in general are, an explanation has been offered already, partly in our work on Sense-perception, and partly in that on The Soul[4].' Again further down: 'With regard to voice, let this suffice for the information not definitely given already in the works on sense-perception and on the soul[5].'

[margin note:] Probable loss of a portion of the tract *de Sensu* treating of sound. The missing treatise cannot be that Περὶ Ἀκουστῶν

---

[1] *Prob.* xix. 35. 920ᵃ 27–38. The Didot punctuation after μελῳδίας (ᵃ38) is here adopted; also Bekker's τ' ἐκεῖνο for τεμεῖν ὅ (ᵃ36).

[2] xix. 38. 921ᵃ 2–4 συμφωνίᾳ δὲ χαίρομεν ὅτι κρᾶσίς ἐστι λόγον ἐχόντων ἐναντίων πρὸς ἄλληλα· ὁ μὲν οὖν λόγος τάξις, ὃ ἦν φύσει ἡδύ.

[3] *Probl.* xix. 50. 922ᵇ 35–9.

[4] Cf. 786ᵃ 23 τίνος μὲν οὖν ἕνεκα φωνὴν ἔχει τὰ ζῷα καὶ τί ἐστι φωνὴ καὶ ὅλως ὁ ψόφος, τὰ μὲν ἐν τοῖς περὶ αἰσθήσεως, τὰ δ' ἐν τοῖς περὶ ψυχῆς εἴρηται.

[5] Cf. 788ᵃ 34 περὶ μὲν οὖν φωνῆς ὅσα μὴ πρότερον ἐν τοῖς περὶ αἰσθήσεως διώρισται καὶ ἐν τοῖς περὶ ψυχῆς, τοσαῦτ' εἰρήσθω.

In the *de Sensu*, however, while the physical properties of
the objects of seeing, smelling, and tasting are examined
and described, those of hearing and touching are entirely
omitted. There, for the psychological import of the five
senses, we are referred back to the work *de Anima* : while
as to the physical character of the objects of all five, we are
promised a discussion to follow ; yet while three of these
are discussed two are passed over. There is no formal
or set treatment of them in that little tract[1]. The fragment Περὶ 'Ακουστῶν is un-Aristotelean. Its opening words
agree with the views of sound-transmission ascribed by
Alexander[2] to Strato, whom therefore Brandis (too hastily
as Zeller thinks) regards as the author. 'According to
the Περὶ 'Ακουστῶν (803ᵇ 34 seqq.), every sound is composed
of particular vibrations (πληγαί) which we cannot distinguish
as such, but perceive as one unbroken sound : high tones,
whose movement is quicker, consist of more vibrations, and
low tones of fewer. Several tones vibrating and ceasing
at the same time are heard by us as one tone. The height
or depth, harshness or softness, in fact every quality of
a tone, depends (803ᵇ 26) on the quality of the motion
originally created in the air by the body that gave out the
tone. This motion propagates itself unchanged, inasmuch
as each portion of the air sets the next portion of air in
motion with the same movement as it has itself.' (Zeller,
Arist. ii. pp. 465–6 nn., E. Tr.)

---

[1] Cf. *de Sens.* iii. 439ᵃ 6–17 τί ποτε δεῖ λέγειν ὁτιοῦν αὐτῶν οἷον . . . ἢ τί
ψόφον . . . ὁμοίως δὲ καὶ περὶ ἀφῆς.

[2] Ad Arist. *de Sens.* (p. 126, Wendland). von Jan, pp. 55 seqq., 135,
ascribes the περὶ 'Ακουστῶν to Heraclides.

THE ANCIENT GREEK PSYCHOLOGY
OF SMELLING

Alcmaeon.

§ 1. WE have little direct information respecting Alc- *Function* maeon's psychological theory of the sense of smell. All *and organ of smelling.* that remains is the following, contained in two passages *Smelling effected by* which I extract, the one from Theophrastus, the other from *air inhaled through* the late compilation of Aëtius. *nostrils* 'He taught that a person smells by means of the nostrils, *and carried to brain.* drawing the inhaled air upwards to the brain, in the respiratory process[1].' Not the nostrils alone, therefore, but these in connexion with the brain form the olfactory apparatus.

'He held that the authoritative principle—the intelligence —has its seat in the brain; that, therefore, animals smell by means of this organ which draws in the various odours[2] to itself in the process of respiration[3].' Besides these two direct references to Alcmaeon, there is a probable allusion to him bearing on the same subject. Socrates in the *Phaedo*, reviewing the history of his own mental development, tells his friends that in his youth he had been interested in psychological questions, and that of these one which presented itself was 'whether it is the brain that furnishes us with the senses of hearing and seeing and smelling[4].' The various theories referred to by Plato in this passage are sufficiently distinctive to show that in mentioning each he is thinking of some particular philosopher. The theory which referred sensation to the opera-

---

[1] Theophr. *de Sens.* § 25 ; Diels, *Vors.*, p. 104 ὀσφραίνεσθαι δὲ ῥισὶν ἅμα τῷ ἀναπνεῖν ἀνάγοντα τὸ πνεῦμα πρὸς τὸν ἐγκέφαλον.

[2] In the following paragraphs the terms 'smell' and 'odour' are sometimes used indifferently for the object of the olfactory sense. So, too, 'taste' is sometimes used for 'savour.'

[3] Aët. iv. 17. 1, Diels, *Dox.*, p. 407, *Vors.*, p. 104 ἐν τῷ ἐγκεφάλῳ εἶναι τὸ ἡγεμονικόν· τούτῳ οὖν ὀσφραίνεσθαι ἕλκοντι διὰ τῶν ἀναπνοῶν τὰς ὀσμάς.

[4] Plato, *Phaedo* 96 B, Diels, *Vors.*, p. 105 πότερον . . . ὁ ἐγκέφαλός ἐστιν ὁ τὰς αἰσθήσεις παρέχων τοῦ ἀκούειν καὶ ὁρᾶν καὶ ὀσφραίνεσθαι.

tion of the brain was characteristic of Alcmaeon. The expression τὸ ἡγεμονικόν in Aëtius betrays the lateness of the writer; for it only came into vogue with the Stoic school. We have, however, the authority of Theophrastus for the statement that Alcmaeon regarded the brain as the great organizing centre of sensation. 'All the senses he regarded as somehow connected with the brain [1].'

What is the internal apparatus with which the breath is brought in contact for the purpose of olfactory sensation? The object of smell, odour, not discussed in the remains of Alcmaeon. Modern physiology helpless over this sensory function: modern physics, over its object.

§ 2. In these meagre statements is contained all that we know of Alcmaeon's psychology of smelling. They amount only to an expression of what ordinary observation might suggest respecting it. Yet even in this short flight of speculation there was room for divergence of opinion. Every one felt convinced that the process of respiration is largely instrumental to the olfactory sense, and also that it is so in virtue of its connexion with some internal apparatus. Thinkers disagreed as to what the latter was. Alcmaeon, for what reasons we are not informed, supposed it to be the brain. Aristotle, as we shall see, firmly held the contrary opinion, that the internal seat of the olfactory sense (as well as the other senses) was not the brain, but the heart—or the region of the heart. We have no information as to Alcmaeon's views respecting the object of this sense, odour, or the manner of its generation as a physical fact. But before we express our disappointment with Alcmaeon's shortcomings on this subject, let us reflect that even now very little more, of any essential import, is known than the brief statements he has given us contain. Anatomy has, of course, enabled modern psychologists to speak with a fullness impossible to the Greeks of the structure of the olfactory apparatus, but as regards the olfactory function itself, and the exact manner of its performance, it has little to teach. Experiments have shown that sensations of smell, like other sensations, may be excited in us without the presence of odorous objects in the ordinary way, by means of other stimuli. But for the explanation of this sense itself, we are still left with such

---

[1] Theophr. *de Sens.* 26 ἁπάσας δὲ τὰς αἰσθήσεις συνηρτῆσθαί πως πρὸς τὸν ἐγκέφαλον.

statements, as that 'particles of odoriferous matters present in the inspired air, passing through the lower nasal chambers, diffuse into the upper nasal chambers, and falling on the olfactory epithelium produce sensory impulses, which ascending to the brain, give rise to sensations of smell.' In this sentence, from the pen of Sir Michael Foster, introducing the subject, it is curious to observe how much might pass for a mere expansion of the brief description of the same facts left us by Alcmaeon [1]. Modern physics is as helpless to explain odour as physiology to explain olfactory function.

Empedocles.

§ 3. The remains of Empedocles, except as regards his theory of ἀπορροαί, show us little more than those of Alcmaeon to elucidate the psychology of smelling.

Organ and function of smelling.

'The act of smelling (he said) takes place by means of the respiration; hence those persons have the keenest sense of smell in whom the movement of inhalation is most energetic [2].' 'Empedocles holds that the sense of odour is introduced with and by the respiration actuated from the lungs; that accordingly, when the respiratory process is laboured, at such times, owing to its roughness, we do not perceive smells when we inhale, as happens with persons suffering from catarrhs [3].' Respiration, on which the introduction of odour and smelling depends, is a process in which the mouth and lungs and also the pores of the skin operate alternately [4]; smelling being incidental to that part of the process in which the mouth and lungs are agents.

Who have keenest olfactory sense? 'Colds' interfere with the keenness of it, as it is dependent on respiration.

[1] Cf. Foster, *Text Book of Physiology*, § 859, p. 1388.

[2] Theophr. *de Sens.* § 9; Diels, *Vors.*, p. 177; Karsten, *Emped.*, pp. 480–3 ὄσφρησιν δὲ γίνεσθαι τῇ ἀναπνοῇ· διὸ καὶ μάλιστα ὀσφραίνεσθαι τούτους οἷς σφοδροτάτη τοῦ ἄσθματος ἡ κίνησις.

[3] Aëtius, iv. 17. 2, Diels, *Dox.*, p. 407, *Vors.*, p. 181 Ἐμπεδοκλῆς ταῖς ἀναπνοαῖς ταῖς ἀπὸ τοῦ πνεύμονος συνεισκρίνεσθαι τὴν ὀσμήν· ὅταν γοῦν ἡ ἀναπνοὴ βαρεῖα γένηται, κατὰ τραχύτητα (sc. τῆς ἀναπνοῆς) μὴ συναισθάνεσθαι, ὡς ἐπὶ τῶν ῥευματιζομένων.

[4] Empedocles illustrated by the filling and emptying of the *clepsydra*. Cf. the verses in Karsten, 275–99, and Burnet's version, *Early Greek Philosophy*, p. 230. Plato in principle adopts Empedocles' theory of respiration, *Tim.* 79 A–E.

Theophrastus criticizes Empedocles' principle of similia similibus *as applied to olfactory sense. Empedocles does not explain the fact that creatures smell which do not respire. Some absurdities would follow if the theory of Empedocles were true. Respiration only indirectly the cause of smelling —not directly, as Empedocles thought.*

§ 4. 'As regards the other senses, how are we to apply the principle " that like is discerned by like " ? . . . For it is not by sound that we discern sound, nor by odour that we discern odour, and so on. . . . When sound is ringing in the ears, when savours are already affecting the taste, when an odour is already occupying the olfactory sense—at such times the senses each and all are dulled, and the more so the greater the quantity of the cognate objects which happen to be in their organs [1].' 'His (sc. Empedocles') explanation of the sense of smelling is absurd. For, in the first place, the cause he has assigned for it is not sufficiently general (οὐ κοινήν), since there are some creatures which possess the sense of smell, but do not respire at all. Again, it is childish to say, as he does, that persons smell most acutely who inhale the breath in greatest amount (τοὺς πλεῖστον ἐπισπωμένους); for respiring is of no avail for this purpose if the sense is not in a healthy condition (μὴ ὑγιαινούσης), or is not, so to speak, (ἀνεῳγμένης πως) open. There are many persons who (no matter how much they inhale) are incapacitated (πεπηρῶσθαι) for smelling, and have no perception whatever of odour. Moreover, those whose (οἱ δύσπνοοι) breathing is distressed, or who are ill (πονοῦντες), or sleeping (καθεύδοντες), should, on Empedocles' theory, perceive odours more keenly than others, as they inhale most air. The contrary, however, is the case. That the act of respiration is not directly (καθ' αὑτό) the cause of smelling, but only indirectly (κατὰ συμβεβηκός), is both evident from the case of the other animals ⟨i.e. those which do not respire yet have this sense⟩, and is further proved by the pathological states just referred to [2].'

Odour, according

§ 5. 'Most odour emanates,' says Empedocles, 'from

---

[1] Theophr. *de Sens.* § 19; Diels, *Vors.*, p. 179 τὰ δὲ περὶ τὰς ἄλλας αἰσθήσεις πῶς κρίνωμεν τῷ ὁμοίῳ; . . . οὔτε γὰρ ψόφῳ τὸν ψόφον, οὔτ' ὀσμῇ τὴν ὀσμὴν οὔτε τοῖς ἄλλοις τοῖς ὁμογενέσιν . . . ἦχον δὲ ἐνόντος ἐν ὠσὶν ἢ χυλῶν ἐν γεύσει καὶ ὀσμῆς ἐν ὀσφρήσει κωφότεραι πᾶσαι γίνονται καὶ μᾶλλον ὅσῳ ἂν πλήρεις ὦσι τῶν ὁμοίων.

[2] The above, as also the following, criticism is determined by the Aristotelean theory of smelling. Theophr. *de Sens.* §§ 21-2; Diels, *Vors.*, p. 179.

bodies that are fine in texture and of light weight'
(Theophr. *de Sens*. § 9). In reply to this Theophrastus
denies that light bodies are especially odorous. 'It is
not true, either, that the bodies which most affect the
sense of smell are the light bodies; the truth is that
if we are to smell them, there must be odour in them to
begin with; for air and fire are the lightest of all, but yet
do not excite the sense of odour[1].' The objective odour
comes, according to Empedocles, in the form of ἀπορροαί
from the odoriferous bodies. Such is the scent which dogs
follow. The hound 'searches with his nostrils for the
particles from the limbs and bodies of the beasts, and for
such whiffs of scent from their feet as they leave on the
tender grass[2].' 'But,' replies Theophrastus, 'if wasting
is a consequence of emanation from a substance (and
Empedocles uses this very fact of the wasting of things

---

[1] Theophr. *de Sens*. § 22; Diels, *Vors*., p. 179.

[2] Plut. *de Curios*. 11, *Quaest. Nat*. 23; Diels, *Vors*., p. 211;
Karsten, *Emped*. p. 253:

κέρματα θηρείων μελέων μυκτῆρσιν ἐρευνῶν
⟨πνεύματά θ'⟩ ὅσσ' ἀπέλειπε ποδῶν ἁπαλῇ περὶ ποίᾳ.

This is Diels' reading. He adopts Buttmann's κέρματα for the τέρματα of Plut. *de Curios*., the κέμματα of *Quaest. Nat*.—the inconsistency
and obscurity of which show the text to be corrupt. By κέρματα
Empedocles denotes not 'fissa ferarum ungula' as Lucretius (*vide
infra*) seems to render, but the ἀπόρροιαι—the material particles which
are the proximate object of, and which stimulate, the sense of smell.
This seems better than (*a*) to read with Karsten τέρματα λεχέων='cubilia
extrema, ultimi ferarum recessus'; or (*b*), with Sturz, to interpret
τέρματα μελέων as='extremitates membrorum,' i. e. 'pedes,' i. e. 'pedum
vestigia'; or (*c*) to accept, with Schneider, κέμματα as a derivative of
κεῖμαι (which would be impossible)='cubilia'; or finally (*d*) to follow
Stein (*Emped*., p. 70) in adopting πέλματα (Duebn.)='the soles of the
feet,' or 'vestigia.' Plutarch, *Quaest. Nat*., explains the meaning to be
that the dogs τὰς ἀπορροὰς ἀναλαμβάνουσιν, ἃς ἐναπολείπει τὰ θηρία τῇ ὕλῃ.
Lucretius had the lines before him when he wrote: 'tum fissa ferarum
ungula quo tulerit gressum promissa canum vis ducit,' *de Rer. Nat*. iv.
680: which reads as if he translated κέρματα (κείρω) by '*fissa* ungula.'
⟨πνεύματά θ'⟩ is Diels' supplement of the words quoted from Empedocles
by Alexander, who denies Empedocles' theory of odours being ἀπορροαί,
asserting that neither odour nor colour can be dispersed (διασπᾶσθαι) in
material particles, as Empedocles' line of reasoning would imply.

as the most general proof of his theory of emanation), and if it is true that odours result from such emanation, the most odorous substances should perish most quickly. But the contrary is the fact, for the most odorous plants are more lasting than any others.'

Function of smelling by the *pores* and *emanations*.

§ 6. The ἀπορροαί of odour find their way into the πόροι of the olfactory organ. If the ἀπορροαί are symmetrical with the πόροι, the sense is stimulated ; if not, no perception occurs.

'Empedocles lays it down, with regard to all the senses alike, that sensation is due to their respective ἀπορροαί fitting into the "pores" of each sense-organ ; whence it is that the several senses cannot discern one another's objects, because the pores of the organs, as compared with the ἀπορροαί of an object other than their own, are in some cases too wide, in other cases too narrow, to admit them ; for he asserts that these ἀπορροαί in the former case pass unchecked straight on, without touching the sides of the pores; while in the latter case, they cannot find ingress at all [1].'

Democritus.

Smelling, like the other senses, is for Democritus a mode of touch. Yet he does not assign the atomic figures on which the various kinds of

§ 7. Democritus has left us considerable information as to his theories respecting sight, hearing, tasting, and touching, but what we know of his views on the sense of smell can be stated very briefly.

He reduced it (as he did all the other senses) to a mode of touch [2]. 'Why is it that Democritus, while he explained the various objective tastes in conformity with the sense of taste, omitted to explain objective odours and colours in conformity with their subjective senses? He ought, if consistent, to have explained these sensibles too by his theory

[1] Theophr. *de Sens.* § 7; Diels, *Vors.*, p. 176 Ἐμπεδοκλῆς δὲ περὶ ἁπασῶν ὁμοίως λέγει καί φησι τῷ ἐναρμόττειν εἰς τοὺς πόρους τοὺς ἑκάστης αἰσθάνεσθαι· διὸ καὶ οὐ δύνασθαι τὰ ἀλλήλων κρίνειν, ὅτι τῶν μὲν (sc. αἰσθήσεων = αἰσθητηρίων) εὐρύτεροί πως, τῶν δὲ στενώτεροι τυγχάνουσιν οἱ πόροι πρὸς τὸ αἰσθητόν, ὡς τὰ μὲν οὐχ ἁπτόμενα διευτονεῖν (= 'pristinum in permeando impetum servare,' Diels, *Dox.*, p. 500, 22 n.), τὰ δ' ὅλως εἰσελθεῖν οὐ δύνασθαι.

[2] Arist. *de Sens.* iv. 442ᵃ 29; Mullach, *Democr.*, p. 405.

of "figures"[1].' Theophrastus tells us that in his theories respecting smelling, touching, and tasting, Democritus 'resembled most other philosophers[2].' For him, as for most of the other φυσιολόγοι, all the several senses were ultimately modifications of the sense of touch. So with the objects of these senses: they too were but variations of the tangible, their qualitative distinctness being merely subjective—due to φαντασία[3]. Having explained in detail the various sensations and objects of tasting, he probably thought that those of smelling—closely related as they are to those of tasting—could be easily explained on the analogy of these, as deducible from the figures of the atoms which caused them. However this may be, 'he neglected to add a definite account of odour; all he tells us respecting it is that the finer matter, passing by emanation from the heavy, produces odour. What the particular natures of the agent and patient in this sensory operation are he did not go on to inform us, though this was the main point[4].'

odour depend; nor give any definite theory of odour at all, except by stating that it is a fine sort of matter emanating from odorous bodies and borne to the nostrils.

Anaxagoras.

§ 8. 'Anaxagoras asserts that we exercise the sense of smell in connexion with the respiratory process[5].' 'Large animals (according to Anaxagoras) hear loud sounds, and at great distances ... small animals low sounds and those close by. And it is likewise as regards the sense of smell; for air

Function and organ of smelling. Smelling connected with inhalation.

[1] Theophr. *de Odor.* § 64; Diels, *Vors.*, p. 390 τί δή ποτε Δημόκριτος τοὺς μὲν χυμοὺς πρὸς τὴν γεῦσιν ἀποδίδωσι, τὰς δ' ὀσμὰς καὶ τὰς χρόας οὐχ ὁμοίως πρὸς τὰς ὑποκειμένας αἰσθήσεις; ἔδει γὰρ ἐκ τῶν σχημάτων.

[2] Theophr. *de Sens.* § 57 περὶ μὲν ὄψεως καὶ ἀκοῆς οὕτως ἀποδίδωσι, τὰς δὲ ἄλλας αἰσθήσεις σχεδὸν ὁμοίως ποιεῖ τοῖς πλείστοις.

[3] Cf. Theophr. *de Sens.* § 63 τῶν δὲ ἄλλων αἰσθητῶν οὐδενὸς εἶναι φύσιν, ἀλλὰ πάντα πάθη τῆς αἰσθήσεως ἀλλοιουμένης, ἐξ ἧς γίνεσθαι τὴν φαντασίαν.

[4] Theophr. *de Sens.* § 82; Diels, *Vors.*, p. 396; *Dox.*, p. 524 περὶ δὲ ὀσμῆς προσαφορίζειν παρῆκεν πλὴν τοσοῦτον, ὅτι τὸ λεπτὸν ἀπορρέον ἀπὸ τῶν βαρέων ποιεῖ τὴν ὀδμήν· ποῖον δέ τι τὴν φύσιν ὂν ὑπὸ τίνος πάσχει, οὐκέτι προσέθηκεν, ὅπερ ἴσως ἦν κυριώτατον. Of ὀδμήν Diels (*Dox.* l. c.) says 'servavi ut Democriteum.' For the Epicurean and probably Democritean theory of smelling, cf. further, Lucret. iv. 673–86 with Giussani's notes.

[5] Theophr. *de Sens.* § 28; Diels, *Vors.*, p. 323 ὡσαύτως δὲ καὶ ὀσφραίνεσθαι ... ἅμα τῇ ἀναπνοῇ.

Large
animals
compared
with small
as regards
olfactory
power.

when thin (he says) is more odorous, since in proportion as it is heated and rarefied its odorousness is increased. A large animal, as it respires, while inhaling the rare air inhales the dense also, but the small animal draws in the rare air by itself; wherefore large animals are more perfect in this form of sense. For odour is more pronounced (μᾶλλον εἶναι) when near than when far off, on account of its greater density (in the former case), and its being weakened by dispersion (in the latter case). He states that as a rule large animals are insensible to the finer sort of odour, while small animals fail to perceive the denser kind [1].' According to Theophrastus, Anaxagoras held that the larger animals had a more perfect sense of odour, as well as of other sensibles, than small animals possess. The general reason for this is that, while the former inhale both the dense and the rare, the latter inhale the rare alone. On this Theophrastus observes that ' it exposes Anaxagoras to a peculiar difficulty. Anaxagoras asserts that the rare air is the more odorous, yet that a more

---

[1] Theophr. de Sens. § 30; Diels, Vors., p. 323, Dox., pp. 507-8 καὶ ἐπὶ τῆς ὀσφρήσεως ὁμοίως· ὄζειν μὲν γὰρ μᾶλλον τὸν λεπτὸν ἀέρα, θερμαινόμενον μὲν γὰρ καὶ μανούμενον ὄζειν. Ἀναπνέον δὲ τὸ μὲν μέγα ζῷον ἅμα τῷ μανῷ καὶ τὸν πυκνὸν ἕλκειν, τὸ δὲ μικρὸν αὐτὸν τὸν μανόν· διὸ καὶ τὰ μεγάλα μᾶλλον αἰσθάνεσθαι. καὶ γὰρ τὴν ὀσμὴν ἐγγὺς (sc. οὖσαν) εἶναι μᾶλλον ἢ πόρρω διὰ τὸ πυκνοτέραν εἶναι, σκεδαννυμένην δὲ ἀσθενῆ. σχεδὸν δὲ ὡς εἰπεῖν οὐκ αἰσθάνεσθαι τὰ μὲν μεγάλα τῆς λεπτῆς [ἀέρος], τὰ δὲ μικρὰ τῆς πυκνῆς. I have thought it better to read, according to Diels' former suggestions Dox., p. 507, 33 n., τὸν πυκνόν for τὸ π., and αὐτὸν τὸν μανόν for αὐτὸ τὸν μ. Though τὸ πυκνόν (=τὸν πυκν. ἀέρα) would serve, yet αὐτὸ τὸν μ. certainly perverts or ignores the reasoning. Also with Diels, Dox., p. 508, 4 n., I reject ἀέρος (after τῆς λεπτῆς) as a ' glossema,' and understand ὀσμῆς with the adjectives λεπτῆς and πυκνῆς. In his Vorsokratiker he does not give effect to all these suggestions, printing τὸν πυκνόν indeed, but keeping the αὐτό, and printing τῆς λεπτῆς ἀέρος in open type, as if to mark a quotation, and to assume that Anaxagoras made ἀήρ feminine. But the τῆς πυκνῆς is not so printed by Diels, nor is it likely that Theophrastus would have thus once retained the Homeric and Hesiodic gender of this word, even if we assume Anaxagoras to have used it in the passage of which Theophrastus was here thinking. Besides ἕλκειν is the verb used of taking in the mere ἀήρ both just above, and later, Th. § 35 ad fin. (τὸν μανὸν ἕλκει): while αἰσθάνεσθαι seems properly to require ὀσμή, as object of the sense of smell.

acute sense of smell is possessed by the animals which
inhale the dense air than by those which inhale the rare[1].'
We can, however, find at least a partial solution in the
fact that while the smaller animals are *confined* to inhaling
the rare air, the larger inhale both the rare and the dense.
But a difficulty remains. In the next sentence, we read,
as a further reason for the superiority in this respect of the
larger animals, that odour is more pronounced (μᾶλλον
εἶναι) when close at hand than when at a distance, on
account of its being more condensed when near, and becoming weakened through dispersion when at a distance;
and that the smaller animals are defective in their perception of the more condensed form of odour, while the
larger fail in that of the rarer form. How these are
reasons for the proposition that the larger animals μᾶλλον
αἰσθάνεσθαι—have the more perfect sense of smell—is not
easy to understand. We may, however, suppose that the
larger animals receive into their larger olfactory organ
a greater quantity of the enfeebled odour from a distant
object, and thus perceive it, while the smaller, receiving
only a small quantity, fail to notice it. But there seems to
be an incoherency in the argument, arising from confusion
and interchange between *air* (rare or dense) as object of
smelling, and *odour proper*, with air merely as its vehicle.
That Theophrastus was perplexed by the argument is plain
from what he says of it in connexion with the other senses
(cf. *supra* 'HEARING,' Anaxagoras,§ 11). Theophrastus finds
in the position thus taken up by Anaxagoras a resemblance
to that of Empedocles, who held that perception is effected
by means of emanations fitting into the pores of the sensory
organs. 'Anaxagoras in explaining the superior sense-
perception of the larger animals by a proportionateness
between the objects which they perceive and their larger
organs of sense, seems to adopt the view of Empedocles;

---

[1] Theophr. *de Sens.* § 35 ; Diels, *Dox.*, p. 509. 1 πλὴν ἐπὶ τῆς ὀσφρήσεως
ἴδιον (i. e. affecting Anaxagoras peculiarly as compared with Empedocles)
συμβαίνει δυσχερές· ὄζειν μὲν γάρ φησι τὸν λεπτὸν ἀέρα μᾶλλον, ὀσφραίνεσθαι
δὲ ἀκριβέστερον ὅσα τὸν πυκνὸν ἢ τὸν μανὸν ἕλκει.

for he represents sense-perception as due to a fitting of something into the pores [1].' It is possible that Anaxagoras merely meant that the larger animals with their larger organs receive a larger amount of stimulus : not that they perceive *fine* distinctions, auditory or olfactory, better than small animals do. Their superiority of sense to the latter would thus be only a qualified superiority, having its drawbacks as we have suggested. Theophrastus may have misunderstood, and then misstated, the intention and effect of his comparison between larger and smaller animals.

Diogenes of Apollonia.

Organ and *Function* of smelling. The air round the brain should be 'symmetrical with odour.' Length and fineness of olfactory passages. Man's inferiority to certain other

§ 9. 'Diogenes held that the sense of smell is effected by the air around the brain, for this is compact and symmetrical with odour. The brain itself is porous, and its veins are fine, but the air around it, in creatures in which its diathesis is unsymmetrical, does not mix with odours; since if a person were assumed to have the temperament of the air within him symmetrical with the temperament of these, he would certainly also have the sensation of them [2].'

'Smelling is most acute in those creatures that have least air in the head, for it (the air) then most quickly blends (with the odoriferous stimulus). Moreover, if one draws in the odour through a smaller and narrower

---

[1] Theophr. *de Sens.* § 35 ; Diels, *Dox.*, p. 509. 12 τὸ δὲ πρὸς τὰ μεγέθη τὴν συμμετρίαν ἀποδιδόναι τῶν αἰσθητῶν ἔοικεν ὁμοίως λέγειν Ἐμπεδοκλεῖ· τῷ γὰρ ἐναρμόττειν τοῖς πόροις ποιεῖ τὴν αἴσθησιν.

[2] Theophr. *de Sens.* § 39. I give the text suggested by Diels, *Vors.*, p. 344 τὴν μὲν ὄσφρησιν τῷ περὶ τὸν ἐγκέφαλον ἀέρι· τοῦτον γὰρ ἄθρουν εἶναι καὶ σύμμετρον τῇ ὀσμῇ· τὸν γὰρ ἐγκέφαλον αὐτὸν μανὸν καὶ ⟨τὰ⟩ φλέβια λεπτά, τὸν δ' ἐν οἷς ἂν ἡ διάθεσις ἀσύμμετρος ᾖ οὐ μείγνυσθαι ταῖς ὀσμαῖς· ὡς εἴ τις εἴη τῇ κράσει σύμμετρος, δῆλον ὡς αἰσθανόμενον ἄν. The suggestion formerly made by Diels (*Dox.*, p. 510, 16 n.) to read ⟨τὰ⟩ φλέβια λεπτά, ἧσσον δὲ οἷς, comparing Arist. *de Sens.* 458ᵃ 7—ἡ λεπτότης καὶ ἡ στενότης τῶν περὶ τὸν ἐγκέφαλον φλεβῶν—gave at all events the required sense, so far as it went ; but the difficult καὶ οὐ remained. The MSS. λεπτότατον δ' ἐν οἷς ἡ διάθεσις ἀσύμμετρος, καὶ οὐ μείγνυσθαι cannot stand. Diogenes could not have said that the air or the brain is λεπτότατον in those whose sense of smell is defective, for according to him the greater the thinness of the air in the brain, and the greater the fineness of its ducts, the more excellent is the faculty of smelling.

passage (he smells more acutely), for thus it is more animals in olfactory
quickly discerned. Wherefore in some of the other power.
animals the sense of smell is more perfect than in man.
Not but that man, too, if the given odour were symmetrical, so as to blend duly, with the (intra-organic) air,
would have this faculty in its highest perfection [1].' In
Diogenes, all the elements which were mixed to form
man's body, and all elements whatever, are reducible to
ἀήρ—the one substance from which all phenomenal substances are differentiated.

Of the physical nature of ὀσμή Diogenes has left no
account that survives. The medium by which it was
conveyed from the odoriferous object to the olfactory organ
was, of course, air.

Plato.

§ 10. 'With regard to smelling, tasting, and touching, Plato does
as sensory *functions*, Plato (says Theophrastus) has told us not attempt an explana-
nothing whatever, nor even whether there are any other tion of the
senses besides these (i.e. the *five*), but he bestows particular *function* of smelling.
care on his theory of the *objects* of the various senses [2].' As regards its *organ*,
While he developed psychological as well as physical theories he merely
of seeing and hearing, his theories of the other senses, being assumes it to be the
confined to their objects, are mainly if not wholly physical. nostrils.
To turn to Plato himself. The *object* does not

' As regards the faculty of the nostrils, no classification admit of division

---

[1] Theophr. *de Sens.* § 41 ; Diels, *Vors.*, p. 344, *Dox.*, pp. 510–11
ὄσφρησιν μὲν οὖν ὀξυτάτην οἷς ἐλάχιστος ἀὴρ ἐν τῇ κεφαλῇ· τάχιστα γὰρ
μείγνυσθαι· καὶ πρὸς τούτοις ἐὰν ἕλκῃ διὰ μακροτέρου (μικροτέρου ? Diels)
καὶ στενωτέρου· θᾶττον γὰρ οὕτω κρίνεσθαι· διόπερ ἔνια τῶν ζῴων ὀσφραντι-
κώτερα τῶν ἀνθρώπων εἶναι· οὐ μὴν ἀλλά, συμμέτρου γε οὔσης τῆς ὀσμῆς τῷ
ἀέρι πρὸς τὴν κρᾶσιν, μάλιστα ἂν αἰσθάνεσθαι τὸν ἄνθρωπον. Diels' suggestion μικροτέρου is supported by the sense. Perhaps μακροτέρου was
a correction of some one who remembered what Aristotle says (*de Gen.
An.* v. 2. 781ᵇ 10) about the more acute sense of distant sounds and
odours being connected with longer tubes inwards from the orifices of
the ear and nose.

[2] Theophr. *de Sens.* § 6 ; Diels, *Dox.*, p. 500 περὶ δὲ ὀσφρήσεως καὶ
γεύσεως καὶ ἁφῆς ὅλως οὐδὲν εἴρηκεν, οὐδ' εἰ παρὰ ταύτας ἄλλαι τινές εἰσιν,
ἀλλὰ μᾶλλον ἀκριβολογεῖται περὶ τῶν αἰσθητῶν. Cf. ibid. *supra* οὐ μὴν
εἴρηκέ γε περὶ ἁπασῶν ἀλλὰ μόνον περὶ ἀκοῆς καὶ ὄψεως.

into genera of its objects can be made (εἴδη μὲν οὐκ ἔνι)[1]. For smells
and species. are of a half-formed nature [2] (τὸ τῶν ὀσμῶν πᾶν ἡμιγενές), and
The four
elements no class of figure has the adaptation requisite for producing
inodorous.
All odours any smell [3], but our veins in this part are formed too narrow
καπνός or for earth and water, and too wide for fire and air: for which
ὀμίχλη,
i. e. water cause no one ever perceived any smell of these bodies; but
passing smells arise from substances which are being either liquefied
into air,
or air or decomposed, or dissolved, or evaporated [4]. For when
passing
into water. water is changing into air and air into water, odours arise
They be- in the intermediate condition; and all odours are *vapour*
long to an
inter- or *mist*, mist being the conversion of air into water, and
mediate vapour the conversion of water into air [5]; whence all smells
condition
of these are subtler than water, and coarser than air. This is
elements.
Odours are proved when any obstacle is placed before the passages
finer than of respiration, and then one forcibly inhales the air; for
water,
coarser then no smell filters through with it, but the air bereft
than air. of all scent alone follows the inhalation. For this reason
Only two
kinds of the complex varieties of odour are unnamed, and are
odour, the ranked in classes neither numerous nor yet simple [6]; only
pleasant
and the *un*-two conspicuous kinds are in fact here distinguished,
pleasant.
Physiologi-*pleasant* and *unpleasant.* The latter roughens and irritates
cal cause all the cavity of the body that is between the head and the
of this
distinction. navel; the former soothes this same region and restores it
with contentment to its own natural condition [7].'

[1] ' Distinctions of kinds of smell are here denied because smell
always has to do with an incomplete and undetermined Becoming, and
because it belongs, as is said in what follows, only to a transient moment,'
Zeller, *Plat.* p. 275 n., E. Tr.

[2] Cf. Aristotle, *infra* § 13.

[3] Mr. Archer-Hind, whose translation I give, observes on this:
' That is, odour does not possess the structure of any of the four—*fire,
air, water,* and *earth.*'

[4] βρεχομένων ἢ σηπομένων ἢ τηκομένων ἢ θυμιωμένων.

[5] εἰσὶ δὲ ὀσμαὶ ξύμπασαι καπνὸς ἢ ὀμίχλη· τούτων δὲ τὸ μὲν ἐξ ἀέρος εἰς
ὕδωρ ἰὸν ὀμίχλη, τὸ δὲ ἐξ ὕδατος εἰς ἀέρα καπνός. Cf. Arist. *Meteor.*
346[b] 32; *de Sens.* 443[a] 26–30.

[6] οὐκ ἐκ πολλῶν οὐδ' ἁπλῶν εἰδῶν ὄντα. 'Smells are not ἁπλᾶ because they do not proceed from any definite single substance, nor
πολλά, because we can only classify them as agreeable or the reverse.'
Archer-Hind, ad loc.

[7] Plato, *Tim.* 66 D–67 A (Archer-Hind's translation). For Aristotle's

§ 11. Plato's theory that smells cannot be classified is controverted by Aristotle, but ineffectually. The theory itself is confirmed by modern psychologists and physiologists. 'Though we may recognize certain odours as more like to each other than to other odours, or can even make a rough classification of odours, we cannot, as we can in the case of visual colour sensations, reduce our multifarious olfactory sensations to a smaller number of primary sensations mixed in various proportions. Nor have we at present any satisfactory guide to connect the characters of an olfactory sensation with the chemical constitution of the body giving rise to it[1].' For a similar judgment from the psychologist's point of view cf. Wundt, *Human and Animal Psychology* (E. Tr.), p. 65.

According to Plato, then, with whom Aristotle here agrees, each of the four elements *per se* is inodorous[2]. Theophrastus re-states the matter thus. 'Plato holds that odours cannot be classified into species, but differ only as they are painful or pleasant. Odour is, he says, a thing more subtle than water, but more gross than air. A proof of this is that when persons inhale the breath through some obstacle it enters without odour. Wherefore it is like vapour or mist from bodies, but invisible. Vapour is the result of a change from water into air, but mist of one from air into water[3].'

§ 12. The pleasures arising from sweet odours are reckoned by Plato among the purer kinds of pleasure.

'Those things which suffer a gradual withdrawing and emptying, but have their replenishment sudden and on a large scale, are insensible to the emptying, but sensible of the replenishment; so that while they cause no pain to the mortal part of the soul, they produce very intense pleasure. This is to be observed in the case of sweet smells[4].' In the *Republic*, Plato tells us that the pleasures

[side notes:] Aristotle's criticism of Plato's statement that odours cannot be classified. It does not affect Plato's contention, which is confirmed by modern physiology and psychology.

Pleasures arising from sweet odours ethically more valuable than those of touch and taste: not merely negative, as the latter are,

criticism of the theory that no classification of odours is possible cf. § 23 *infra*, Arist. *de Sens.* v. 443[b] 17 seqq.

[1] Foster, *Text Book of Physiology*, § 860, p. 1389.

[2] Cf. Arist. *de Sens.* v. 443[a] 10 τά τε γὰρ στοιχεῖα ἄοσμα οἷον πῦρ ἀὴρ ὕδωρ γῆ.   [3] Theophr. *de Sens.* § 85 ; Diels, *Dox.*, p. 525.

[4] *Tim.* 65 A (Archer-Hind's trans.).

nor followed by pain. Not so valuable, however, as those of colour and sound.

of smell are not merely negative, i. e. depending on the removal of a pain; nor are they followed by any pain. They are instances, therefore, of καθαραὶ ἡδοναί—*pure* pleasures [1].

In the *Philebus* also he grants that there are true pleasures arising from the sense of smell. They depend on wants which are not felt *as* wants, or as painful, while the supply of them is felt, and felt as pleasurable [2]. These pleasures are, however, of a less exalted kind than those of colours and sounds.

Aristotle.

Difficulty presented by the olfactory sense. Inferiority of this sense in man. All sensations of odour are for man mixed with pleasure or pain. We distinguish *odours* as obscurely as hard-eyed creatures do *colours*, which to them are only significant of the presence of danger or the contrary.

§ 13. Aristotle recognizes the difficulty of treating satisfactorily of the sense of smell, its objects and their classification, and accounts for it by the fact, as he states it, that this sense is in man comparatively imperfect. 'Savours as a class display their natures more clearly to us than odours, the cause of this being that the olfactory sense of man is inferior in acuteness to that of the lower animals, and that this, compared even with man's other senses, is the least perfect of all. Man's sense of touch, on the contrary, excels that of all other animals in fineness, and taste is a form of touch [3].' 'It is less easy to form definite conceptions on the subject of odour—the object of the sense of smell—than on the subjects hitherto dealt with, seeing, hearing, and their objects. It is not as clear what the physical nature of odour is as what the natures of colour and sound are. The ground of this is, that *our* olfactory sense is not exact in its perceptions, but inferior to that of many other animals. Mankind have but an imperfect sense of smell; they perceive none of the objects of this sense, except in connexion with their pleasurableness or unpleasantness, which at once betrays the imper-

---

[1] *Rep.* 584 B–C εἰ ᾽θέλεις ἐννοῆσαι τὰς περὶ τὰς ὀσμὰς ἡδονάς· αὗται γὰρ οὐ προλυπηθέντι ἐξαίφνης ἀμήχανοι τὸ μέγεθος γίγνονται, παυσάμεναί τε λύπην οὐδεμίαν καταλείπουσιν. Μὴ ἄρα πειθώμεθα καθαρὰν ἡδονὴν εἶναι τῆς λύπης ἀπαλλαγήν.

[2] *Phil.* 51 B ὅσα τὰς ἐνδείας ἀναισθήτους ἔχοντα καὶ ἀλύπους τὰς πληρώσεις αἰσθητὰς καὶ ἡδείας καθαρὰς λυπῶν παραδίδωσιν.

[3] Cf. *supra* § 9 ; *de Sens.* iv. 440ᵇ 30–441ᵃ 3.

fection of our olfactory organ. The case of hard-eyed animals, with regard to seeing colours, resembles that of man in relation to odours : the distinctive qualities of colours are not apparent to them except as indicating the presence or absence of something terrifying. With the same vagueness one may suppose that human beings perceive odours. The sense of smell appears analogous to that of taste, and the various kinds of odours to those of tastes ; and yet our sense of taste is more perfect, which appears due to its being a mode of touch—the sense in which man is superior to all other animals [1].'

§ 14. There is a sensible analogy between smells and tastes. 'Smells are, like tastes, distinguished as *sweet* and *bitter*. In some objects, however, the smell is analogous to the taste; in them, for example, both taste and smell are sweet. In others the taste and the smell are of opposite sorts. Odours, as well as tastes, are likewise distinguished as *pungent* (δριμεῖαι), *harsh* (αὐστηραί), *acid* (ὀξεῖαι), and *succulent* (λιπαραί). But since odours are not as clearly discernible as tastes, it is from the latter that odour has derived these distinguishing names, in virtue of the sensible resemblance between the things. For example, the smell of saffron is *sweet*, and so is the smell of honey; while that of thyme and such things is *pungent*, and so on in like cases [2].' But the analogy of smells to tastes must not be pressed too far. Many things have an agreeable odour, yet a most disagreeable taste, and conversely [3]. 'From the physical analogy between the object of smell and that of taste, there should be an analogy between their effects on sense. This is certainly the case with some odours and tastes. There are odours called *pungent, sweet, harsh, sour* (στρυφναί), and *succulent*, and one might speak of *fetid* smells as analogous to *bitter* tastes ; wherefore the former make inhalation as offensive as the latter make swallowing [4].' The sense of smell occupies a place midway

Sensible analogy between odours and tastes, marked by community of names. It is from tastes that odours have by analogy derived their names. Also a physical analogy between smells and tastes in virtue of the common origin of their objects. Smell stands midway between touch and taste on one hand, and sight and hearing, on the other.

---

[1] Arist. *de An.* ii. 9. 421ᵃ 7-20.   [2] Arist. l. c. 421ᵃ 26–421ᵇ 3.
[3] 421ᵃ 27 ἀλλὰ τὰ μὲν ἔχουσι ... τὰ δὲ τοὐναντίον.
[4] *De Sens.* v. 443ᵇ 6-12. For the above analogies see also § 19 *infra.*

between the two senses which are modes of touch (i. e. ἀφή and γεῦσις), and the other two which perceive through an external medium [1].

Organ of smelling in animals generally. In birds and serpents. In non-respiring animals.

§ 15. The organ of smelling is (as Aristotle thinks, contrary to the opinion of previous psychologists, who held it to be of fire) constituted of air in animals which respire, of water in the case of aquatic animals. In the former class it is, perhaps, furnished with a πῶμα, or cover, analogous to the lid which covers the eye (see *infra* § 18, p. 151). The veins or pores of this covering must be opened by the breath inhaled, before smelling can take place [2]. This explains why it is that we perceive odour only when inhaling, not when exhaling or holding the breath, and that under water we cannot smell, since inhalation is there impossible. Aquatic animals can smell under water just because probably they are without this covering of the organ of smell (*vide infra*, § 18). 'The organs of smell are placed with good reason between the eyes. For as the body consists of two parts, a right half and a left, so also each organ of sense is double.' This is not so obvious in the cases of taste and touch as in the senses of hearing, seeing, and smelling. 'There are two nostrils, though these are combined together. Were they otherwise disposed, and separated from each other as are the ears, neither they nor the nose in which they are placed would be able to perform their office. For in such animals as have nostrils olfaction is effected by means of inhalation, and the organ of inhalation is placed in front, and in the middle line. This is the reason why nature has brought the two nostrils together, and placed them as the central of the three sense-organs, setting them, as it were, on either side of a single line, in a direction parallel to the inhalatory motion [3].' 'In the generality of quadrupeds and viviparous animals there is no great variety in the forms of the organ of smell. . . . In no animal is this so peculiar

[1] 445ᵃ 5–8.

[2] *De An.* ii. 9. 421ᵇ 14 seqq.; *de Sens.* v. 444ᵇ 22 seqq.

[3] Arist. *de Part. An.* ii. 10. 656ᵇ 31–657ᵃ 11 (Dr. Ogle's Transl. with a few changes).

as in the elephant, where it attains an extraordinary size and strength, for the elephant uses its nostril as a hand. ... Just as divers are sometimes provided with instruments for respiration, through which they can draw air from above the water, and thus may remain for a long while under the sea, so also have elephants been furnished by nature with their lengthened nostril ; and when they have to traverse the water, they lift this up above the surface, and breathe through it. ... A nostril is given to the elephant for respiration as to every animal that has a lung, and its proboscis is its nostril. ... In birds and serpents there is nothing which can be called a nostril, except from a functional point of view. ... A bird, at any rate, has nothing which can be properly called a nose. In its beak, however, are olfactory passages, but no nostrils. ... As for those animals that have no respiration, it has been already explained why it is that they are without nostrils, and perceive odours either through *gills*, or through a *blow-hole*, or, if they are insects, by the *hypozoma* ; and how their power of smelling depends, like their motions, upon the innate spirit of their bodies which in all of them is implanted by nature and not introduced from without [1].' 'Another part of the face is the nose, which forms the passage for the breath. ... Through this part is performed respiration. It is, indeed, possible to live without breathing through the nose, but through this alone *smelling*, i. e. the sense by which we perceive odour, is effected. Its parts—for it is bipartite— are the septum, which is of cartilage, and an empty duct on either side of this [2].' 'Nature, as it were *en passant*, employs the respiratory process, in the case of certain animals, for the purpose of the sense of smelling. Hence, almost all animals have the sense of smell, though all have not the same sort of olfactory organ [3].'

§ 16. The sense of smelling operates through a medium— *Medium* of smelling :

---

[1] Arist. *de Part. An.* ii. 16. 658ᵇ 27–659ᵇ 19 (Dr. Ogle).
[2] Arist. *Hist. An.* i. 11. 492ᵇ 5–17.
[3] Arist. *de Respir.* 7. 473ᵃ 23–7 ; cf. *de Sens.* v. 444ᵃ 25–8 for similar words.

air or water[1]. Aquatic animals appear to have a sense of odour. This sense is possessed alike by sanguineous and by bloodless animals, and generally by all which live in the air (τὰ ἐν ἀέρι); for some of the last come from great distances directly to their food when they have got the scent of it[2]. What the organ of smelling (or hearing) is in the case of fishes and other animals that live beneath the water is not known[3]. But the medium is in general the same as that of seeing, viz. the diaphanous : only it is not *qua* diaphanous that it serves as medium of smelling, but (§ 19 *infra*) *qua* having the power of washing or rinsing its native quality out of the sapid dryness (p. 152, n. 1). How the medium acts, or how odour is conveyed through or by it from the odorous object to the organ, had been considered before Aristotle's time. Older writers took the essential constituent of the organ of smelling to be fire[4], and regarded odour itself as a fumid exhalation (καπνώδης ἀναθυμίασις) consisting (according to Aristotle) of the elements earth and air[5]. 'Indeed,' says Aristotle, 'all are inclined to this

air or water. The latter is the medium of odour for aquatic creatures. The *general* medium, viz. the 'diaphanous' includes both. Not however *qua* diaphanous is this a medium of odour, but *qua* capable of extracting and absorbing the quality of the *sapid dry*. Former writers

---

[1] 419[a] 32, 443[a] 2, 419[a] 35, 421[b] 9-11, 533[b] 4, 444[a] 21.

[2] 421[b] 12. 'The old hypothesis that vultures find their prey by the aid of this sense (smell) has been abundantly disproved.' Romanes, *Mental Evolution in Animals*, p. 92.

[3] 444[b] 15, 656[a] 36.   [4] 438[b] 20-22.

[5] 443[a] 21 seqq. In *de Sens.* ii. 438[b] 20-25 Aristotle himself appears to adopt these very views of the organ and object of smelling. Bäumker, however, with whom Zeller (*Arist.* ii. 63 n., E. Tr.) agrees, on the strength of the reading εἰ before δεῖ, asserts (*Arist. op. cit.* p. 31) that Aristotle there speaks not from his own but from an alien point of view with which he does not agree. Kampe, *Erkenntniss-theorie des Arist.*, p. 77, accepts the statements of *de Sens.* ii as containing Aristotle's own opinions, notwithstanding the inconsistencies which thus emerge. The health-theory of ὀσμή, propounded in *de Sens.* v (where the statements of ch. ii that ὀσμή is καπνώδης ἀναθυμίασις is energetically contradicted) requires this very assumption of ὀσμή being ἐκ πυρός ; for the wholesome effect of ὀσμή on the brain is derived from the heat of the former. Cf. 444[b] 1 σύμμετρος γὰρ αὐτῶν (sc. τῶν ὀσμῶν) ἡ θερμότης, and also 444[a] 22-4 ἡ γὰρ τῆς ὀσμῆς δύναμις θερμὴ τὴν φύσιν. Though ἀήρ is *hot* and moist, I cannot think that it is to air and not fire that the heating effect of ὀσμή is intended to be ascribed in these passages. How the inconsistency is to be explained is another matter. See *infra*, § 22.

exhalation-theory.' It furnished them with the analogy which they sought for to explain the transmission of the odorous particles through the medium. Heraclitus implied his acceptance of it when he asserted that 'if all existing things were reduced to "smoke"[1] (i. e. the above fumid exhalation) the nose would be the organ which would perceive or discern all things.' Aristotle (*de Sens.* v) though he regards odour as naturally 'hot,' rejects this theory of its being καπνώδης ἀναθυμίασις, for other reasons but particularly because (*a*) since fumid exhalation does not occur under water, it leaves inexplicable the fact that fishes have the olfactory sense; and because (*b*) this theory is analogous to, and must stand or fall with, the theory of emanations, which he has already declared to be untenable. All that has been urged against the theory of ἀπορροαί in relation to the other senses, may be used in argument against it in relation to the sense of smell. Aristotle probably intends here to confute Plato, who regarded all odour as either καπνός or ὀμίχλη[2]. Perfection of the sense of smelling, as of the senses of seeing and hearing, involves two things, viz. (*a*) perception of its object at a long distance; and (*b*) nice discrimination of differences of quality in the object. The latter element of perfection depends on the purity of the organ, and the freedom from alien matter of the membrane which covers it. The former element depends on the length of the passages in the organ which convey the external stimulus inwards to the 'point of sense.' These rules of perfection hold alike, indeed, for the three organs which have external media, viz. those of seeing, hearing, and smelling[3]. We are led to infer that the operation of smelling is ultimately effected by the σύμφυτον πνεῦμα, or connatural spirit, with which the olfactory channel is filled. This spirit conveys the ὀσμή, or stimulus of ὄσφρησις, to the blood vessels around the brain, and thence to the heart. The case is analogous with that of hearing[4].

Marginal notes: had made the essential element of the organ of smell to be fire, and the object to be humid, or fumid, evaporation. Heraclitus held the ἀναθυμίασις theory of odour. Conditions of perfection in the olfactory sense, (*a*) distant perception, (*b*) nice discrimination. The former depends on having long tubes or passages connected with the organ. The latter on the purity of the constitution of the organ. Dependence of the olfactory function on the σύμφυτον πνεῦμα.

---

[1] καπνός. It must be remembered that by words like this and ἀήρ the Greeks denoted what we, after van Helmont, speak of as 'gases.' The word 'air' did duty for the idea of 'gas' in English until about 100 years ago.

[2] Cf. § 10 *supra*.   [3] 781[a] 17-[b] 29.   [4] 744[a] 3 seqq.

Inhalation a condition of smelling. How creatures which do not respire perceive odours is a mystery. Insects have smell. Proof that they do possess this sense.

§ 17. In mankind and other creatures which have lungs and respire, the power of smelling is suspended while the breath is held or exhaled[1]. Only while inhaling can a person smell, as may be ascertained by experiment[2]. 'Since bloodless animals do not respire, and yet possess olfactory sense, some one may doubt whether it is really this sense which they possess, and not some other over and above the common five senses. To this we reply that, if what at such times they perceive is *odour*, it cannot be that they perceive it by any other than the olfactory sense; for the sense which discerns odour, pleasant or unpleasant, *is* the olfactory sense and nothing else[3].' 'That creatures which do not respire possess this sense is evident. Fishes and all insects have, thanks to the species of odour correlated to nutrition (*vide infra* § 23), a keen sense of their proper food from even a very great distance; e.g. bees as regards honey, and also ants, of the small kind called κνῖπες. Among marine animals, too, the purple-fish and many other similar creatures have an acute perception of their food by its odour[4].' 'Further, they are deleteriously affected by strong odours of the kind by which human beings are injured, e.g. those of *bitumen, brimstone*, &c. These animals, therefore, must possess the sense of smell even without the faculty of respiration[5].'

Difficult to determine

§ 18. 'It is not so easy to be confident as to the organ by

---

[1] 421[b] 18 ἀλλὰ τὸ ἄνευ τοῦ ἀναπνεῖν μὴ αἰσθάνεσθαι ἴδιον ἐπὶ τῶν ἀνθρώπων. This sentence is, as Hayduck (*Observationes criticae in Arist.*, Greifswald 1873) pronounces, corrupt: it states what is both false *per se* and contradictory of 419[b] 1–2 ὁ μὲν ἄνθρωπος καὶ τῶν πεζῶν ὅσα ἀναπνεῖ ἀδύνατα ὀσμᾶσθαι μὴ ἀναπνέοντα: as also of 444[b] 16–24 and 473[a] 15–27. He also finds ἀνθρώπων in 421[b] 18 wrongly opposed to πάντων (αἰσθητῶν) just before. He therefore reads (instead of ἀνθρώπων) ὀσφραντῶν, thus getting rid of an extraordinary proposition, and making perfect sense.

[2] *De An.* ii. 9. 421[b] 13–19. While the breath is being held or exhaled no odorous object can be smelled—not even if placed within the nose on the very nostril. But (adds Aristotle) *contact* between object and organ defeats perception in the cases of all the mediated senses.   [3] 421[b] 19–23: cf. 444[b] 19–21.

[4] 444[b] 7–15.   [5] 421[b] 23–6.

which they smell.   Though they have the olfactory sense, the organ of this sense in them cannot be like that in man and creatures which respire.   In the latter, this organ, as compared with the analogous organs in the other creatures, seems to differ from them much as man's eyes differ from those of hard-eyed animals.   The eyes of man have, in their lids, a kind of shelter or envelope, whence a person cannot see without first raising and removing the eyelids.   But hard-eyed creatures are without anything of this sort ; they see at once whatever presents itself to them in the diaphanous medium of vision [1].'   'They do not need, besides eyes, an eye-opening apparatus, but see directly, once there is anything to be seen [2].'   'In the selfsame way in the non-respiring animals the olfactory organ seems to stand uncovered, like the eye in the case described ; while in creatures which respire this organ seems to have upon it a sort of lid ($\pi\hat{\omega}\mu\alpha$) or curtain ($\epsilon\pi\iota\kappa\dot{\alpha}\lambda\upsilon\mu\mu\alpha$), which the breath inhaled lifts off and removes, the veins and pores being then dilated ; hence they can smell only when inhaling.   In creatures which do not respire, this lid may be regarded as permanently removed [3].'   'The reason why animals which respire cannot smell under water is now manifest.   To smell they should inhale air, and for them to do this under water would be impossible [4].'   The connexion, therefore, between the sense of smell and respiration is not, as Empedocles thought, necessary, but merely contingent '(§ 15 *supra*)[5].

[margin: the organ of smell in such creatures. Our eyes have lids needing to be lifted for vision. So our olfactory organ may have some sort of lid, while that of non-respiring creatures is without it. Respiration is only contingently by the economy of nature, joined with smelling in certain animals.]

§ 19.   Physically regarded, odour consists of the Dry, just as taste consists of the Moist, and as the object of smell is *actually*, such is the organ *potentially*[6].   As, therefore, there is a sensible analogy between tastes and smells, so there is a physical analogy also, resting on their origin respectively.   'Our physical conception of odours must be analogous to that of savours, inasmuch as the sapid moist (see note 1, p. 152) effects, in water and air alike, in the sphere of another sense, what the (nutrient) dry effects in

[margin: Object of smell—odour—regarded physically. Analogy between odour and taste, so regarded. Effects of cold on odours: it destroys]

---

[1] Arist. *de An.* ii. 9. 421^b 26–32.　　　[2] 444^b 27–8.

[3] 421^b 32–422^a 3 and 444^b 21–8.　　　[4] 422^a 3–6.

[5] Theophr. *de Sens.* § 21 ; Diels, *Vors.*, p. 179.　　　[6] 422^a 6–7.

the 'scent.' the water (moist) only [1]. We attribute diaphanousness to
Odour ex-
ists under both water and air; but it is not in virtue of this quality
water. Def. that either of these is a vehicle of odour, but in virtue of
of *odour.*
the power which the so-called diaphanous has of rinsing
out, and so contracting, the quality of sapid dryness from
objects which possess it. Again, if the dry produces in
water and air an effect as of something washed out into
these, there must be an analogy between savours and
odours. . . . Plainly, odour is, in water and air, what savour
is in water. This explains why excessive cold, as of frost,
dulls the odour and taste of things; as it destroys the
kinetic heat by which sapidity—the base of odour—is
wrought into the substance of the moist. That the object
of smell—odour—exists not only in air, but also in water, is
proved by the case of fishes and testacea, which are seen to
possess the faculty of smelling, in spite of the fact that
water does not contain air (since air generated under water
always rises to the surface and escapes), and though these
creatures do not respire. Hence, if we grant that air and
water are both moist, it follows that we may define *odour*
as *the natural substance of the sapid dry in a moist medium*[2];
and whatever is of this nature is an object of smell [3].

Odour   § 20. We may see by comparing the things which have
originates
in taste,   odour with the things which have it not, that the property
physically of odorousness originates in that of sapidity. Simple
regarded.
Substances substances (viz. the elements earth, air, fire, water) are
which do
not possess tasteless, and hence they are inodorous [4]. The elements

[1] 442[b] 27–443[a] 2. The nutrient dry produces sapidity in water: the
sapid moist produces odorousness in air and water. The quality of
sapidity is derived from τὸ ξηρόν, which, however, to be tasted, has to
be presented in a moist vehicle, or medium. In this medium it can
be called the sapid moist, and as such it is the foundation of odour.
The ἔγχυμον ξηρόν is the *ultimate*, the ἔγχυμον ὑγρόν the *proximate*
cause of odour. Hence Aristotle uses either expression—sapid dry
(443[a] 2) or sapid moist (442[b] 29)—in this connexion, and Torstrik's
ξηρόν for ὑγρόν in 442[b] 29 is needless.

[2] In air or water; air is hot and moist as water is cold and moist.

[3] Arist. *de Sens.* v. 442[b] 30–443[a] 8 and 443[b] 6–16.

[4] Cf. Theophr. Περὶ Ὀσμῶν, i. 1 αἱ ὀσμαὶ τὸ μὲν ὅλον ἐκ μείξεώς εἰσι καθάπερ
οἱ χυμοί· τὸ γὰρ ἄμεικτον ἅπαν ἄοσμον, ὥσπερ ἄχυμον, διὸ καὶ τὰ ἁπλᾶ ἄοδμα,
οἷον ὕδωρ ἀὴρ πῦρ· ἡ δὲ γῆ μάλιστα ἢ μόνη ὀδμὴν ἔχει, διὸ μάλιστα μεικτή.

are inodorous because in them the moist and the dry are taste have without sapidity, until some added ingredient introduces it. Sea-water, on the other hand, possessing savour as well as dryness [1], possesses *odour* also. Various other substances are found to vary in odorousness directly in proportion to their *sapidity*. Such are *salt* as compared with *soda, wood* as compared with *stone*; *bronze* and *iron* as compared with gold [2]. 'In fact *odour* and *savour* are physically almost the same affection, though each is realized for sense under different conditions from the other [3]. Odour is in its nature possessed of heating power [4], a property which, as we shall see, makes it conducive to the health of the brain.

taste have no odour : odorousness varies directly as sapidity. Odour and savour physically the same.

§ 21. Odour is transferred from the odorous object to the olfactory organ in a medium which, as we have seen, may be *air* or *water*. Its passage through the medium is not instantaneous; unlike light, it requires time to travel. A person who is nearer to an odorous object perceives its odour sooner than one who is farther off [5]. Odour is wafted to us in the air, so that we can smell distant objects. So savour is propagated through water, and, no doubt, if we were denizens of the water, we should be able to taste things, as we now smell them, from a distance [6]. The stimulus of smell like that of hearing takes time to reach us. The only object of sense which involves no time of transit is the object of vision, colour, which depends on light: for light has no transit-time. Its diffusion is co-instantaneous in diverse places.

Odour travels through its medium. Colour is the only mediated object which takes no time in transitu.

In reading this account of odour travelling through a

---

[1] ξηρότητα : sea-water, according to Aristotle, contains earth, the distinctive characteristic of which is dryness, *de Gen. An.* iii. 11. 761ᵇ 8-12 ; *Meteor.* iv. 4. 382ᵃ 3 λέγεται δὲ τῶν στοιχείων ἰδιαίτατα ξηροῦ μὲν γῆ, ᵇ 3 τιθέμεθα δὲ ὑγροῦ σῶμα ὕδωρ, ξηροῦ δὲ γῆν.

[2] 443ᵃ 8-21. Aristotle's theory of odours depends on his theory of tastes, hence a good deal of the above must, to be understood, be read in the light of what will follow in the section on Tasting.

[3] 440ᵇ 29-30. Πάθος = the effect of the (ἔγχυμον) ξηρόν in the ὑγρόν —of air and water, or of water only.

[4] 444ᵃ 24-5.   [5] 446ᵃ 23.

[6] 422ᵃ 11-14, 447ᵃ 6-9. Taste, for Aristotle, is, however, a mode of Touch, 434ᵇ 18-24.

medium one should not forget that Aristotle steadfastly opposed the theory of ἀπορροαί, or particles floating from the object to the organ. What he believed was that the object caused a *change* (κίνησις or πάθος) in the adjacent part of the medium, which change, propagated onwards to the point where medium and organ meet, became the stimulus of perception. (See *de An*. iii. 12. 434ᵇ 27 seqq.)

Odour is not 'evaporation' either fumid or humid. Reasons. Apparent incongruity between views of Aristotle on this point in different parts of *de Sensu*.

§ 22. 'Odour is not fumid evaporation [1], consisting of earth and air. Popular though this idea of it has been, we must reject it. Yet all writers incline to take odour as evaporation in some form, whether fumid or humid [2], or either indifferently [3]. The humid is mere moisture, but fumid evaporation is, as we have said, composed of air and earth. The former, when condensed, forms water ; the latter, a species of earth. Odour is not either of these. The one, too, consisting as it does of water, is tasteless, and therefore without odour; while the other evaporation cannot occur in water, and would not, as physical basis of odour, account for the fact that subaqueous or aquatic creatures possess a sense of this [4].'

It causes much surprise when, on turning from the chapter in which we read as above to an earlier chapter of the *de Sensu*, we find it stated that odour, the object of smell, is (καπνώδης ἀναθυμίασις) fumid evaporation : the proposition denied so energetically three chapters later. 'The olfactory organ is essentially composed of fire' (we read in ch. ii) ; 'for the olfactory organ is *potentially* what the olfactory sense (as actualized) [5] is *actually*. The object is that which causes the actualization of each sense ; so that the sense itself must, to begin with, have the corresponding potentiality. Now odour, the object of this sense, is fumid evaporation, which arises from fire ; hence the

---

[1] Cf. 341ᵇ 6 seqq., 357ᵇ 24 seqq. καπνώδης ἀναθυμίασις is, in plain English, a form of *smoke*, καπνός. [2] 'Mistlike evaporation,' ἀτμίς.

[3] It will be remembered that Plato reduced ὀσμή in all forms to either καπνός or ὁμίχλη, i. e. to the καπνώδης ἀναθυμίασις or the ἀτμίς of our passage. [4] *De Sens*. v. 443ᵃ 21–31.

[5] ὃ γὰρ ἐνεργείᾳ ἡ ὄσφρησις, τοῦτο δυνάμει τὸ ὀσφραντικόν, where ὄσφρησις —the actualized sense—is awkwardly put for ὀσμή—its actualizing object.

organ that is brought to actuality by this object is
potentially fire.'

Is is not easy to explain this discrepancy or to explain
it away. To assert (see p. 148, n. 5) that in the earlier
passage Aristotle speaks from an alien point of view is not
sufficient. Aristotle himself adopts and everywhere maintains all the points there laid down respecting the nature
of the other organs. The thermic property of the object
of smell is plainly asserted[1] even in ch. v, in the argument
which expounds the wholesome effect of odours upon the
brain of man. This effect they owe to their thermic
properties. Thus, notwithstanding the denial in ch. v
that odour is καπνώδης ἀναθυμίασις, it is there made to
retain the property of heat which, in ch. ii, forms the
ground of the assertion that it *is* καπνώδης ἀναθυμίασις.
We may perhaps assume, that, despite the proximity in
which chapters ii and v of the *de Sensu* now stand, they
were written at some considerable interval of time from one
another, which would render explicable a change of view
on the writer's part. We cannot suppose that in the earlier
chapter, where ὀσμή is said to be fumid evaporation,
Aristotle merely uses the current terminology and adopts
the current opinion, which he corrects afterwards when he
comes to deal directly, at close quarters, with this opinion
itself. In the *Meteorologica*, indeed, he adopts respecting
ὄψις (the light ray) a view opposed to his own theory of
vision, but one which was and had long been current.
There, however, he was not concerned with psychology
but with optics, and the current view was good enough for
his purpose; which could not be said here. We have to
fall back upon the *patchwork* character of even some of
the indisputably Aristotelean writings (however it came
about) to explain many such apparent incongruities.

§ 23. 'Despite statements to the contrary[2], odours are Odours can
be classi-

---

[1] 444ª 22–4, 444ᵇ 1. See, however, Neuhäuser, *Arist. Erkenntnissvermögen*, pp. 20–26.

[2] 443ᵇ 17 seqq. Aristotle here seems to censure Plato, *Tim.*: *vide
supra* §§ 10–11. Plato held that odours are incapable of division and

divisible into species. They have an aspect in which they run parallel to tastes. In this aspect their pleasant or unpleasant quality belongs to them only as a consequence of their relation to savour.' Plato, rejecting all classification of odours, except into pleasant and unpleasant, overlooked the distinction between the pleasantness of certain odours *per se* and that of others which depends on appetite for the food from which they arise. But there is a close connexion between the taste of things and the nutrient faculty of the soul, and animals find the odour of food pleasant when they have an appetite for the food itself. When they are satisfied and want no more food, they cease to feel the odour of it pleasant. Their agreeable or disagreeable quality belongs to such odours only incidentally, i.e. as a result of their relationship to food ; but just because of this relationship, all animals without exception perceive them. But there is a different class, viz. that of odours which are *per se* agreeable or disagreeable, as for example, those of flowers, which have nothing to do with appetite (though they preserve *health*, as below explained) either as stimulating or as dulling it. Odours of the former class are divisible into as many sub-classes as there are different classes of savours. Those of the latter class are not divisible in the same way.

These latter odours are perceptible to man, and man only, as agreeable or disagreeable. Other animals perceive only those of the former kind. If they perceive such odours as those of sweet flowers, they are not in the least degree attracted by them. If they perceive the odours which to man are essentially disagreeable, they evince not the slightest repugnance to them, unless, indeed, besides being disagreeable, they are noxious or pernicious, like the fumes of charcoal and brimstone. By the latter animals and men alike are affected, and animals, like men, shun them on account of their effects. But certain plants, which to us smell offensively, seem no way offensive to the lower animals, nor do they concern themselves with them, except as affecting their food.

subdivision into genera and species, and can only be classed· as either pleasant or unpleasant.

§ 24. The reason why the perception of such odours is confined to man is to be found in the comparative size and coldness of man's brain, which is, in proportion to his bulk, larger and moister than that of any other species of animal. Now odour is naturally akin to the hot, and being introduced through the act of respiration, in the case of all animals which respire, it mounts up to the brain, and tempers with its heat the coldness of that organ which might otherwise be excessive. The heat which odour contains renders it light, so that it naturally ascends into the region of the brain, and thus produces in the latter a healthy tone and temperature[1]. While this is true of odour in all animals alike, man, for the reason above given, has, in his perception of odours essentially pleasant or unpleasant, an additional provision for the same purpose. It was nature's own device for counteracting the dangers arising from the greater size and coldness of the human brain. Man's richer endowment in this sense, evidenced by his perception of pleasures and pains of odour in which other animals have no share, is thus and thus only to be explained. This is the sole purpose of his perception of such odours. That they effect this purpose is manifest enough, for odours sweet *per se* are (unlike sweet tastes, which often mislead) universally found to be beneficial, irrespectively of particular states of health or appetite[2]. In

[1] For medicinal effects of ὀσμή cf. Theophrastus, Περὶ Ὀσμῶν, §§ 42 seqq.; Athenaeus 687 D (Kock, *Com. Att.* ii. p. 368) οὐκ οἶδας ὅτι αἱ ἐν τῷ ἐγκεφάλῳ ἡμῶν αἰσθήσεις ὀδμαῖς ἡδείαις παρηγοροῦνται προσέτι τε θεραπεύονται, καθὰ καὶ Ἄλεξίς φησιν ἐν Πονήρᾳ οὕτως—

ὑγιείας μέρος
μέγιστον, ὀσμὰς ἐγκεφάλῳ χρηστὰς ποιεῖν.

In what follows Athenaeus dilates at great length on the wholesome efficacy of odours sweet *per se*.

[2] Arist. *de Sens.* v. 443ᵇ 17–445ᵃ 16. The passage in which the writer expounds his theory of the classification of odours is very confused and ill-composed. It digresses frequently into other matters; but, worst of all, it leaves obscure the precise point on which the difference between man and other animals consists. At one time (444ᵃ 3, 8, 29) the writer says, man alone *perceives* the second class of odours. Later on (444ᵃ 31–3) he seems to qualify this, as if his

smells sweet *per se* never betray. general, however, what taste is for nutrition, this smell is for health [1].

Position of the olfactory among the other senses, and that of the object of this among other objects of sense: smelling comes midway between the tactual and the externally mediated senses: odour midway between the objects of the two classes respectively. § 25. It has been already observed (§ 14 *supra*) that the *sense* of smell occupies a middle position between the senses which perceive by contact and those which perceive through an external medium. The senses are five, that is, they form an *odd* number; and an odd number has a middle unit, which answers to the position of smelling among the other five senses. Hence the *object* of smell, too, has an analogous place among those of the other senses. It is an effect (§ 19 *supra*) produced *in* water or air by the ἔγχυμον ξηρόν (or ὑγρόν), and therefore involves at once affinities for the nutrient objects, which come within the provinces of taste and touch, and also for the objects of seeing and hearing, whence it is that water and air—the media of seeing and hearing—are its vehicles. Accordingly, odour is something belonging to both spheres in common. It has its more material side in the provinces of touch and taste, its less material in the provinces of seeing and hearing. From this fanciful position Aristotle deduces a justification of the figure, by which he described odour as a sort of 'dyeing' (cf. Neuhäuser *op. cit.* p. 24, and Arist. 441[b] 16) or 'washing' of 'dryness' in the moist and fluid [2].

Pythagorean theory that odour is nutrient § 26. 'The theory held by certain Pythagoreans [3] that certain animals are nourished by odour alone is untenable. For food must be composite, as the animal structure

meaning was that man alone *feels pleasure* in their perception. We must suppose that this *pleasurable perception* by man is the distinguishing feature in his case, and that it implies a keenness of scent for odours of this class surpassing that of other animals; so that while they may or may not (ὡς εἰπεῖν, 444[a] 32, seems to indicate uncertainty on this point) perceive them objectively, or in their effects, at all events they do not feel pleasure or pain in these odours as such. Their sense of them lacks the vividness and force with which they impress the consciousness and benefit the health of man.

[1] 445[a] 30.

[2] οἷον βαφή (' Abfärbung ') τις καὶ πλύσις, 445[a] 4–14, 443[a] 1.

[3] On the ground of Alexander's stating that certain physicians held this opinion, Zeller doubtfully refers it to Alcmaeon.

nourished by it is composite. Even water, when unmixed, does not suffice for food; that which is to form part of the animal system must itself be corporeal; but air is even less capable than water of assuming the required corporeal form.

Besides, food passes into the stomach, whence the body derives and assimilates it. The organ by which odour is perceived is in the head, and thither—to the respiratory tract—odour goes in the process of inhaling.' But, not going to the stomach, it is impossible that odour should act as food[1].

mistaken and false. Odour a κίνησις of air, not capable of forming food, which must be solid. Besides, odour goes upwards to the brain; food downwards to the stomach.

[1] *De Sens.* v. 445ᵃ 16–29; *de An.* ii. 3. 414ᵇ 10.

THE ANCIENT GREEK PSYCHOLOGY
OF TASTING

Alcmaeon.

Organ and function of tasting. The tongue is porous like a sponge, and so absorbs the sapid particles which it dissolves by its warmth and moisture. Helplessness of psychology to explain taste.

§ 1. ALCMAEON says 'it is with the tongue that we discern tastes. For this being warm and soft dissolves the sapid particles by its heat, while by its porousness and delicacy of structure it admits them into its substance and transmits them to the sensorium[2].' In the *Placita* he is reported as teaching 'that tastes are discerned by the moisture and warmth in the tongue, in addition to its softness[3].' Diogenes of Apollonia compares the tongue to a sponge, and Alcmaeon seems to have had the same idea. It absorbs the sapid juices of food, and then transmits them to what Alcmaeon regarded as the sensorium—the brain. This very·popular and superficial view of the matter may be compared with that which has still to serve for the psychology of tasting, little though it helps us as regards the essential point, viz. how it comes to pass that the sapid particles are *perceived* as tastes. 'In the ordinary course of things these sensations are excited by the contact of specific sapid substances with the mucous membrane of the mouth, the substances acting in some way or other, by virtue of their chemical constitution, on the endings of the gustatory fibres[4].' Anatomy, Physiology, and Chemistry, despite the enormous advantage they give the psychologist of to-day, have been able to advance the psychology of taste little beyond the popular and superficial stage at which Alcmaeon left it. Here, as in Touching, Psychology tends to merge itself in Physiology.

[1] Theophr. *de Sens.* 25; Diels, *Vors.*, p. 104 γλώττῃ δὲ τοὺς χυμοὺς κρίνειν· χλιαρὰν γὰρ οὖσαν καὶ μαλακὴν τήκειν τῇ θερμότητι· δέχεσθαι δὲ καὶ διαδιδόναι διὰ τὴν μανότητα καὶ ἀπαλότητα. So Wimmer reads for MSS. τὴν μ. τῆς ἀπαλότητος.

[2] Plut. *Epit.* iv. 18, Diels, *Dox.*, p. 407; *Vors.*, p. 104 Ἀλκμαίων τῷ ὑγρῷ καὶ τῷ χλιαρῷ τῷ ἐν τῇ γλώττῃ πρὸς τῇ μαλακότητι διακρίνεσθαι τοὺς χυμούς.

[3] Foster, *Text-Book of Physiology*, § 865, p. 1398.

Empedocles.

§ 2. 'As to tasting and touching, Empedocles says nothing definite respecting either of them, not stating the mode in which or the causes by which they are effected, except merely to enunciate his general principle that all sensation whatever is due to the fitting of emanations into the pores [1].' 'Parmenides, *Empedocles*, Anaxagoras, Democritus, Epicurus, and Heraclides held that the particular sensations are produced in us by the symmetrical relations between the pores of the sense-organ and the object of sense, i.e. when each sense has its proper object of perception fitting into its pores [2].' Theophrastus observes that the theory of ἀπορροαί is, notwithstanding objections, a possible theory regarding the other senses, but is met with difficulties of a special sort as regards those of tasting and touching [3]. It may be that this difficulty prevented Empedocles from developing his theory of emanation with reference to the sense of tasting and touching.

Taste: its function performed by the fitting of symmetrical emanations into the pores of the organ.

§ 3. But though, except for this vague doctrine, he teaches nothing respecting the function of tasting, he gives certain opinions on the physical nature of tastes, objectively regarded, i.e. the sapid substances which cause the sensations of taste. The following we learn from Aristotle: 'Taste is a mode of touch. Now the natural substance water tends to be tasteless, but it is necessary *either* that the water should have in itself the various genera of sapid qualities, though imperceptible owing to their minuteness, as Empedocles holds, *or* &c.[4]' In accordance with this is the view ascribed to Empedocles by Aelian that the sea contains particles of sweet water among the

Taste, objectively regarded, according to Empedocles. All its various kinds exist primarily in water, but in particles of infinitesimally small size, and therefore not perceptible

---

[1] Theophr. *de Sens.* § 9; Diels, *Vors.*, p. 177 περὶ δὲ γεύσεως καὶ ἁφῆς οὐ διορίζεται καθ᾽ ἑκατέραν οὔτε πῶς οὔτε δι᾽ ἃ γίγνονται, πλὴν τὸ κοινὸν ὅτι τῷ ἐναρμόττειν τοῖς πόροις αἴσθησίς ἐστιν.

[2] Aët. iv. 9, Diels, *Dox.*, p. 397; *Vors.*, p. 180 Παρμενίδης, Ἐμπεδοκλῆς, Ἀναξαγόρας, Δημόκριτος, Ἐπίκουρος, Ἡρακλείδης παρὰ τὰς συμμετρίας τῶν πόρων τὰς κατὰ μέρος αἰσθήσεις γίνεσθαι τοῦ οἰκείου τῶν αἰσθητῶν ἑκάστου ἑκάστῃ ἐναρμόττοντος.

[3] Theophr. *de Sens.* § 20 τὸ περὶ τὴν ἀπορροὴν . . . περὶ δὲ τὴν ἁφὴν καὶ γεῦσιν οὐ ῥάδιον.   [4] Arist. *de Sens.* iv. 441ᵃ 3.

162 THE FIVE SENSES

severally.
The tastes
of plants
and fruits,
whence
derived.

predominating salt. 'Empedocles of Agrigentum says that there is a certain portion of sweet water in the sea, though not perceptible to all creatures, and that it serves for the nourishment of the fishes. He declares that the cause of this sweetness which is produced amidst the brine is a natural one [1].' Unfortunately Aelian omits to state what natural cause Empedocles assigned for the sweetness of sea-water; yet we may connect his view of this with what Aristotle tells us above, that Empedocles regarded all genera of taste as existing in water, but in particles too small to be separately perceptible. The several sorts of particles might combine according to their affinities, and when enough of them come together, and are combined like with like, the perceptibly *sweet, bitter, harsh, acid,* and other tastes appear [2]. We must further connect with this view the statement attributed to Empedocles that wine is water which has undergone fermentation [3]. 'The differences of taste in plants correspond to the variations in the manifold of their nutrient particles, and hence in the plants themselves, since they assimilate the kindred particles, from that which nourishes them, differently ⟨in different soils⟩, as we see in the case of vines. It is not differences in the vines that make the wine good or bad, but differences in the soil which nourishes them [4].' The nourishment of

[1] Aelian, *Hist. An.* ix. 64 Ἐμπεδοκλῆς ὁ Ἀκραγαντῖνος λέγει τι εἶναι γλυκὺ ἐν τῇ θαλάσσῃ ὕδωρ, οὐ πᾶσι δῆλον, τρόφιμον δὲ τῶν ἰχθύων· καὶ τὴν αἰτίαν τοῦδε τοῦ ἐν τῇ ἅλμῃ γλυκαινομένου λέγει φυσικήν.

[2] Karsten, *Emped.*, pp. 439 and 482. Cf. Arist. 357^b 24; Diels, *Dox.*, p. 381.

[3] Arist. *Top.* Δ 5. 127^a 17 ὁμοίως δ' οὐδ' ὁ οἶνός ἐστιν ὕδωρ σεσηπός, καθάπερ Ἐμπεδοκλῆς φησί; Diels, *Vors.*, p. 205

οἶνος ἀπὸ φλοιοῦ πέλεται σαπὲν ἐν ξύλῳ ὕδωρ.

Wine is water that has penetrated from the rind of the vine inwards, and undergone decomposition or fermentation within the wood.

[4] The version is from the text of Galenus, *Hist. Phil.*, with Diels' ⟨παρὰ⟩: τὰς διαφορὰς τῶν χυμῶν ⟨παρὰ⟩ παραλλαγὰς γίγνεσθαι τῆς πολυμερείας καὶ τῶν φυτῶν διαφόρως ἑλκόντων τὰς ἀπὸ τοῦ τρέφοντος ὁμοιομερείας. The τῆς ⟨γῆς⟩ πολυμερείας of Diels (*Vors.*) is unfortunate, as Empedocles held not γῆ but ὕδωρ for the source of χυμοί. Cf. Diels, *Dox.*, p. 439; *Vors.*, p. 172.

plants, according to Empedocles, is effected by the attraction of kindred elements into them through their pores from the earth in which they grow.

Democritus.

§ 4. According to Democritus, 'The atomic figure has absolute existence (καθ' αὑτό ἐστι), but the sweet, like objects of sense in general, is relative and dependent on extraneous things' (πρὸς ἄλλο καὶ ἐν ἄλλοις)[1]. ' He does not specify the atomic shapes (μορφάς) which generate *all* objects of sense, but rather those which form tastes (χυλῶν) and colours ; of these he treats definitely and in detail those that are the objective condition of tastes (τὰ περὶ τοὺς χυλούς), explaining how they present themselves as purely relative to us (ἀναφέρων τὴν φαντασίαν πρὸς ἄνθρωπον). The *acid* taste (ὀξύν) he declares to be formed from atomic shapes that are angular, winding, small, and thin (γωνοειδῆ[2] τῷ σχήματι καὶ πολυκαμπῆ καὶ μικρὸν καὶ λεπτόν). ... The *sweet* taste (γλυκύν) is composed of shapes which are spherical and not too (ἄγαν) small. ... The astringently *sour* (στρυφνόν) is composed of shapes large and with many angles, and having very little rotundity. ... The *bitter* (πικρόν) consists of shapes small, smooth, and spherical, having got a spherical surface which actually has hooks attached to it (τὴν περιφέρειαν εἰληχότα καὶ καμπὰς ἔχουσαν). ... The *saline* is composed of large shapes, not spherical, but in some cases also not scalene[3], and therefore without many flexures. . . . The *pungent* (δριμύς) is small, spherical, and regular, but not scalene. . . . In the same way he explains the other "powers" (δυνάμεις) of each taste-stimulus, reducing them all to their atomic figures (ἀνάγων εἰς τὰ σχήματα). Of all these shapes he says that none is simple or unmixed with the others, but that in each taste there are combined many shapes, and that each one and the same taste involves somewhat of the smooth, the rough, the spherical, the sharp, and the rest. But of the shapes that which is chiefly involved determines

[Marginal note: The *object* of taste : only subjectively real. Differences of taste depend on the differences of shape in the atoms of sapid things. *Acid, sweet, sour, bitter, saline, pungent, succulent;* explained according to the particular shapes of the atoms affecting the organs in each case. But the bodily state of the person has to be also taken into account.]

---

[1] Theophr. *de Sens.* § 69. [2] So Diels, 'ut ex γῶνος,' *Dox.*, p. 517 n.

[3] Diels, *Vors.*, p. 393 ἀλλ' ἐπ' ἐνίων καὶ ⟨οὐ⟩ σκαληνῶν. See next page, note 3.

the effect upon sensation, and the sensible "power" of the whole. It makes much difference also what the bodily state is with which the shapes come into relation; for from this it happens sometimes that the same stimulus (τὸ αὐτό) produces contrary subjective effects, and that contrary stimuli produce the same subjective effect[1].'

§ 5. 'Democritus investing each taste with its characteristic figure makes the *sweet* that which is round and large in its atoms; the astringently *sour* that which is large in its atoms, but rough, angular, and not spherical; the *acid*, as its name imports, that which is sharp in its bodily shape (ὀξὺν τῷ ὄγκῳ), angular, and curving, thin, and not spherical; the *pungent* that which is spherical, thin, angular, and curving; the *saline*, that of which the atoms are angular, and large, and crooked (σκολιόν) and isosceles; the *bitter*, that which is spherical, smooth, scalene[2], and small. The *succulent* (λιπαρόν) is that which is thin, spherical, and small[3].' We need not here endeavour to reproduce the reasons given, on the authority of Theophrastus, for the assignment of the particular shapes to the production of the respective tastes. To us the whole theory seems almost a play of fancy; yet we must not forget that to its author it was a serious attempt, on the most scientific and common-sense lines at that time known, to account physically for these sensations. Our interest in it is mainly and primarily historical. Except for the general idea of atomism, this theory of 'atomic shapes' has little affinity to any modern scientific theory of taste, physiological or psychological.

Democritus, as sufficiently appears from what precedes,

Theophr.
de Caus.
Plant.
restates this
theory of
tastes. For
Democritus
tasting, like
every other
sense, a
mode of
touching.

[1] Theophr. *de Sens.* §§ 64-7; Diels, *Vors.*, p. 393; Mullach, *Democ.*, p. 219.

[2] Mullach reads ἔχοντα σκαληνίαν; Diels keeps the MSS. σκολιότητα, 'crookedness.'

[3] Theophr. *de Caus. Pl.* vi. 1. 6. I have given this extract for comparison with the preceding. It shows that some degree of consistency was observed in the respective descriptions of the corpuscular shapes which according to Democritus go to form the various stimuli of taste. It may be noted that here the atoms of the saline are described as ἰσοσκελῆ. This confirms the insertion of οὐ before σκαληνῶν Theophr. *de Sens.* § 66.

reduced the sensations of taste to modifications of the sense of touch. This was not peculiar to his system. It was, says Aristotle, a doctrine shared by him with most of the natural philosophers [1] who tried to explain the sensory functions. They all conceived the objects which affect the senses generally as being *tangible*.

§ 6. Theophrastus, having stated that Democritus' opinions as regards the sensory operations of smelling, tasting, and touching were much like those of most other writers [2], criticizes as follows his theory of tastes, and the physical account he gives of them. 'There is this strange feature too in the theory of those who advocate the atomic shape doctrine, viz. the different kind of sensory effect which they ascribe to atoms alike in shape, and differing only in smallness or largeness. For this would imply that their powers as affecting sense depend not only upon their shapes, but on their bulks. But though one might assign atomic bulk as cause of the greater force or impressiveness of a sensory stimulus, or of the amount or degree of sensory effect produced, it is not reasonable to explain in this way differences in the quality or kind of sensory effect. Democritus' leading hypothesis is that the sensory powers depend on the figures [3] of the atoms; since, if the figures of different stimuli were homogeneous, their effects on sense would be homogeneous in the sphere of taste, as in other spheres; just as a triangle of sides a foot long agrees with one with sides of ten thousand feet in having its three angles together equal to two right angles [4].'

'One might, as against Democritus, well ask how it is that the different tastes are generated from or succeed one another. For either the atomic figures must be altered so as, for instance, from scalene and angular to become spherical; or, assuming that all the various shapes which give rise to certain tastes are in ⟨the moist founda-

(marginal note:) Democritus ascribes different *kinds* of taste to atoms alike in shape but different in size. Theophrastus criticizes this. Again, how are *alterations* of taste produced? are the atomic shapes and bulks altered? or are some removed from, some introduced into, the former aggregate? If the latter be true, what is the *efficient cause* of the removal or introduction?

---

[1] Cf. Arist. *de Sens.* iv. 442ᵃ 29.   [2] Theophr. *de Sens.* § 57.

[3] Theophrastus argues as if Democritus had asserted σχήματα *alone* to be the cause of the perception of sensible qualities.

[4] Theophr. *de Caus. Pl.* vi. 2. 3; Diels, *Vors.*, p. 390. 13; Mullach, *Democr.*, p. 350.

tion⟩, e.g. those of the sour, the acid, and the sweet, some must be separated from the rest—those, that is, which determined the previous tastes in each case respectively, and were proper to them severally—while the others should hold their ground ; or else, in the third place, some must go out from the mass and others must come in. Now since alteration in the atomic figures is out of the question, the atom being incapable of change, it remains either that some must leave and others must enter, or else, simply, that some must stay, while some leave. Both these latter hypotheses are untenable, however, unless it can be shown further what it is that produces these movements— what is their efficient cause [1].' Democritus held that the moist—τὸ ὑγρόν—is, as it were, a πανσπερμία of tastes [2]. This moist is in every case the foundation of taste ; the element in which the taste atoms are, so to speak, suspended. If now a change takes place in a given taste, so that, e.g., from στρυφνός it becomes γλυκύς, either the atoms *proper* to στρυφνότης, in some given moist medium, alter their shape (which is impossible) to suit γλυκύτης ; or else from the portion of the moist medium which is, in the given case, the vehicle of στρυφνότης, those atomic shapes depart on which this quality depended, leaving behind them those proper for γλυκύτης (as there must have been some such, since tastes are never composed of atomic shapes of one single kind, but *all*, or many, are associated in each case, the predominating kind fixing the quality of the whole) ; or else from that portion of the moist medium which yielded στρυφνότης all the atomic shapes which character- ized the taste before depart, while other shapes, suitable to γλυκύτης, are then imported from somewhere in the wider

---

[1] Theophr. *de Caus. Pl.* vi. 7. 2 ; Diels, *Vors.*, p. 390. 20.

[2] Cf. Arist. *de Sens.* iv. 441ᵃ 6 ἢ ὕλην τοιαύτην εἶναι [τὸ ὕδωρ] οἷον πανσπερμίαν χυμῶν, καὶ ἅπαντα μὲν ἐξ ὕδατος γίνεσθαι, ἄλλα δ' ἐξ ἄλλου μέρους, which words must, as Alexander states, apply to Democritus. The Empedoclean theory had been stated in the preceding line, while that of Aristotle himself (which was also that of Theophrastus) comes in the following lines. πανσπερμία is used of the Democritean theory by Arist. 203ᵃ 20.

moist medium outside the given portion. The first supposition contradicts the fundamental hypothesis of atomism ; the two latter require an efficient cause which Democritus neglected to supply. Aristotle and Theophrastus regard water—the moist medium—as tasteless *per se*, but capable of being qualified to sapidity by τὸ ξηρόν, which produces its effect in the medium by the force or efficiency of τὸ θερμόν [1].

Theophrastus states that the different species of tastes were popularly regarded as seven in number, or eight if the saline is separated from the bitter. Thus the number of these would correspond with those of the different species of odours and of colours [2].

Anaxagoras.

§ 7. 'Anaxagoras held that touching and tasting discern their objects in the same fashion (sc. by contraries). For that which is equally hot or cold with the organ of sense affects it with the feeling neither of heat nor of coldness when it comes in contact with it, nor do they perceive the *sweet* or the *acid* by means of these themselves, but they discern the *cold* by contrast with the *hot*, and the *drinkable* (sc. sweet, of water) by contrast with the *saline*, the *sweet* (generally) by contrast with the *acid*, according to the deficiency of each of these respectively, as compared with its opposite : since all alike, he says, exist within us [3].' According to the Anaxagorean theory of πᾶν ἐν παντί, all qualities—those of taste as well as others—are found together : where one is, there are all the rest. But some

(marginal note:) Tasting like other sensory functions involves the operation of contraries, or of *unlike* upon *unlike*. The cold hand feels water warm and vice versa ; and so in tasting, it is by the bitter within us that we

---

[1] Cf. Theophr. *de Caus. Pl.* vi. 1–7, for an exposition of his own (which is probably a more detailed Aristotelean) account of taste, and a criticism of that of Democritus.

[2] Theophr. *de Caus. Pl.* vi. 4. 1–2 (he concludes : ὁ δὲ ἀριθμὸς ὁ τῶν ἑπτὰ καιριώτατος καὶ φυσικώτατος) ; Arist. *de Sens.* iv. 442ᵃ 19–29. For Democritus' theory of tasting cf. further Lucret. iv. 615–32, with Giussani's notes.

[3] Theophr. *de Sens.* § 28 ; Diels, *Vors.*, p. 323. 8 τὸν αὐτὸν δὲ τρόπον καὶ τὴν ἀφὴν καὶ τὴν γεῦσιν κρίνειν· τὸ γὰρ ὁμοίως θερμὸν καὶ ψυχρὸν οὔτε θερμαίνειν οὔτε ψύχειν πλησιάζον, οὐδὲ δὴ τὸ γλυκὺ καὶ τὸ ὀξὺ δι' αὐτῶν γνωρίζειν, ἀλλὰ τῷ μὲν θερμῷ τὸ ψυχρόν, τῷ δ' ἁλμυρῷ τὸ πότιμον, τῷ δ' ὀξεῖ τὸ γλυκὺ κατὰ τὴν ἔλλειψιν τὴν ἑκάστου· πάντα γὰρ ἐνυπάρχειν φησὶν ἐν ἡμῖν.

perceive the sweet, &c. Πᾶν ἐν παντί, therefore, where one taste is, all are; only some one predominates and characterizes the total. Thus in our organisms too; so that the required contrariety between organ and stimulus is always present. The saline taste of the sea.

preponderate, others are comparatively deficient in certain cases. 'This being so, in all composite substances we must conceive many sorts of matter with all sorts of qualities to be inherent, and germs of all things, possessing forms and colours and *savours* of all kinds. Thus, too, human beings are constructed, and all other animals—all things that possess a soul[1].' Thus in the human body and in the organs of sense are found these infinitesimal specimens of all sorts of qualities; and the senses as above explained owe their discriminating power to the opposition between the qualities of the sense-organ and its object in each case. With regard to the physical nature of the *saline* taste, as exhibited in sea-water, we have the following: 'Anaxagoras supposed that when the moisture which originally flooded all the earth had been subjected to the scorching heat of the sun in its revolutions, and the finest part of the water had thus been evaporated, the sediment which remained became salt and bitter[2].' 'A third opinion as regards the manner in which the sea became briny is that the water which forms it, being filtered through the earth, and contracting by infiltration the qualities of this, becomes saline, because of the earth containing such tastes within itself; whereof writers produced a proof in the fact that salt and natron are obtained from mines dug into the earth; and they assert that in many places in the earth sharp or acid savours are found[3].'

[1] Simplic. *in Phys. Arist.* (Diels) pp. 34–5; Diels, *Vors.*, p. 327. 29; Schaubach, *Anax.*, p. 85 τούτων δὲ οὕτως ἐχόντων χρὴ δοκεῖν ἐνεῖναι πολλά τε καὶ παντοῖα ἐν πᾶσι τοῖς συγκρινομένοις καὶ σπέρματα πάντων χρημάτων καὶ ἰδέας παντοίας ἔχοντα καὶ χροιὰς καὶ ἡδονάς. Diels renders this last word here *Gerüche*: in Diogenes (see *infra*, p. 170 n. 1) he renders ἡδονῆς *Geschmack*. But there seems to be no reason for regarding the meaning as different in the two cases. Probably the ideas of smell and taste are united in ἡδονή, here and in Diogenes, very much as they both enter into the meanings and associations of our words *savour* and *savoury*, ἡδονή thus being to χυμός what *nidor* is to *odor*.

[2] Aëtius, iii. 16. 2, Diels, *Dox.*, p. 381, *Vors.*, p. 322. 32 Ἀναξαγόρας τοῦ κατ' ἀρχὴν λιμνάζοντος ὑγροῦ περικαέντος ὑπὸ τῆς ἡλιακῆς περιφορᾶς καὶ τοῦ λεπτοτάτου ἐξατμισθέντος εἰς ἀλυκίδα καὶ πικρίαν τὸ λοιπὸν ὑποστῆναι.

[3] Alexander, *in Arist. Meteor.*, p. 67 (Hayduck); Diels, *Vors.*, p. 322. 35 τρίτη δὲ δόξα περὶ θαλάσσης ἐστὶν ὡς ἄρα τὸ ὕδωρ τὸ διὰ τῆς γῆς διηθού-

Diogenes of Apollonia.

§ 8. 'Diogenes held that, owing to the porousness of the tongue and its softness, as well as to the fact that the vessels from the body converge into it, the various sapid juices are diffused from it, being drawn to the sensorium and the intelligent governing power, as if squeezed from a sponge[1].' Theophrastus also states that, according to Diogenes, tasting is effected by the tongue owing to its porosity and softness or delicacy of structure[2]. On the same authority we learn that, according to Diogenes, the tongue is in the highest degree capable of discerning 'pleasure (see note),' inasmuch as it is most delicate in structure and porous, and, moreover, all the vessels extend into it; whence, too, its great significance as indicating the condition of persons who are ill[3]. 'For it (the air) is various in character, exhibiting varying degrees of heat and cold, of dryness and moisture,

Organ and function of tasting: the tongue porous and absorbent like a sponge; the blood vessels of the body all converge towards it. Significance of the tongue for diagnosis of illness. Diogenes (like Anaxagoras) uses

μενον καὶ διαπλῦνον (cf. Arist. 445ᵃ 14) αὐτὴν ἀλμυρὸν γίνεται τῷ ἔχειν τὴν γῆν τοιούτους χυμοὺς ἐν αὐτῇ· οὗ σημεῖον ἐποιοῦντο τὸ καὶ ἅλας ὀρύττεσθαι ἐν αὐτῇ καὶ νίτρα· εἶναι δὲ καὶ ὀξεῖς χυμοὺς πολλαχοῦ τῆς γῆς. Theophrastus says that Anaximander and Diogenes of Apollonia were of this opinion, which Alexander, l. c., ascribes to Anaxagoras and Metrodorus. Cf. Diels, *Dox.*, p. 494, who quotes Arist. *Meteor.* ii. 2. 355ᵃ 21 seqq. and 353ᵇ 5 seqq. Empedocles (Diels, *Dox.*, p. 381) spoke of the sea as ἱδρὼς τῆς γῆς ἐκκαιομένης ὑπὸ τοῦ ἡλίου, as if suggesting by analogy an explanation of its *saline* quality. Olympiodorus refers to Heraclitus for the same figure, which Aristotle allows as a poetic metaphor, but dismisses with contempt as a scientific dictum.

[1] Aëtius, iv. 18, Diels, *Dox.*, p. 407, *Vors.*, p. 345. 40 Διογένης τῇ ἀραιότητι (here = μανότητι) τῆς γλώττης καὶ τῇ μαλακότητι καὶ διὰ τὸ συνάπτειν τὰς ἀπὸ τοῦ σώματος εἰς αὐτὴν φλέβας διαχεῖσθαι τοὺς χυμοὺς ἑλκομένους ἐπὶ τὴν αἴσθησιν καὶ τὸ ἡγεμονικὸν καθάπερ ἀπὸ σπογγιᾶς. The use of the Stoic term τὸ ἡγεμονικόν shows us how far we are in this from the actual words of Diogenes, and how much reason there is to regard with suspicion even the substance of such information ; cf. Diels, *Dox.* proll., p. 223.

[2] Theophr. *de Sens.* § 40 τὴν δὲ γεῦσιν τῇ γλώττῃ διὰ τὸ μανὸν καὶ ἁπαλόν.

[3] Theophr. *de Sens.* § 43 κριτικώτατον δὲ ἡδονῆς τὴν γλῶτταν· ἁπαλώτατον γὰρ εἶναι καὶ μανὸν καὶ τὰς φλέβας ἁπάσας ἀνήκειν εἰς αὐτήν· διὸ σημεῖά τε πλεῖστα τοῖς κάμνουσιν ἐπ' αὐτῆς εἶναι κτέ. I cannot help thinking that Theophrastus here misunderstood the word ἡδονή, used by Diogenes (and also by Anaxagoras) in the traditionally limited sense of 'the pleasure of taste,' or even of 'taste' itself, as an objective thing—savour.

ἡδονή in sense of 'savour' or 'taste.'

rest and movement; and undergoes besides many qualitative changes infinite in variety of savour and colour[1].'

Plato.

Function and *organ* of tasting. This sense effected by contractions and dilatations of the parts of the organ; according to the qualities—the roughness or smoothness, e.g.— of the stimulating particles. Ducts reach from tongue to heart.

§ 9. As to the general way in which the stimuli of taste affect the gustatory organ we have some information—not much—from Plato in the *Timaeus*. 'It appears that these —sc. sensations of taste—like most other sensations are effected through certain contractions and dilatations (διὰ συγκρίσεών τέ τινων καὶ διακρίσεων γίγνεσθαι), but, besides these, they employ, more than other sensations do, the qualities of roughness and smoothness in their stimuli. Earthy particles (γήϊνα μέρη) enter in the region of the ducts (φλέβια), which are as it were the test tubes or feelers (δοκιμεῖα) of the tongue, reaching from this to the heart[2], and, entering, strike upon the moist and tender parts of the flesh. These particles, as they are dissolved, cause the ducts to contract and to become dry[3].' In this we have the general explanation of the manner in which the sapid particles work upon the organ of taste in order to give rise to the sensation. In the *Locrian Timaeus* (which is not by Plato, but Platonic enough perhaps to be received in evidence of Plato's theory of sense) we read: 'The objects of taste resemble those of touch, for it is by dilatation and contraction, and by the way in which particles enter into

---

[1] Panzerbieter, *Diogenes*, p. 64; Diels, *Vors.*, p. 349. 10 ἔστι γὰρ πολύτροπος [ὁ ἀήρ], καὶ θερμότερος καὶ ψυχρότερος καὶ ξηρότερος και ὑγρότερος καὶ στασιμώτερος καὶ ὀξυτέρην κίνησιν ἔχων, καὶ ἄλλαι πολλαὶ ἑτεροιώσιες ἔνεισι καὶ ἡδονῆς καὶ χροιῆς ἄπειροι. By Anaxagoras ἡδονή (Schaubach, *Fr.* 3, p. 86, *supra* § 7) is used in the same way to signify 'savour' or 'taste.' Panzerbieter in his excellent note shows that the word means taste here, and Diels translates 'noch viele andere Abänderungen und unendliche Abstufungen von *Geschmack* und Farbe.' Cf. Aristot. *de An.* ii. 3. 414[b] 13 πεῖνα δὲ καὶ δίψα ἐπιθυμία, καὶ ἡ μὲν πεῖνα ξηροῦ καὶ θερμοῦ, ἡ δὲ δίψα ψυχροῦ καὶ ὑγροῦ· ὁ δὲ χυμὸς οἷον ἥδυσμά τι τούτων ἐστίν : cf. Xen. *Anab.* ii. 3. 16 τοῦ φοίνικος . . . οἱ πολλοὶ . . . ἐθαύμασαν . . . τὴν ἰδιότητα τῆς ἡδονῆς. In a fragment of Heraclitus ap. Hippol. *Ref. Haer.* ix. 10 ἡδονή='smell' (Bywater, *Fr.* xxxvi) ἀλλοιοῦται δὲ ὅκωσπερ ὁκόταν συμμιγῇ ⟨θύωμα⟩ θυώμασι· ὀνομάζεται καθ' ἡδονὴν ἑκάστου.

[2] Such teaching may have determined, to some degree, Aristotle's theory of the heart as sensorium.   [3] Plato, *Tim.* 65 C-D.

the pores (τᾷ ἐς τὼς πόρως διαδύσει), and by their figures
(σχημάτεσσι), that tastes are either *astringent* or *smooth*
(στρυφνὰ ἢ λεῖα); they are presented as astringent when
they dissolve (ἀποτάκοντα) and rinse (ῥύπτοντα) the tongue;
the contrary are smooth and sweet[1].'

§ 10. 'With regard to savours (χυμῶν), Plato, in treating
of water, mentions four species of water. Among *saps*
(χυλοῖς) he places wine, verjuice (ὀπόν), oil, honey, while
among the affections (πάθεσι) which water undergoes, he
places the earthy taste (τὸν γεώδη χυμόν). And it is by
these particles[2] compressing and contracting the pores[3]
that ⟨tastes are generated⟩[4]. The rougher particles
are the *astringent* tastes, those less rough[5] are the *harsh*.
That which acts as a detergent or kathartic on the pores
(τὸ δὲ ῥυπτικὸν τῶν πόρων καὶ ἀποκαθαρτικόν) is the *saline*.
That which is detergent in an extreme degree, so as actually
to dissolve (ὥστε καὶ ἐκτήκειν) their tissues, is *bitter*. Those
particles which are warmed by the heat of the mouth, and,
ascending, dilate the pores are *pungent*. Those which
cause fermentation[6] are *acid*; those which together with
the moisture that is in the tongue tend to relax (διαχυτικά)
and restore it to its normal state (συστατικὰ εἰς τὴν φύσιν) are
sweet[7].' The part of the *Timaeus* which Theophrastus had
in view here is the following: 'These (earthy particles)
if they are very rough (τραχύτερα) are *astringent* (στρυφνά)
in taste, if less rough, they are *harsh* (αὐστηρά). Those of
them which are detergent (ῥυπτικά) and rinse (ἀποπλύνοντα)[8]
the whole environment (πᾶν τὸ περὶ τὴν γλῶτταν) of the

Plato's theory of objective tastes. Four species of 'water' (the element). Astringent, harsh, bitter, pungent, acid, sweet tastes, explained. Plato had the idea of taste as a chemical sense clearly before his mind, so far as this was possible at the time. See his explanation of acid in particular.

---

[1] *Tim. Locr.* 100 E.   [2] The γήινα μέρη of *Tim.* 65 D.

[3] I read πόρους after Philippson for the, to me, unintelligible χυμούς. Plato has φλέβια in the corresponding place in the *Timaeus*, and πόρων here occurs farther on.

[4] In spite of Diels' remark on the condensation and brevity of Theophrastus in quoting Plato, it seems that there must have been—as Wimmer held—something lost here. I supply the sense as above.

[5] Cf. ἧσσον τραχύνοντα, Plato, *Tim.* 65 D.

[6] κυκῶντα: cf. ζέσιν τε καὶ ζύμωσιν, *Tim.* 66 B.

[7] Diels, *Dox.*, p. 525. 4; Theophr. *de Sens.* § 84.

[8] Similar terms are used by Aristotle in connexion with the physical stimulus of taste.

tongue, if they do this immoderately, and fasten upon it so as to dissolve some of its very tissues, as is the power of alkalies (ἡ τῶν λίτρων δύναμις), all under such circumstances are named *bitter* ; those which come short of the character of the aforesaid alkalies, and have the rinsing effect in but a moderate degree, are called *saline* (ἀλυκά), being without rough bitterness, and appear rather agreeable than otherwise. Those which go into partnership (κοινωνήσαντα) with, and are soothed (λεαινόμενα) by, the warmth of the mouth, being both set aglow themselves and, in turn, acting as counter-caustics (ἀντικάοντα) on that which caused their heat, being borne upwards by their lightness towards the senses of the head (πρὸς τὰς τῆς κεφαλῆς αἰσθήσεις), and cutting through all that they come in contact with—on account of these powers all such are called *pungent* (δριμέα). But when these same earthy particles have been progressively fined down by decomposition, and insinuate themselves into the narrow veins (sc. of the tongue), being as they are symmetrical with such particles of earth and air as are already in these, so that, setting these particles in motion, they cause them to be mixed together (περὶ ἄλληλα), and, as they are mixed, to tumble about, and, entering severally into different places, to produce concavities which envelop the things that enter them, and which, being but hollow globules of water, become dewy vessels of air, when the dewy cellule of each, whether earthy or pure, has enveloped a particle of air; so that those of them which are of pure moisture form transparent encinctures for the air, and are called bubbles, while those which are made of the earthy moisture, that sways and rises in all parts alike, exhibit what is called *seething* or *fermentation*: then that which is the cause of all these affections is denominated *acid*. An affection the opposite of all those thus described is that arising from an opposite cause, when the collocation of the entering particles in the moist environment, being naturally akin to the normal condition of the tongue, glazes and smoothes over the roughened parts, while, as for those abnormally contracted or dilated, it contracts the latter and

relaxes the former, and re-establishes all as far as possible
in their normal state. Every such remedy of the violent
affections being, when it takes place, pleasant and agreeable
to every one, is called *sweet*[1].' In this passage Plato,
largely by the aid of a vivid and not unscientific imagination, attempts to describe what would now be called
a *chemical* process. In thus explaining the effect of the
stimuli of taste upon the organs, he has taken a considerable
step beyond his predecessors, so far as they have left us
any knowledge of their views on this subject. Modern
empirical psychologists have at command more perfect
knowledge of the gustatory tissues and structures, but the
conception which still vaguely dominates theories of tasting,
is that of chemical changes set up by the sapid particles in
the gustatory apparatus. Chemistry as a science did
not exist in Plato's time, or for many centuries afterwards,
and it is, therefore, the more surprising that he should
have had recourse to an idea which is purely chemical for
his explanation of at least one of the objects of taste—the
acid. In this he shows a conception far in advance of all
predecessors, and more developed than that of Aristotle.

§ 11. 'Most forms of waters intermingled with one
another are, taken as a whole class, called *saps*[2] when they
have been filtered through the plants that grow out of
the earth[3]; but having, owing to their various mixtures,
severally acquired dissimilar natures, they present, for the
rest, many nameless kinds; yet there are four of them which
are of a fiery nature, and which, being most transparent, have
received special names:—(1) That which warms the soul
together with the body is *wine*. (2) That which is smooth
and dilates the visual current (διακριτικὸν ὄψεως), and therefore presents itself as bright in appearance, and glistening
and oily—a thing of oily species—such is *resin*, or *castor-
oil* (κίκι), or common *olive-oil* (ἔλαιον) itself, or other things

The theory of 'saps' (χυμοί). These are in their origin modifications of water, produced by filtering through the tissues of plants. Four special sorts: (1) wine; (2) oil; (3) honey; (4) verjuice. The nature of each and its effects

---

[1] Plat. *Tim.* 65 D–66 C.

[2] χυμοί is here used by Plato in the sense in which χυλοί is regularly
used by Theophrastus.

[3] ὑδάτων εἴδη . . . ξύμπαν μὲν τὸ γένος διὰ τῶν ἐκ γῆς φυτῶν ἠθημένα χυμοὶ
λεγόμενοι.

upon the organ—the tongue. of the same power. (3) That which relieves the tension of the passages in the mouth and restores their natural condition (διαχυτικὸν μέχρι φύσεως τῶν περὶ τὸ στόμα ξυνόδων), producing by this property sweetness to the taste—this has received the name of *honey* as its most general appellation. (4) That which dissolves the flesh (διαλυτικὸν τῆς σαρκός) by burning, a frothy kind of substance (ἀφρῶδες γένος), is, when singled out from all the other saps and taken by itself, what has been named *verjuice*[1].' For Plato the organ of tasting is 'the tongue'; he (like Aristotle) does not speak of 'the palate' as concerned. Plato does not probe into questions (*a*) respecting the proper organ of this sense, or (*b*) regarding its relationship to touch or smell.

Aristotle.

Object and *function* of tasting. Taste a variety of touch. Its medium not external to the body. The tongue or its flesh is, properly, medium of taste. The moist is the vehicle of taste. Water and taste. Water *per se tasteless.* Derives sapidity from earth when filtered through this. Phy-sical *defi-nition* of taste. The tongue is § 12. Tasting is the variety of touching which peculiarly subserves nutrition. The object of taste, viz. the *gustable*, is something tangible[2]: this explains why it is not per-ceptible through a foreign body interposed as a medium; for the sense of touch acts through no foreign (i. e. extra-organic) medium. The tongue is, however, itself a medium, though internal, i. e. belonging to the body. It is related to the organ of taste proper, as e. g. air is to the organ of hearing[3]. Moreover χυμός, the object of taste, is con-veyed in the moist as its vehicle, and the moist is a tangible: which again exhibits the object of taste as tangible. The object of taste, being conveyed thus in the moist vehicle, is naturally regarded as connected in its physical origin with water. Views have differed as to the nature of this connexion. Empedocles held that the water already as such contains fully developed within itself all sorts of savours, which, however, are so infinitesimally small as to be imper-ceptible; others again have held water for the material out of which, as out of a seminary (πανσπερμία) of all kinds of seeds, tastes of all kinds are developed—one from this part of the water, another from that, and so on. Neither of

---

[1] Plato, *Tim.* §§ 59 E-60 B.
[2] 422ᵃ 8 τὸ δὲ γευστὸν ἁπτόν τι.   [3] 423ᵇ 17 seqq.

these views commends itself to Aristotle.   Water contains, potentially moist.
he thinks, *per se* none of the διαφοραί of taste, as Empedocles Taste
held.   Without any contributory activity on the part of the perceives the gust-
water, such διαφοραί are wrought into it by an extraneous able and
cause, which affects it as agent affects patient.   Just so one the non-gustable.
can impart a taste to water by washing something sapid in Two mean-
it.   Such is the way in which nature produces all savours— ings of latter.
χυμοί—by sifting or straining the moist element (of water)
through the dry (of earth), and so imparting to the former
its sapid quality[1].   Hence the gustable—χυμός or τὸ γευστόν—
may be physically defined as *the affection produced in the
moist by the dry*[2], and *capable of converting the faculty of taste
from potentiality to actuality*[3].   Were we creatures living in
water instead of air[4], we should indeed perceive the sweet
if infused into this water; yet our perception would still
be one of touch : not even then would it be perceived
through the water as external medium.   It would be perceived immediately, owing to the sweet being blended with
the particular moisture with which we happened to be in
contact, just as in the case of the water which we drink and
find sweet.   It is not thus, i.e. by mixing with the medium,
that colour is perceived.   Taste has no medium externally
to the organ: its medium *is* the so-called organ (the tongue)
itself when moistened.   Nothing produces the sense of
taste without moisture ; everything which excites this sense
has moisture actually or potentially ; as for example, the
saline, which is in itself easily liquefied, and by its lique-
faction tends to actualize the potential liquidity or moisture
of the tongue.   The sense of taste, like the others, has for
its object a genus embracing contraries.   It perceives the
gustable and the non-gustable, meaning by the latter either
that which is sapid but only in an infra-sensible degree, or
else that the taste of which is destructive of the sense.   The
difference between the palatable and unpalatable in drinks
seems the foundation of the matter.   Both are objects
of taste, but while the former is natural and normal, the

[1] 441ᵃ 4–441ᵇ 14.   [2] Sc. τὸ τρόφιμον ξηρόν, 441ᵇ 24.
[3] 441ᵇ 19.   [4] 422ᵃ 11.

latter is in its tendency destructive. The 'drinkable,' too, as an object is perceptible by touch as well as taste.

§ 13. Since the object of taste is moist [1], the tongue, *qua* organ of taste [2], must be neither actually moist nor incapable of becoming moist. The sense of taste is passively affected by the object. Hence the part of the body which is to be the organ of this sense should be something capable of being moistened, while yet preserving its distinctive nature, not something actually and always moist [3]. A proof that the organ should be thus capable of being moistened, yet not actually moist, is found in the fact that tasting is impossible, or difficult, when the tongue is either quite dry, or excessively moist. In the latter case, when we attempt to taste something, what ensues is merely a tactual perception of the moisture of the tongue, in which the sense of taste proper is merged and disappears. With this tactual perception the organ is preoccupied, as it might be with a previous taste, if a person after tasting something of very strong savour were immediately to try to taste some other savour. So it is that sick persons find sweet things bitter, because the tongue is full of bitter moisture. The tongue is an organ of touch as well as of taste [4]. With this same part wherewith we taste, we can perceive any given object of touch [5].

§ 14. None of the elements—not even water—has a taste *per se*. All tastes arise from some sort of mixture in the

The tongue, qua organ, must not be actually moist: only potentially so, i. e. capable of being moistened. Tasting impeded by excessive dryness or excessive moisture of tongue. Tongue, an organ of touch also.

The elements all per se taste-

---

[1] 422ᵃ 34 seqq.

[2] Sc. the tongue (533ᵃ 26 τὸ τῶν χυμῶν αἰσθητήριον τὴν γλῶτταν), popularly regarded as the organ of taste: all this has to be considered in the fuller light of Aristotle's discussion of the organs of touch and taste.

[3] σῳζόμενον: preserving its distinctive nature *as an organ of taste.* The moistening which the organ has to undergo is only subsidiary to its gustatory function, which primarily depends on something else than the moisture, viz. upon the sapid stimulus of which the moisture is but the solvent or vehicle. The moisture is a means—something secondary— employed by the organ for its proper purpose; thus were the organ to become actually moist, it would forsake its distinctive and proper character.

[4] Aristotle, notwithstanding what he says 423ᵇ 17, often speaks of the tongue as organ—instead of intra-organic medium—of taste. Cf. § 12 *supra.*

[5] 423ᵃ 17-18.

moist medium. Wine and all sapid substances, which, from a state of vapour, are condensed into moisture, become water. Others are affections of water itself caused by something mixed with it. The taste ensuing corresponds to that which is thus mixed with the water[1]. Moreover no simple element—only a mixture of elements—can effect the purpose of nutrition. Hence there is a fundamental connexion between taste and nutrition[2]. The object or final cause of this sense is nutrition[3]. Yet only the sweet actually nourishes: all other varieties of taste are, like the saline and the acid, merely ways in which nature seasons the sweet to make it the more suitable for its purpose[4]. In the case of objective tastes, as of colours, the contraries are relatively simple, i. e. the *sweet* and the *bitter*. These are the elements of the other tastes[5]. Next to the *sweet*, and perhaps as a variety of this, comes the *succulent* (λιπαρός); the *saline* and the *bitter* are closely akin; while between the sweet and bitter come the *harsh* (αὐστηρός), the *pungent* (δριμύς), the *astringent* (στρυφνός), and the *acid* (ὀξύς). If the succulent is a kind of sweet, there appear to be seven leading varieties of tastes, as there are of colours[6]. The *faculty* of taste is that which is potentially such as each of these objective tastes is; while the object of taste is that which in each case makes the faculty actually such[7].

Marginal note: less. All tastes involve a mixture in moist vehicle. Taste (objectively) is nutriment; and this is always a composition of moist and dry. Only the sweet, however, actually nourishes. Between the two extremes of sweet and bitter fall saline, harsh, pungent, astringent, acid. There are seven species of taste, as of colour and odour.

§ 15. Taste is a sort of touch, if only because it has to do with nutrition. Nutriment must be something tangible. *Sound, colour,* and *odour* do not nourish, nor do they cause either growth or decay. Hence tasting must be (as we have said) a mode of touching, as it is that which perceives the nutrient tangible. All animals with the sense of touch possess ἐπιθυμία, or the impulse towards what is pleasant. Moreover they have a discriminating perception of their

Marginal note: With faculty of touch and its modification taste necessarily arises desire (ἐπιθυμία).

---

[1] 358ᵇ 18, 443ᵃ 26 seqq.
[2] 441ᵇ 24 seqq., 442ᵃ 1 seqq.
[3] 436ᵃ 15 ἡ δὲ γεῦσις διὰ τὴν τροφήν, 435ᵇ 22, 434ᵇ 18 ἡ γεῦσις ὥσπερ ἀφή τις· τροφῆς γάρ ἐστιν.
[4] 442ᵃ 8.   [5] 442ᵃ 12.   [6] 442ᵃ 19 seqq.
[7] For the original of §§ 12–14 cf. Arist. 422ᵃ 8–ᵇ 16, 414ᵇ 1–16.

food ; for touch gives them this (viz. through its modification, taste). All are nourished by things dry and moist,
hot and cold, i. e. by the objects of touch. The objects
of other senses nourish only incidentally; just as sound,
colour, smell may put an animal on the track of food,
but they cannot in themselves feed it. χυμός is a variety

Hunger and thirst. then of the ἁπτόν or tangible. Hunger and thirst constitute
ἐπιθυμία in relation to food and drink. Hunger is (ἐπιθυμία)
for the dry and hot ; thirst for the cold and moist, and χυμός
is a sort of seasoning (ἥδυσμα) of these objects.

Touch and taste essential to the being of an animal. § 16. Touching and tasting, then, are essential to the
very *being* of an animal. The others are subservient rather
to its *well*-being, and do not belong to all species of
Use of taste to distinguish the pleasant and unpleasant in food and drink. The heart is the true organ of touch and taste; these manifestly connect themselves with the heart. Man's excellence in touch and taste. animals, but only to some; especially to those which
have the power of locomotion [1]. Animals have the sense
of sight in order that they may be able to see objects
while yet distant through the medium of the διαφανές.
They have hearing in order that they may be able to
apprehend significant sounds conveyed through the air to
their ears ; and they possess in the tongue an organ
wherewith to convey such sounds to others. But they
possess taste on account of the difference between the agreeable and the disagreeable in food and drink ; in order that
they may be able to apprehend this difference, and according to such apprehension, may direct their movements
to the seizure or avoidance of certain things as food.
Serpents and saurians have a peculiarly delicate and keen
sense of taste, nature having endowed them with tongues
long and forked, with a fine extremity furnished with hairs.
This formation of the tongue doubles the pleasure which
such creatures feel in agreeable tastes, since the sense itself
is thus possessed of twofold power [2]. The organ of taste
like that of touch is connected with the vital organs. The
region of the heart is the foundation of the senses, of which
two—those of touch and taste—are manifestly connected
with the heart [3]. Of all animals man is the most finely sensi-

---

[1] Arist. *de An.* iii. 12. 434ᵇ 18–26.
[2] *De part. An.* 660ᵇ 6–10.   [3] 469ᵃ 12–16, 656ᵃ 27–31.

tive as regards touch. Man's tongue, too, is soft[1], which makes it particularly sensitive in touching; and tasting, the tongue's proper function, is a kind of touching. Man's sense of touching is the most perfect, and in it he excels all other animals. Next comes his sense of tasting. In the other senses he has no superiority to the lower animals, many of which, on the contrary, have better sight and hearing, and a keener olfactory sense[2]. As to the way in which the organ of taste discharges its function, Aristotle has made no real advance beyond the positions taken up by Alcmaeon or Diogenes.

[1] 660ª 20–22 reading ἡ γλῶττα μαλακή, instead of Bekker's ἡ μ. γλ.
[2] 494ᵇ 16–18, 421ª 17–26.

THE ANCIENT GREEK PSYCHOLOGY
OF TOUCHING

Alcmaeon—Empedocles.

§ I. THE pre-Aristotelean psychologists have left comparatively little on record respecting this sense, although
it was, according to the opinion of several of them, the
fundamental sense — that from which the others are
developed, or at least in some way derived. Not indeed
until we come to Aristotle himself do we find a real or
business-like attempt to treat of touching. True, Plato
gives a detailed account of the objects of the sense, as he
conceived them ; but of the organ, or its operation, we read
little in his remains or those of his predecessors. That
little has, however, in accordance with the plan hitherto
followed, to be here set forth in its entirety.

According to Theophrastus[1] Alcmaeon altogether omitted
to treat, at least in his writings, of the sense of touching—
its organ or mode of operation. Theophrastus makes a
similar statement of Empedocles, with this difference that
while, according to him, the former seems to have omitted
all reference to touching; the latter, though not indeed
treating it with complete neglect, failed to give a distinct
and detailed theory of touch. He merely threw out the
general suggestion that this, like the other senses, is to be
explained by the operation of 'emanations' entering into
and fitting the 'pores' of the organ[2]. Theophrastus is of
opinion that the Empedoclean theory of perception by
'emanations' is even less plausible with regard to touching
(and tasting) than in reference to the other senses. 'How,'
he asks, 'are we to conceive sensible distinctions of taste
or touch as made by means of emanation (ἀπορροῇ)? how

Marginal notes:
Touching, though the fundamental sense, most scantily treated.

Alcmaeon.

Empedocles.

Theophrastus' criticism of Empedocles' account of the function of touching.

---

[1] Theophr. *de Sens.* § 26.

[2] περὶ δὲ γεύσεως καὶ ἀφῆς οὐ διορίζεται καθ' ἑκατέραν οὔτε πῶς οὔτε δι' ἃ
γίγνονται, πλὴν τὸ κοινὸν ὅτι τῷ ἐναρμόττειν τοῖς πόροις αἴσθησίς ἐστιν,
Theophr. *de Sens.* §§ 7, 9. Also Arist. *de Gen. et Corr.* A. 8. 324[b]
26 seqq.

are we to discriminate "the rough" or "the smooth" by its
fitting into "the pores [1]"?' Yet Empedocles seems to bring
all the other sensations under the sense of touch. ' He says
of all alike that they are caused ultimately by "emanations"
entering and fitting into the pores of the respective organs.
Whence it is that one sense-organ is not susceptible of
the sensations proper to another; since the "emanations"
which fit the pores of one are too large or too small for
those of another, and therefore are not followed by the
due sensory effect. Those that are too small pass right
through the pores *without touching* (οὐχ ἁπτόμενα) its sides;
those that are too large cannot enter at all [2].' Thus the
primary condition of the proper exercise of each and every
sense-organ is found to consist in a fact of touch—the due
contact between the 'emanation' and the inner surface of
the pore; yet of the *sense* of touching he has propounded
no special theory. No idea of the sensory function of
nerves existed till long after Empedocles; and the seeming
'immediacy' of touch was, perhaps, what debarred it in his
opinion from being easily explained in detail by the theory
of ἀπορροαί, which operate at a distance and through a
medium [3]. The difficulty felt in applying his general
theory to touching was of course felt also in reference to
the kindred sense of tasting. Accordingly we have from
Empedocles no particular information as to either the *objects*
or the *organs* and *functions* of touching and tasting.

Democritus.

§ 2. Here, too, we are disappointed. The whole tenor of Demo-
the physics and psychology of Democritus himself, as well as critus referred all
the assertions of Aristotle, make it perfectly clear that for other
senses to
Democritus the sense of touching was the primary sense. that of
' Democritus and most of the " physiologi " who treat of touch, yet
fails to
sense do a very extraordinary thing: they represent all give a par-
objects of sense as objects of touch. If, however, this ticular or
detailed
is true it plainly follows that each of the other senses is account

---

[1] *De Sens.* § 20.   [2] Theophr. *de Sens.* § 7.
[3] By ἀπορροαί too he explains the properties of the magnet. Cf.
Alex. *Quaest.* ii. 23, p. 72. 9 (Bruns).

a kind of touch, which is manifestly impossible [1].' This was not only a biological but a physical conclusion. It was the opinion of Democritus that we see, hear, smell, taste, and touch by the agency of atoms, which are the sole ultimately real; the ultimate 'things.' We must distinguish carefully between *res naturae*, i.e. such 'things' as we perceive, and the atoms, or real things, which reason alone reveals. The physical qualities of each atom are *weight* and *solidity*. To these must be added local *motion*, which in each and every atom goes on eternally. It has also *geometrical* qualities—*figure* and *magnitude*. The primary physical qualities of *res naturae* are also weight and solidity. Their weight depends on the number and size of the atoms in them; their solidity (which is only comparative) on the density of the atoms. The differences of the atoms compared *inter se* when forming *sensibilia* consist of *order*, *figure*, and *position*. A H differ from H A in order; A differs from H in figure; I from H in position [2]. Besides atoms, void was postulated to explain the possibility of movement. The principal 'distinction' (διαφορά) for Democritus seems to have been that of figure: hence the name 'figure' is frequently employed to designate the atom. Thus the only ultimate properties or qualities of sensible things are *tangibilia*, and from the physical point of view we see how all the objects of sense had to be reduced to those of touch. Only the above-named qualities are objectively real; the rest are subjective, due to our sensibility.

§ 3. Such are our sensations of taste, colour, smell, sound, and (among tangibles) temperature. It would seem then

---

[1] Arist. *de Sens.* iv. 442ᵃ 29. This criticism appears to exhibit Aristotle as incapable of profoundly apprehending the idea of biological development. Yet, strangely, he himself most firmly held the theory that Touch is the original sense from which all others have been differentiated. *Vide* SENSATION IN GENERAL, § 23, and SENSUS COMMUNIS, § 49.

[2] Cf. VISION, § 19, p. 37 n. 2 *supra*. Theophrastus (*de Sens.* §§ 61 seqq.), in stating the physical qualities of the atoms, seems to use σκληρότης loosely for πυκνότης—*hardness* for *solidity*. Plato (§ 6 *infra*) did not confound these.

as if the desirability of a full investigation of the sense of
touching should have impressed itself upon Democritus[1].
But we are told he left this part of his subject without any
attempt at originality of treatment. The fact of his not
having attempted such investigation may perhaps be
explained (a) by his ignorance of the nerve-system, and
(b) by assuming that he felt the difficulty of satisfying himself with any explanation of the way in which the *merely*
physical, conceived as such without original reference to
mind, could 'pass into' the mental. This difficulty
confronted him—as it must confront every one—most
formidably, just at the point where the ultimate analysis
of sense (or what seemed to him to be so) is reached.
To this may be traced the half-heartedness, barrenness, or
absence of early physiological psychology with reference
to the organ and functions of touching. To this also is
due the fact that even modern physiological psychologists,
when they come to deal with the sense of touching, have
to be content with conclusions which scarcely take us
outside the province of anatomy. It is chiefly, if not
solely, in that province that real advances have been
made beyond the position in which this sense was left
by the ancients. True, modern psychologists have distinguished, as the ancient Greeks failed to do, between
cutaneous sensations (of touch proper, and of pressure),
sensations of *temperature*, and *muscular* sensations; and
attempts have been made, not very successfully, to connect
each of these with their proper nerves or nerve-endings.
But these are small matters. The biological question as
to the differentiation of touch into the other senses remains
now as it was then—a mystery only vaguely soluble by
reference to a long process of evolution. And—to say
nothing of the metaphysical difficulty of accepting touch
as the ultimate authority for objective reality—there was
yet another biological question, viz. that of the history of
this parent-sense. How did touch itself, with all its implicit
powers of development, arise? Democritus could not answer.

Why Democritus did not examine the sense of touch psychologically.

[1] σχεδὸν ὁμοίως ποιεῖ τοῖς πλείστοις, Theophr. *de Sens.* § 57.

This question we, too, must still either shelve, or slur over in the best way we can. All attempts at explaining a 'transition' from the physico-physiological to the psychical or conscious fact have been futile. Most moderns prefer to speak or think of the so-called two facts as really one, but with two (or more) different aspects. We hesitate even to think of such 'transition.'

Anaxagoras.

Touch (like the other senses) perceives by contraries. The cold hand feels the water hot, &c. If the water be of the temperature of the hand, the latter feels it neither hot nor cold—feels no temperature.

§ 4. Anaxagoras teaches that sensation is effected by the interaction of opposites ; for like is incapable of being affected by its like. This principle he tries to carry out with reference to each particular sense. Touching (and tasting) distinguish their objects as seeing and hearing do, i. e. by interaction of opposites. That which is of like temperature with the hand does not by its contact give us the sense either of coldness or of heat. By the warm we cognize the cold, as by the saline we cognize the 'potable'[1]. Except for this we have scarcely any record of Anaxagoras' teaching regarding the sense of touch. As Theophrastus informs us, Anaxagoras has not left on record his views of the more corporeal senses[2]. Diogenes also having left no opinions on record concerning the sense of touching, we pass on to Plato.

Plato.

Organ and function of touching: treated with little regard by Plato. Notices that the tactile

§ 5. Plato, too, has treated this sense with comparatively slight care[3]. He has given little to determine the nature of the organ and function of touching. It is distinguished, he says, from the other senses in that it is not confined to some particular part, but diffused all over the body. He reckons the sensations of touch among the κοινὰ παθήματα— those belonging to the whole body as pleasant or painful[4]—

---

[1] Theophr. *de Sens.* §§ 27–8.

[2] § 37 οὐ δηλοῖ δὲ τὰς σωματικωτέρας αἰσθήσεις.

[3] Theophr. *de Sens.* § 5 Πλάτων . . . οὐ μὴν εἴρηκέ γε περὶ ἀπασῶν ἀλλὰ μόνον περὶ ἀκοῆς καὶ ὄψεως.

[4] *Tim.* 64 A. Here Plato comes near recognizing the *sensus communis* of modern parlance, i. e. a 'general feeling' such as that of comfort or discomfort, nausea, faintness—a totally different thing from Aristotle's *sensus communis*.

among which he names *hot* and *cold, hard* and *soft, heavy* and *light, rough* and *smooth*. In the *Timaeus*, 61 D seqq., he drafts an explanation of some of these objects of touching. ' First then,' he says, ' let us see what we mean by calling fire hot. We must consider the matter as follows, remembering the power of dividing and cutting which fire possesses and exercises upon our body. That the sensation is a sharp one, we are all well enough aware ; and we must take into account the fineness of its edges and sharpness of its angles[1], besides the smallness of its particles and the swiftness of its motion, all of which qualities combine to render it so vehement and piercing as keenly to cut whatever meets it, remembering the genesis of its figure, that this more than any other substance separates our bodies and minutely divides them, whence the sensation that we now call *heat* justly derives its quality and name. The opposite condition, though obvious enough, still must not lack an explanation. When the larger particles of moisture which surround the body enter into it, they displace the smaller, and because they are not able to pass into their places, they compress the moisture within us ; and, whereas it was irregular and mobile, they render it immovable owing to uniformity and contraction, and so it becomes rigid. And what is against nature contracted struggles in obedience to nature and thrusts itself apart ; and to this struggling and quaking has been given the name of trembling and shivering ; and both the affection and the cause of it are in all cases termed ' cold.'

§ 6. *Hard* is the name given to all things to which our flesh yields ; and *soft* to those which yield to the flesh ; and so also they are termed in their relation to each other[2]. Those which yield are such as have only a small

sense is distributed all over the surface of the body, not, like the others, confined to certain parts. He therefore calls the affections of touch κοινὰ παθήματα. Names chief distinctions made by touch hot-cold, heavy-light, hard-soft, rough-smooth. Explains objective nature of hot-cold physically.

Explanation of hard-soft: exactly anticipates Locke's account of hardness.

---

[1] For an account of the elementary structure of fire in accordance with Plato's geometrical physics, see *Timaeus* 53 C seqq.

[2] Cf. Locke, *Essay concerning Human Understanding*, ii. 4. 4 ' And, indeed, hard and soft are names that we give to things only in relation to the constitutions of our own bodies ; that being generally called hard by us, which will put us to pain sooner than change figure by the pressure of any part of our bodies ; and that, on the contrary,

base of support; and the figure with square surfaces, as
it is most firmly based, is the most stubborn form; so, too,
is whatever from the intensity of its compression offers the
strongest resistance to external force.

Heavy-
light: must
be investi-
gated to-
gether with
the notions
of *above*
and *below.*
These
directions
only *rela-*
tive. The
universe as
a whole,
being
spherical,
really con-
tains no
such dis-
tinction.
Heaviness
of a body
is its ten-
dency
towards its
kindred
element.
Thus earth
tends to
earth. The
direction
of this
tendency
is called
'down-
ward' or
below. The
contrary
direction is
'upward.'
Fire is light
because it
tends away
from the
earth.
But if we
were ten-
ants of the
empyrean,
and tried

§ 7. Of 'heavy' and 'light' we shall find the clearest
explanation if we examine them together with the so-called
'below' and 'above.' Here follows an argument showing
that the popular notion of the universe being divided into
an *upper* and a *lower* portion, to the latter of which all
bodies naturally tend, is false; the truth being that, as the
universe is a *sphere*, there is really no such thing as an
upper and a lower region in it. 'Whence (Plato goes on 63 A)
these names ("upper" and "lower") were derived and under
what conditions we use them to express this division of
the entire universe we may explain on the following
hypothesis. If one were in that region of the universe
which is specially allotted to the element of fire, the region
wherein is to be found collected in greatest mass the fiery
element to which our earthly fire is attracted; and if
he, possessing the requisite power, takes his stand on
this mass and separates from it portions of the fire and
weighs them in scales, when he raises the balance and
forcibly drags the fire into the alien air, evidently he
overpowers the smaller portions more easily than the
larger; for when two masses are raised at once by the
same force, necessarily the smaller yields more readily to
the force, the larger, owing to its resistance, less readily;
hence the larger mass is said to be heavy and to tend downwards; the smaller to be light and to tend upwards. This
is exactly what we ought to detect ourselves doing in our
own region. Standing as we do on the earth, we separate
portions of earthy substances, or sometimes earth itself,
and drag them into the alien air with unnatural force, for
each portion clings to its own kind. Now the smaller
mass yields more readily to our force than the larger,
and follows quicker into the alien element; therefore we

soft which changes the situation of its parts upon an easy and unpainful
touch.'

call it "light," and the place into which we force it "above"; to detach a piece of fire, we should find it heavy as earth is here, and our notions of *up* and *down* would be reversed. while to the opposite conditions we apply the terms "heavy" and "below"... In every case it is the tendency towards its kindred element that makes us call the moving body "heavy," and the place to which it moves "below"; while to the reverse relations we apply the opposite names.... Of the affection "smooth" and "rough" any one could perceive the cause and explain it to another: the latter *Smooth-rough*, explained. is produced by a combination of hardness and irregularity; the former by a combination of uniformity and density[1].'

§ 8. For Plato the *organ* of touching was undoubtedly The *function* and *organ* of touching. what he called flesh—σάρξ. In the *Timaeus*, 61 C, having explained σώματα by geometrical figures in various combinations, he says we must assume that all these 'bodies' Plato thinks the object must be explained first; but his explanation of this is not followed by an account of the former. are perceptible to sense, but of σάρξ and its concomitants, as well as of the soul in its mortal nature, he has, as yet, given no account. These, however, cannot be really explained apart from the sensible qualities of body, nor can the latter be explained apart from the former. Nor can they be dealt with together. He has, therefore, to assume provisionally the several distinct sensory faculties, to a particular account of which he purposes afterwards In the *Timaeus Locrus* we find Aristotle's doctrine that the qualities of body *qua* body are all tangibles. The tangible to return[2]. The promised account is, however, nowhere satisfactorily rendered. In what follows the organ and function of touching remain almost without an attempt at explanation. In the *Timaeus Locrus*[3], however, we have a few remarks bearing on this subject. Though not by Plato, they deserve to appear here for comparison with Plato's views. 'All the sensible affections (πάθεα)

[1] Plato, *Tim.* 61 C–64 A. Mr. Archer-Hind's translation has been for the most part adopted.

[2] *Tim.* 61 C–D πρῶτον μὲν οὖν ὑπάρχειν αἴσθησιν δεῖ τοῖς λεγομένοις ἀεί· σαρκὸς δὲ καὶ τῶν περὶ σάρκα γένεσιν, ψυχῆς τε ὅσον θνητόν, οὔπω διεληλύθαμεν. τυγχάνει δὲ οὔτε ταῦτα χωρὶς τῶν περὶ τὰ παθήματα ὅσα αἰσθητὰ οὔτ' ἐκεῖνα ἄνευ τούτων δυνατὰ ἱκανῶς λεχθῆναι, τὸ δὲ ἅμα σχεδὸν οὐ δυνατόν. ὑποθετέον δὴ πρότερον θάτερα, τὰ δ' ὑποτεθέντα ἐπάνιμεν αὖθις· ἵνα οὖν ἑξῆς τὰ παθήματα λέγηται τοῖς γένεσιν, ἔστω πρότερα ἡμῖν τὰ περὶ σῶμα καὶ ψυχὴν ὄντα. I adopt here Mr. Archer-Hind's αἰσθητὰ for αἰσθητικὰ of MSS.

[3] *Tim. Locr.* 100 D–E.

and the visible were the first created properties of body: without earth no tangible, however; without fire, no visible.

of body, as they are called, are named in relation to the sense of touching [1] (ποτὶ τὰν ἀφὰν κληίζεται), while some of them are denominated from their tendency towards the earth (ῥοπᾷ ποτὶ τὰν χώραν). It is touch that distinguishes the vital properties (τὰς ζωτικὰς δυνάμιας)—*heat, coldness*; *dryness, moistness*; *smoothness, roughness*; things *yielding* to the touch (τὰ εἴκοντα); things *resisting* the touch (τὰ ἀντίτυπα); *soft* things, *hard* things. It is touch that primarily distinguishes (προκρίνει) heavy and light, but it is reason (λόγος) that defines them (ὁρίζει) by their inclination to the centre or from the centre (τᾷ εἰς τὸ μέσον καὶ ἀπὸ τῶ μέσω νεύσει). Motion 'downwards' and 'towards the centre' are identical. . . . The 'hot' is held to be composed of fine parts (λεπτομερές) and to have a tendency to dilate or separate the parts of bodies (διαστατικὸν τῶν σωμάτων), whereas the 'cold' is thought to consist of grosser parts (παχυμερέστερον) and to tend to compress and close their pores (συμπιλατικὸν πόρων).

Created matter must be both *visible* and *tangible*. But without fire nothing could ever be visible; and nothing could be tangible without something solid in it, i.e. without earth (see Arist. § 12 *infra*). Hence when God framed the body of the universe He formed this of fire and earth. These, however, required a bond to unite them. The best bond is that which makes itself and the things bound by it as much one as possible; and the agency which is best fitted for such a bond is proportion (ἀναλογία). . . . God accordingly set air and water between fire and earth, making them as far as possible proportional; in such a way that fire is to air as air to water, and air is to water as water is to earth. Thus He constructed a universe both visible and tangible [2].

Aristotle.

The *organ* of touch: is it σάρξ

§ 9. Nowhere is the advance made by Aristotle in the psychology of the senses more evident than in the intro-

[1] Cf. Arist. § 10 *infra*: he also made the qualities of body *qua* body *tangibles*.

[2] Plato, *Tim.* 31 B–32 B, with Mr. Archer-Hind's notes.

ductory words of the chapter in which he treats of the sense of touching and its objects. He raises the question whether σάρξ is the real organ of touch, or whether the real organ is not rather something internal, to which σάρξ only serves as a medium. This question initiates an inquiry which could be satisfied only by a minute examination of the bodily structures concerned in touching, and which was destined in later times to lead to important results for physiological psychology. These results were not, however, reached by Aristotle, who may be considered nevertheless as a pilot of research. A second question here also raised by him, viz. whether this sense usually considered *one* is not really *several*, is of equal importance. To these questions he gives answers which correct the popular views. He concludes that the 'flesh' is not the true organ of touching; and he indicates his conviction that this sense is really a combination of several senses, prominent among which are the senses of temperature and resistance. The σάρξ and γλῶττα, popularly looked on as the organs of touch and its modification taste, are related to the true organs of these, as air and water are to the organs of seeing, hearing, and smelling [1].

§ 10. The sense of touching, like the other senses, is best explained if its object be first analysed and examined. (a) If touching be one sense, its object should be one (i. e. should fall under one conception bounded by contrary poles, as *colour* is a province lying between the contraries white and black). But if it have several objects it must be not one but several senses. (b) Again; what exactly is the organ which perceives the *tangible*? Is it the flesh—in creatures possessing flesh—and, in other creatures, that which is analogous to flesh? Or is this merely the medium, while the organ proper is something different, situated within? As regards the former question (a), every other sense is regarded as related in its object to one pair of opposites. Such is the case, for example, with seeing. This, as above remarked, is related to the opposition of

Margin notes: (as is generally supposed) or 'something within'? Is the sense of touch one sense, or a group of senses? The flesh (σάρξ) not the true organ, but only the medium, of touch. The sense of touch is not one sense but a combination of several senses.

Touch not one single sense, for object of sense of touch cannot be brought under a single pair of contraries like the objects of every other sense. Here we have (1) hot-cold; (2) fluid-solid (or wet-dry); this pair of con-

[1] Cf. 422ᵇ 17-424ᵃ 16 with Trendelenburg-Belger, pp. 329–337.

trarieties is not reducible to one contrariety. Therefore the sense which perceives them is more than one sense. These two pairs contain the qualities of body *qua* body, and form the ultimate tangibles.

white and *black.* So hearing, too, is related to *acute* and *grave* tones; tasting, to *sweet* and *bitter.* But within the *tangible* many kinds of opposition are included[1], all or most of which are reducible to the two of *hot* and *cold, fluid* and *solid*[2]. These two, however, are not further reducible[3]. A sort of answer to this question may be given by saying that there are several oppositions in the case of certain of the other senses also; for instance, in the case of sound, there is not merely the *high* and *low*, but also the *loud* and *faint*, the *soft* and the *harsh.* In regard to colour also there are corresponding kinds of opposition. But as Themistius observes, this answer is not satisfactory. It could not have been so to Aristotle himself[4]. It contradicts his frequent declaration that each special sense has a single ἐναντίωσις. Besides, what is the one conception sufficient to embrace all the *tangibles* in their various oppositions, in the way in which the notion of sound embraces all the *audibles*? There is no one obvious generic conception capable of containing under it the various, or the two chief, oppositions which come under touching[5]. All that can be said is that the tangible qualities are those of body *qua* body[6], and that their four above-named irreducible varieties determine the four elements of all bodies[7]. Hence either the sense of touch is one, with the difficulty that there is no one generic concept of its objects, or else it is two senses with two forms of ἐναντίωσις falling under it.

The organ of touch is not the *flesh.*

§ 11. As regards the other question above-raised, viz. whether flesh is the true organ of touch, decisive evidence is not to be found in the fact that the perception of touch

---

[1] 422[b] 25-7, 647[a] 16-20.

[2] These words best represent ὑγρόν and ξηρόν in this connexion. It may be observed that this opposition covers that of soft-hard; see § 16, p. 195, n 6 *infra.*

[3] 330[a] 25 αὗται δ᾽ οὐκέτι εἰς ἐλάττους ⟨ἀνάγονται⟩.

[4] τοῦτο μὲν οὖν ἴσως ἄν τις οὐκ ἀποχρώντως ἀλλὰ πιθανῶς διαλύσειεν, Them. *de An.* ii. 11, p. 72. 21 (Heinze; ii. 130. 20, Spengel).

[5] 422[b] 32.   [6] 423[b] 26.

[7] 330[b] 3 τὸ μὲν γὰρ πῦρ θερμὸν καὶ ξηρόν· ὁ δ᾽ ἀὴρ θερμὸν καὶ ὑγρόν ... τὸ δ᾽ ὕδωρ ψυχρὸν καὶ ὑγρόν· ἡ δὲ γῆ ψυχρὸν καὶ ξηρόν.

occurs simultaneously with contact between the flesh and an object. For if one were to take a thin membrane and strain it close around the flesh, this membrane would, just like the naked flesh, seem to take the impression of touch into consciousness co-instantaneously with the occurrence of contact between it and an object. Yet such a membrane would not, of course, be the organ of touch; though if, instead of being thus placed artificially round the flesh, it were connatural with it, the sensation of touch would pass through it even more quickly, and still more would it seem to be itself sensitive. A decisive argument to the contrary is this: immediate contact between the flesh and an object causes sensations of touch; but no other sense-organ has its specific sensations excited by immediate contact with its object. Hence we must conclude that flesh is only to be looked on as a *medium* of the sense of touch, somewhat as the air would be of the other senses, if it were a natural growth around our bodies. On the latter supposition we should have been thought to perceive sound, colour, and odour by one and the same organ; and seeing, hearing, and smelling would be held to be in a manner one and the same sense. 'As matters stand, however, owing to the separateness from us (i.e. from our bodies) of the medium through which the movements stimulating each of these three senses pass, the difference of their several organs is manifest[1]. But now as regards touching, this remains

Margin note: True this sense acts concurrently with the contact between flesh and an object: but so it would were a fine membrane strained tightly over the skin. The medium of touch and taste, however, is internal, for the flesh is a part of the body itself. It is this fact (of the medium being combined with the organ *in* the body) that makes us uncertain not only what the organ is, but whether the sense is one or several.

[1] 423ᵃ 10. I take δι' οὗ γίγνονται αἱ αἰσθήσεις as Simplicius did, and as Bäumker (*op. cit.*, p. 43) does, referring it to the medium-air, which is not according to the above hypothesis περιπεφυκὼς ἡμῖν, but διωρισμένος. It is hard to see how Wallace's translation (which follows Themistius and Trendelenburg's note) can be acquitted of tautology. 'Now, however, as matters stand, *by reason of the difference in the organs* by which the movements are effected, *the organs* of sense which we have mentioned *are clearly seen to be different* from one another (the italics are mine).' If the air were ἡμῖν περιπεφυκώς, then (according to Aristotle's notion here) the sensibility to colour, sound, and odour would be as widely diffused over the surface of the body as is the sensibility to tangibles. The connatural air, no matter where the κίνησις affected the periphery of the body, would transmit this κίνησις to the sensorium, and the local separateness which marks and distinguishes the organs of seeing, hearing, and smelling would disappear.

uncertain [1].'   Hence those two senses—of touch and temperature—which, according to Aristotle's principle of determining sensory faculties according to their objects, ought to be separated, remain for ordinary consciousness combined in one single sense.

§ 12. There must, however, be such a medium of sense as flesh, notwithstanding its effect in defeating our attempts at analysis of the sense of touching. 'An animate body cannot be composed of air or water singly [2] : it must be something solid. Accordingly it must be composed of a mixture of earth and these two other elements, i.e. it should be such a thing as flesh and what is 'analogous to flesh' tend to be. Hence by implicit necessity the body must be interposed as medium between the organ of touch and its object, and cohering naturally with the former, through which body the varieties of sensation classed under touch all alike pass notwithstanding their severalty and plurality. That touching does comprise several kinds of sensation is proved by the sense of touch immediately connected with the tongue. For in virtue of the tongue, which is one and the same organ, one has the sensation of all the other objects of touching and also that of taste. Now, if the rest of the flesh (as well as that of the tongue) had also been endowed with a sense of taste, touching and tasting would have been regarded as one and the same sense [3]. As it is, however, they are seen to be two, owing to the fact that their organs are not thus each capable of discharging the other's functions.

Notwithstanding this, such a medium as flesh is necessary. In order to perceive the qualities of body qua body, viz. solid-fluid, hot-cold, we require a solid medium. The possibility of several senses being mediated through the same medium is seen in the case of the tongue.

§ 13. One might ask: if every body possesses a third dimension—depth: and if two bodies, between which there is a third, cannot touch one another: and if, further, that which is moist and fluid has, by implication, body, as it

Can things submerged in water touch one another ? Can things

---

[1] 423ᵃ 11. What remains uncertain ? The answer is: *both* the things in question, viz. (1) what is the organ of touching (whether the flesh or something internal) ? and (2) is the sense of touching really not one but a plurality ? This uncertainty arises from the σάρξ being a 'connatural' medium, and therefore obscuring differences between organs otherwise discernible.

[2] 423ᵃ 11 seqq.   [3] 433ᵃ 19 seqq.

necessarily either *is* or *contains* water; and if things which touch one another in water have not (as they cannot have) their tangent extremities dry, and, therefore, necessarily have water between them, the water with which the said extremities are flooded;—if all this is true, it is impossible that in water any one thing should really touch any other. And so, too, in air; since the air is to things in air just as water is to the things in water; though, as regards the question whether one thing touches another, when both are immersed in the fluid air, we (owing to our *living* in air) are less likely to notice the difficulty of it, just as aquatic animals (owing to their living in water) would be as to the question whether one wet thing touches another[1].'

§ 14. 'This being so (i. e. even supposed contact being only close proximity), it is natural to ask: is the sense-perception of all objects whatever effected similarly, or are some objects perceived by sense in a fundamentally different way from others, just as, in fact, the senses of tasting and touching are both held to operate, i. e. by immediate contact with their objects, while the other three senses are supposed to perceive their objects from a distance? Or is this distinction false, and do we perceive the objects of touching, e.g. hard and soft, through media, just as we do the object of hearing, the object of seeing, and the object of smelling, only that while we perceive the objects of these three senses at long distances[2], we perceive objects of touching only near at hand? Owing to this nearness[3] it may well be that the mediation in the second case escapes notice; the truth being that we perceive all alike through a medium, only that in the case of these things (the objects of touch and taste, owing to their proximity) the mediation is not observed. Yet, as we said before, if we were to perceive all objects of touch through a membrane, which separated us from the objects without our knowing that it did so, we should be in the same condition, relatively to it, in which we now are, in fact, relatively

(marginal notes:) in air? All supposed contact in touch and taste is but close proximity.

In requiring mediation between object and organ, touch and taste do not stand apart from the other senses. The only differences are (1) that the objects of touch and taste must come *near* the body: and (2) that the medium in the case of touch and taste is itself part of the body. In touching and tasting we perceive concurrently with the affection of the medium.

---

[1] *De An.* ii. 11. 423ᵃ 21–31. [2] 423ᵇ 6.
[3] It has been shown or suggested (§ 13) that supposed contact is only close proximity.

to water and air when we touch objects in them. For
it is supposed that we touch the very objects themselves,
with nothing between us and them. But the object of
touching differs from the objects of seeing and hearing in
this, that we perceive the latter in virtue of the external
medium producing an effect upon us, while we do not
perceive the tangible by such operation of the object *through*
an external medium, but we perceive it *concurrently*, or *co-
instantaneously, with* the flesh regarded as medium ; just as
when a soldier is struck by a javelin which pierces his shield.
It is not that the shield is driven against and strikes the
man, but that shield and man seem to be struck together [1].

The true
organ of
touching
and tasting
is the heart.
Such is
Aristotle's
real con-
viction.
Yet he em-
ploys the
current ter-
minology,
based on
a partial
truth.

§ 15. On the whole (i.e. except for this last point) it
seems that the flesh in general, in touching, or that
of the tongue, in tasting, is what air or water is with
reference to the function of seeing, hearing, or smelling:
that is to say, it is related to the organ of touch (or taste)
proper as either of these media is to the organ in each
case. Accordingly, just as there would be no sensation
of whiteness if the white object were laid immediately
on the eye, so there would be no sensation of touch if the
tangible object were placed immediately on the veritable
organ of touch, and not on the flesh. Hence it follows
that the latter organ is not the flesh [2]. Thus only would
the facts in the case of touch (and taste) be analogous to
those of the other senses.' The whole matter may be
summed up thus. Aristotle abandoned the theory of
his predecessors, that touch and taste are unmediated
senses, because (*a*) the apparent simultaneity of tactual
perception with contact between σάρξ and the object, regarded as an argument for this, proves nothing ; (*b*) all the
other senses have media ; and (*c*) even between σάρξ and
the object absolute contact is impossible, since water or
air always intervenes. The true organ of touching (and

---

[1] 423[b] 12 seqq. Aristotle had no conception of a 'nerve process'
which takes time to reach the centres of consciousness.

[2] 422[b] 19, 656[b] 35 οὐκ ἔστι τὸ πρῶτον αἰσθητήριον ἡ σὰρξ καὶ τὸ τοιοῦτον
μόριον ἀλλ' ἐντός.

of tasting) is the heart, or the 'region of the heart[1].'
Yet, in spite of all this, we often (cf. p. 198, n. 2) find
Aristotle speaking in terms of the popular view which
makes flesh the organ of touching and tasting. He speaks
of the flesh as organ of touch[2], and of the tongue as organ
of taste[3]. The key to this seeming inconsistency is the
relative truth contained in the popular view. The flesh is
not, indeed, the true organ; yet it is not such a medium
as *air* is, viz. something external to us. It is part of our
organism, and a sort of auxiliary organ; standing to the
true internal organ as τὸ διαφανές (the external medium)
would stand to ἡ κόρη were it naturally united with this,
so as to form part of the whole living organism[4]. Flesh
is a peculiar medium, yet a medium all the same[5].

§ 16. 'It is by touching that the distinctive qualities
(διαφοραί) of body *as* body are discernible, i. e. the qualities
which characterize the different elements respectively, hot
cold, solid fluid, of which we have already treated in our
work on the elements[6]. Now the organ which perceives
these is that of touching, being that part wherein primarily
what we call the sense of touching resides. This is a part
of the body which is *potentially* such as the object which
affects it is *actually*. For to perceive by sense is to be
affected in a way in which the (agent or) object so acts
upon the organ (the patient) as to impart to the latter
actually the quality which the object itself actually has, but
which the organ before had only potentially. This explains

Right margin notes: By touching the qualities which belong to body as such are discerned. The organ which perceives these must be potentially what the objects are actually. Thus alone can the ἀλλοίωσις,

---

[1] 656ᵃ 29 αἱ μὲν δύο φανερῶς ἠρτημέναι πρὸς τὴν καρδίαν εἰσί, ἥ τε τῶν ἁπτῶν καὶ ἡ τῶν χυμῶν : cf. 439ᵃ 1–2.

[2] 647ᵃ 19.   [3] 533ᵃ 26.

[4] 653ᵇ 24 seqq. ὥσπερ ἂν εἴ τις προσλάβοι τῇ κόρῃ τὸ διαφανὲς πᾶν.

[5] Cf. Bäumker, *Arist. op. cit.* pp. 55–6.

[6] 423ᵇ 26 seqq., 329ᵇ 7 seqq. The second class of tangibles is elsewhere referred to as the hard and soft (τὸ σκληρὸν καὶ τὸ μαλακόν) but
remains the same. The ὑγρόν is the soft or fluid or moist : the ξηρόν
is the dry, the solid, the hard : i.e. in a loose and popular mode of
expression. Even now it is not unusual for even men of science to
oppose water to *solids*, as if water were not 'solid' (cf. Locke, *Essay*,
Book II, ch. iv, and p. 185, n. 2 *supra*); what they mean is that water
is soft. But this opposition is traditional from remotest times.

in which consists the physiological condition of all perception, take place. The hand feels cold water *as* cold because, relatively to it, it is itself warm: hot water *as* hot, because relatively to it, it is cold. So with the perception of hard and soft, &c. The organ is a μέσον, and hence κριτικόν of the above distinctions of quality. The sense of touch perceives both the *tangible* and the *intangible.*

why, when an object of touch is at first equally hot or cold, equally hard or soft, with the organ, we do not perceive it *as* hot or cold, hard or soft, when we touch it [1]. It is the tangible qualities in excess or defect of those already actually belonging to the organ that we perceive; since each sensory function results from the organ being in the position of a *mean* between *any two different qualities*, no matter what, in the scale of those which lie between the two opposites determining the province of the sense. This is what gives each sense its discriminating faculty (τὸ κρίνειν). The mean is that which discerns; and it can do so because it presents itself to a pair of different homogeneous qualities, allied each to different extremes, in such a way that when confronted with either it becomes the other. To cold water the hand can be hot: to hot water the same hand can be cold. Accordingly, as the organ which is to discern white and black must be actually neither but potentially both (and so on with the other organs), so the organ of touching must be actually neither hot nor cold.'

There is another analogy between touching and seeing. 'Seeing is, as we have pointed out, related at once to the visible and the invisible, and the three other senses with which we have dealt are similarly each related to opposites; so also the sense of touching is related to the tangible and the intangible. By "intangible" here we mean, on the one hand, those among tangibles which contain only an exceedingly small amount of tangible quality (and so are beneath our tactual capacity)[2], as, for example, is the case with air,

[1] Cf. § 17, p. 198 *infra.* In reference to the sense of touching Aristotle explains his idea of the μεσότης of the sense-organ most fully.

[2] 424ᵃ 12. He wants it to be understood that he is not referring simply to the non-tangible, a wide class which would include objects of all other senses (e. g. *whiteness*), and intellectual and moral conceptions (e. g. *thinking, virtue*), and even nonentities, all of which would be irrelevant to his subject here. His intangible does not involve a μετάβασις εἰς ἄλλο γένος, but a descent to or below the very lowest, or an ascent to or above the very highest, degree of the consciously tangible. τῶν ἁπτῶν is partitive genitive depending on τὸ ἔχον. The extremes here treated of as apprehensible by ἀφή both lie within the class τὰ ἁπτά: the one consists of such ἁπτά as are not actually but only potentially

and, on the other hand, such tangibles as are in excess of our tactual capacity; for example, things like a thunderbolt, which, if touched, would destroy us [1].'

§ 17. 'Among the senses that of touching is fundamental. The attribute which first distinguishes animal from merely living forms is tactual sensibility. Just as the function of nutrition may exist apart from the sense of touching and from sense generally, so the sense of touching may exist apart from all the other senses. Plants or vegetables possess the nutrient function: it is by the possession of the sense of touch that animals first rise above and are distinguished from vegetables [2].' 'If a body is to possess sensory faculty, it must be either simple or compound. But it cannot be simple, for if it were, it would not possess the sense of touching, which it must, however, possess, if it is to possess sensory faculty, or even live, at all, as will be manifest from the following considerations. Since an animal is an animate body, and every body is tangible, and that which is perceived by touch *is* the tangible, it follows that the body of an animal must have the sense of touch, if the animal is to live and preserve itself. For the other senses, smelling, seeing, hearing, perceive their objects through media; but if the animal body comes into contact with some other, but does not possess the sense of touch, it will be lacking in the guidance needful to enable it to shun tangibles of the dangerous sort, and to seize on those desirable for its food. Such an animal would be incapable of preserving its existence [3].'

'It is manifest that the body of an animal cannot be simple, i. e. composed wholly of a single element, e. g. *fire* or *air*. For an animal cannot possess any other sense if it have not that of touching, since this is what distinguishes

Marginal note: The sense of touch the fundamental sense. Its possession first distinguishes animal from vegetable. To possess this sense animal bodies must be composed of as many elements as have qualities corresponding to the oppositions which come under the sense of touch. As earth is needed for the perception of hard-soft (or solid-fluid) so fire is needed for the perception of hot-cold. The organ of touch the most composite

tangible, the other of such as are tangible, but only with an effect destructive of the organ of touch, or even of life and perception generally. Philoponus understood this, but Trendelenburg does not seem to do so, for he misunderstands Philoponus, whose note, he thinks, proves him to have read τοῦ ἁπτικοῦ for τῶν ἁπτῶν.

[1] For the preceding paragraphs see *de An.* ii. 11. 423^b 1–424^a 15.
[2] *De An.* ii. 2. 413^b 4 seqq.   [3] *De An.* iii. 12. 434^b 8–18.

of all organs. No sensibility in parts consisting too exclusively of any one element, e. g. earth. So no feeling in hair or bone *per se*. Plants, being of earth for the most part, have no sensibility.

and defines the animal. Now the other organs of sense might conceivably be formed without [1] earth, since they all effect sensation by some medium or third thing, external to the body, through which each perceives its object. The sense of touch, on the contrary, as its very name shows, acts only by immediate contact between its organ and the tangible object. If the other senses perceive by a sort of contact it is at least a *mediated* contact, one brought to pass by the intervention of a third thing. This sense alone perceives its objects—or is held to do so—immediately [2]. Thus if an animal is to possess touch, its body cannot consist of any one of the elements of which the externally mediated sense-organs might consist (i. e. of air or water alone). Earth is necessary as an element in the apparatus of this sense [3]. Yet earth *alone* without, e.g. fire, is not enough, this sense being a mean between all tangibles, and capable of discerning not only the distinctive qualities of earth, but also the qualities denominated hot and cold [4], and all other tangibles. The organ of touch, in fact, is, or should be, the most composite of all the organs. This is natural to expect, since it discerns a greater variety of objects than other organs, and its objects have more than one form of opposition [5]. We have no sensibility in bone or hair, since such parts are formed too largely of earth alone. Plants, for the same reason, are destitute of sensation [6]. Without touch no other sense can subsist, and its organ consists neither

[1] 435ᵃ 11–15. Here ἔξω γῆς = 'without earth.' Cf. Pind. *Isth.* v. [vi.] 72 where, by a metaphor, γλῶσσα δ' οὐκ ἔξω φρενῶν = 'his word is not without understanding.' The obvious opposition here between τὰ ἄλλα and ἡ ἀφή below makes it certain that by ἄλλα is meant not στοιχεῖα, but αἰσθητήρια.

[2] 435ᵃ 17. Aristotle here adopts the popular view of σάρξ as organ of touch; it is for his present argument as suitable as the other; the medium being in this case part of the body, and the question whether σάρξ is or is not the true organ being irrelevant here.

[3] For the reasons *vide* 423ᵃ 14, § 12 *supra*.

[4] 435ᵃ 23. The need of *fire* is here clearly implied, though not stated.

[5] 647ᵃ 14.

[6] τὰ φυτὰ διὰ τοῦτο οὐδεμίαν ἔχει αἴσθησιν ὅτι γῆς ἐστιν, 435ᵇ 1 : this does not mean that φυτά have γῆ *alone* in their composition. All μεικτὰ σώματα have in them *all* the elements, the only difference being as to the degree in which these predominate in the compound.

of earth nor of any single element alone. The requisite
μεσότης of sense could not subsist in one single uncom-
pounded element.'

§ 18. Touch is the one sense deprivation of which means Destruc-
death to an animal. Nothing can have this sense but an tion or
privation
animal, nor, to be an animal, is any other necessary. of touch
Hence the objects of the other senses—colour, sound, alone
means
odour—do not, when felt in excess, destroy the animal, but death to
an animal.
only the organs: unless, indeed, incidentally, as when with Excess
a sound a thrust or a blow is incidentally associated, or as in the other
sensations
when, by the sights or odours, other things are set in action may de-
which by their contact destroy the animal. Taste, when stroy the
organ or
it destroys an animal, does so only so far as its object is its function ; but
tangible. But all excess of the tangible qualities of the excess of
hot or cold, or the hard, destroys animal life. In every the tangible de-
province of sense, indeed, excessive action in the object stroys the
destroys the organ of the sense : so that this happens also animal's
life.
with regard to the organ of touching. The latter organ,
however, is one on which the animal's life depends, and
without which no animal exists. Hence with destruction
of this organ, not only the organ itself but the living
animal perishes forthwith [1].

§ 19. 'The flesh, or what is "analogous," is *per se* the The organ
principium of the body of animals. An animal is defined of touching
[i. e. using
by having sensation, but particularly that of touching—the the popu-
primary sense. The organ of this sense is a bodily part such lar term,
flesh] is
as has been described, viz. a μόριον ὁμοιομερές, such as σάρξ[2]. a μόριον
ὁμοιομερές.
This is either the essential organ of touching, as the κόρη is Touch is
of vision; or else it has been conjoined with the essential the one
sense in
organ as its auxiliary or instrument ; just as if one were which all
to conceive the whole διαφανές, or external medium of animals
are akin.
vision, joined with and superadded to the pupil. In the Man's
superior
case of the other senses it would have been superfluous for intelligence
nature to produce this fleshy environment, but the sense due to the

---

[1] *De An.* iii. 13. 435[a] 12–[b] 1–19.

[2] 653[b] 19 seqq. The ὁμοιομερῆ (e.g. flesh, bone, hair) no matter how
much subdivided severally yield parts still homogeneous with one
another and the whole. An 'organic' part, e.g. the hand or face,
could not be so divided into hands or faces.

fineness of
his sense
of touch :
not, how-
ever, to
this alone,
but also to
the perfec-
tion of the
way in
which in
his organ-
ism the
elements
are mixed.
Twofold
form of
the organ
of touch
obscured
by the
nature of its
medium.
of touch requires it, this organ being of all others the most corporeal in its character [1]. All animals have one sense in common—touching. Hence the part wherein this is naturally generated is without a common or generic name ; for in some animals this part is the same (viz. σάρξ), in the remainder it is that which is analogous to this [2].' The assertion that touch is common to all animals, and the distinctive mark of animal as compared with vegetable life, is found in passages too numerous to mention in Aristotle. The connexion between this sense and the life of the animal harmonizes at least with the fundamental importance which, as we shall see hereafter, touch assumes for Aristotle as the *basis* of the whole sensory endowment of animals and men : as primary, not merely from a biological but also from a psychological standpoint. His insistence on this everywhere makes it the more surprising that he rejects Democritus' theory that all senses are reducible to that of touch. As this fundamental character of touch is explained or asserted by him in reference to the *sensus communis* (the κοινὴ αἴσθησις and its κύριον αἰσθητήριον or *sensorium commune*), we will postpone the further consideration of it until we come to treat of the latter, in which Aristotle's psychology of the senses culminates [3].

' In the fineness of his sense of touch man excels all other animals, and also in his sense of taste, which is a mode of touch. Owing to the delicacy of his sense of touch it is that man is the most intelligent of all animals. A proof of this is that within the human race itself men show genius, or the lack of it, in a degree parallel with the degree of fineness in their organ of touch, and none other. Those who are hard-fleshed [4] are dull, while the soft-fleshed are the

[1] 653[b] 24 seqq.   [2] *Hist. An.* i. 2. 489[a] 17-19.
[3] In what precedes we have seen the remark often repeated that ἡ ἁφή is the only sense essentially requisite for animal existence. There is no inconsistency between this and the statements found in 436[b] 13, 455[a] 7, that ἡ ἁφή *and* ἡ γεῦσις must accompany animal life, for it is Aristotle's constant doctrine that γεῦσις is a mode of ἁφή, or ἁφή τις.
[4] Cf. our term 'thick-skinned.'

persons of genius [1]. The mental superiority of man, however, according to Aristotle, rests also upon a very different ground—that chosen by Empedocles—the superiority of the mixture of the elements in his bodily organism [2].'

The sense of touching is subject to illusion. 'If we cross the fingers, one object placed between them so as to touch both their adjacent surfaces appears as if two. We do not, indeed, call it two, for the sense of sight, which is superior in authority, pronounces it one; but if we had only the sense of touch, we should actually call it two objects [3].'

'Each of the sensory organs is twofold, except that of the sense of touching, in which the twofold character appears absent; but this appearance is due to the fact that the flesh is not really the organ of touching, and that the true or primary organ is something internal [4].'

---

[1] *De An.* ii. 9. 421^a 22-6.

[2] Cf. 744^a 30 δηλοῖ δὲ τὴν εὐκρασίαν ἡ διάνοια· φρονιμώτατον γάρ ἐστι τῶν ζῴων ἄνθρωπος. Against this complacent opinion of human wisdom may be set a favourite dictum of Polybius (e. g. xviii. 15. § 15 ; 40. § 1), that 'of all animals man is the most foolish, being taken repeatedly in the same traps, political and military.'

[3] Cf. *de Insomn.* 2. 460^b 20-22, 461^b 2.

[4] *De Part. An.* ii. 10. 656^b 32-6.

PART II. SENSATION IN GENERAL

ITS COMMON AND PECULIAR FEATURES

§ 1. In dealing with the Greek psychology of the special senses, we have used the terms 'sensation,' 'sense-perception,' &c., as if their meaning had been already determined. We must hereafter consider how far the Greeks themselves had reached a clear conception of the general and characteristic force of these terms. It has to be remarked that they failed for the most part (*vide*, however, § 6 *infra*) to distinguish between sensation as the elementary fact, and perception as the more complex and developed, implying objective reference. Αἴσθησις for them (when it did not mean *feeling*) usually denoted what we call perception. We have to inquire here what general statement of the meaning of sensation, or sense-perception, served them at once to clear up the intrinsic connotation of these words, and to distinguish—if they did distinguish—between the facts which they denote and others such as those of physical interaction between bodies. How does *seeing*, for example, differ from the reflexion of images in a mirror? How does *touching* differ from mere physical contact? These questions were raised by some of the ancients, and answers were in some few cases attempted. Of their psychological importance there can be no doubt. Having considered in Part II what the Greek writers with whom we have here to do contributed to their settlement, we shall in Part III proceed to the consideration of the *sensus communis*, the faculty of distinguishing and comparing, imagining and remembering, with the synthetic or organizing function which Aristotle, rightly or wrongly, attributed to τὸ αἰσθητικόν.

§ 2. The problem of mind is complicated with that of life. An animal must live if it is to feel and perceive. To live it must be nourished, and the faculty of nutrition is for Aristotle biologically prior to that of sense-perception: indeed, for all Greek writers this empirical relation between

vital and psychical faculty is axiomatic. Aristotle, there- order and inter-rela-tionship.
fore, was not taking a course peculiar to himself, but merely
emphasizing his empirical standpoint, when he in his
psychology discussed the faculties of the soul in this
order—*nutrient* (and *generative*), *sentient* (with *appetitive*
and *locomotive*), *intellectual*[1]. The nutrient faculty can exist
without any of the others ; these cannot exist without the
nutrient. So the sentient can exist without the intellectual,
but the latter cannot exist without the former. The animal
world is distinguished by the super-addition of αἴσθησις to
the lower or nutritive (and generative) faculty. All animals
possess sensation, though some do not possess all the
varieties of sensation. There is, however, one sense which
all possess—that of touching, with its modification tasting.
This is that in which all animals fundamentally agree. If
then one wishes to ascertain Aristotle's views as to the most
general and fundamental characteristics of sensation, one
should understand first what he has to say of this particular
form of sense-perception. We shall deal with it more
particularly in connexion with his theory of the *sensus
communis* with which it is so closely connected. But first
we must consider how much his predecessors had done for
the purpose of clearing up the notion of sensation in general,
and how much Aristotle owed to their efforts in this
direction. We shall find that he owed but little to any
except Plato.

Alcmaeon.

§ 3. We have but scanty information—if indeed we have Alcmaeon had little to say of sensation in general, except that it is due to the inter-action of dissimilars. He distin-guished
any—as to Alcmaeon's views of the common and peculiar
characteristics of sensation. According to Theophrastus[2],
he regarded it as brought about by the interaction of dissimilars ; he distinguished between τὸ αἰσθάνεσθαι and τὸ
φρονεῖν (or τὸ ξυνιέναι), the latter being probably not σωματικόν, and declared that while the lower animals possess
sense-perception, man alone has intelligence. In all this

---

[1] *De An.* ii. 3. 414ᵃ 31 seqq. He varies slightly in his statements,
but generally speaking adheres to this arrangement.
[2] *De Sens.* §§ 25–6.

sensation
from in-
tellect.

we do not discover what we wish to find, namely, how Alcmaeon would have distinguished between the fact of sense-perception in general and merely physical facts, or how he would have stated the fundamental characteristics in which all the varieties of sense-perception agree. He most probably was, however, of opinion that there is even in sensation a peculiarity which distinguishes it from merely physical processes (see Rohde, *Psyche*, ii. p. 171 n.).

Empedocles.

Empe-
docles
thought he
solved the
question by
his theory
of pores and
symmetri-
cal emana-
tions ;
but in
reality he
only ob-
scured it.
Neither
did he help
to answer
the ques-
tion by his
principle
that ' like
perceives
like.'

§ 4. Empedocles, as we may infer from our records, approaches more nearly to an appreciation of these questions. As we have already repeatedly observed, he held that all the particular operations of sense are effected by ἀπορροαί entering the pores of the sensory organ, when each organ has its fitting object supplied, and when relations of symmetry [1] subsist between the ἀπορροαί from the object and the pores of the organ. Here, then, we find a conception of a common characteristic of all varieties of sense-perception : this requisite συμμετρία between the ἀπορροαί and the πόροι. But nevertheless for Empedocles there is in this nothing peculiarly characteristic of sensation. Such agreement between ἀπορροαί and the pores of objects is the universal condition of the interaction of material bodies. Theophrastus, therefore, pertinently asks [2], how animate beings differ, according to Empedocles, from inanimate in this respect ? Shall we have to admit that, when emanations from a body fit the pores of an inanimate body, the latter has sensible experience of the former? or have all things whatever a capacity for sense-perception? If Empedocles' theory were sufficient, says Theophrastus, all substances which naturally blend together should be said to perceive

---

[1] It would be worth while to consider how far in this notion of συμμετρία Empedocles anticipates or paves the way for the Aristotelean doctrine of the μεσότης or λόγος of each αἰσθητήριον, in virtue whereof it grasps the form without the matter of the αἰσθητόν. As regards the composition of σάρξ and ὀστοῦν, Aristotle himself states (642ᵃ 19–24) that Empedocles made these severally to consist of a λόγος τῆς μείξεως τῶν στοιχείων—not of any one or two or three elements, or of all merely put together.   [2] *De Sens.* §§ 7 and 12.

one another[1]. Another point in which, according to Empedocles, all sensory operations agree is that like is perceived by like. We perceive external objects by elements homogeneous, or identical in kind, with them, forming part of our bodily structure and constituting the soul itself. Thus to the former requisite relation of συμμετρία is added the further requirement of ὁμοιότης between object and organ. By this second principle also, Empedocles did but little which could be said to raise psychology above the level of physics. He showed, indeed, or tried to show, in what the various kinds of sense-perception agree, but not that which at the same time distinguishes them from physical processes. Rather he implicitly denied that there is any such fundamental distinction. Perception is for him only interpenetration—a material conception. We shall, indeed, find that philosophers divide themselves, henceforth, on this very point, viz. into (1) those who assert (implicitly or explicitly) that there is no difference at bottom between sense-perception and physical interaction, and (2) those who maintain such fundamental difference.

Democritus.

§ 5. Democritus considered all relations between realities of every kind as reducible to the purely mechanical form. Therefore for him no difference could be admitted ultimately between the kind of interaction involved in sense-perception and that involved in the action of any atomic bodies upon one another. All interaction whatever consists in or involves contact: and this is as true of the interaction between a percipient and a perceived object as of any other. Sensation is due in the last resort to a contact between the objects of sense, or ἀπορροαί from these, all of which are atoms combined in various ways, and the spherical atoms of which the soul is composed. Theophrastus strangely hesitates as to whether for Democritus sense-perception was

For Democritus the difference between sensation and physical interaction is merely apparent; nor can there be a fundamental difference between sensation and intellect. All interaction

---

[1] Theophr. *de Sens.* § 12. Empedocles no doubt would accept the full consequences of his cosmical doctrine. Despite his discrimination of γυίων πίστις from νοεῖν, he did not believe in any *absolute* distinction between sensible and insensible forms of interaction: cf. Rohde, *Psyche*, ii. 171 seqq.

whatever, or was not to be explained by the interaction of like with
that of *per-* like [1]. When we reflect that for Democritus differences of
cipiens and
percipien- kind, being all due to sensory discrimination (which can-
dum in-
cluded, is not be ultimate), must resolve themselves into quantitative
ultimately differences, and that he allowed even physical interaction
mechanical
interaction between similars (a doctrine in which he differs from the
between
atoms in majority), we cannot share such hesitation. It is, therefore,
a void.
manifest that we cannot find in the doctrine of Democritus
anything to *distinguish* sensory facts from physical facts :
the former are but a mode of the larger physical total.
What, then, has he to say on the other side of the question,
viz. as to the common feature in which all sensory facts
agree? We can find no clear statement on this point either.
The facts of sense-perception are reduced to physical facts
of contact between the object and the organ : that is all.

Did Demo-   § 6. On the general subject of sensation, however, it is
critus con-
ceive of interesting to notice a dictum contained in the *Placita*,
actual
αἰσθητά that 'Democritus regarded the αἰσθήσεις as being more
which our numerous than the αἰσθητά, but that owing to want of
senses are
incapable correspondence between the αἰσθητά and the multitude of
of perceiv-
ing? Or of αἰσθήσεις, some of the latter (or the former?) escape observa-
αἰσθήσεις tion [2].' Diels (*Dox.*, p. 399 n.) renders: *sensuum affectiones*
of which
we are *plures sunt perceptis, sed cum percepta multitudini (affec-*
ourselves *tionum) non respondeant, illae non omnes agnoscuntur.* In
uncon-
scious? his lately issued *Vorsokratiker* (p. 388), however, he illustrates by quoting Lucret. iv. 800 *quia tenuia sunt, nisi se*
contendit acute, cernere non potis est animus. Zeller, on
the other hand (*Pre-Socr.* ii. 267 n., E. Tr.), supplies (not
τὰς αἰσθήσεις as Diels, but) τὰ αἰσθητά before λανθάνειν, and
interprets the passage as having in its original form meant
that 'much is perceptible which is not perceived by us,
because it is not adapted to our senses.' This interpretation Siebeck (*Geschichte der Psychologie*, pt. i. p. 114) adopts,
and, as an illustration, mentions our want of 'a sense

---

[1] *De Sens.* § 49.   See p. 24, n. 1 *supra.*

[2] Stob. *Ecl.* i. 51, Diels, *Dox.*, p. 399, *Vors.*, p. 388 (πόσαι εἰσιν αἱ
αἰσθήσεις) Δημόκριτος πλείους μὲν εἶναι τὰς αἰσθήσεις τῶν αἰσθητῶν τῷ δὲ μὴ
ἀναλογίζειν (ἀναλογεῖν, Diels) τὰ αἰσθητὰ τῷ πλήθει (sc. τῶν αἰσθήσεων, Diels)
λανθάνειν.   What does ' correspondence ' or ' analogy ' here mean ?

for the perception of magnetic currents, which we can only conceive by translating them psychologically into phenomena of seeing.' It is true that Democritus was committed to a belief in the infra-sensible qualities of the atoms, which are αἰσθητά, perhaps, *ex hypothesi*, but 'disproportionate' to *our* αἰσθήσεις. Still, in order to get the sense which Zeller and Siebeck find in the words, we should have πλείω τῶν αἰσθήσεων τὰ αἰσθητά, or else take τὰς αἰσθήσεις as equivalent to *possible* sensations, or sensory *powers*, and τῶν αἰσθητῶν as *actualized* percepts, which would be very awkward, even if legitimate. Interesting as it would, no doubt, be to find Democritus (who stood at the head of the 'science' of that time) conceiving tones which our ears cannot hear, colours which our eyes cannot see, and so on, as well as the infra-sensible atoms themselves on which his physical theory rested, yet it is more than questionable whether—on the strength of an excerpt (such as that here under discussion) five hundred years at least later than the writings of Democritus, and of a doubtful reading or interpretation of it—we have any right whatever to attribute such conceptions to him. Besides, such a theory would implicitly objectivize the so-called secondary qualities, contrary to all that we know of his teaching. Adopting Diels' rather than Zeller's construction, we might as well, and with equal justification, find in the words the germ of some such theory as that of so-called 'latent mental modifications,' or that of *perceptions insensibles* afterwards developed by Leibniz. Our αἰσθήσεις are more numerous than our αἰσθητά (Democritus might then seem to say), because we do not notice the former unless when we notice the latter. In modern terms, we do not notice sensations which, not being referred to an object, are not perceptions. There are, in this way, many αἰσθήσεις which pass without being attended to or coming 'into consciousness.' The argument of Arist. *de An.* iii. 1, that 'there are not more senses than the recognized five,' was directed, perhaps, against the very speculation of Democritus (whatever it really was), which is alluded to in

the above words of the *Placita*, but of which unfortunately
we know nothing more [1].

Anaxagoras.

§ 7. According to Anaxagoras νοῦς was the principle of
orderly movement, both in the cosmos and in the individual.
He did not distinguish νοῦς from ψυχή [2], representing both
as absolutely different from any form (or, at least, from
any *other* form) of material things. While he *implies* the
peculiarity of the interaction implied in sensation, we look
in vain to him for an account of it. He does not define
the general features which characterize all sensory activity,
and at the same time distinguish it from other kinds of
activity. The scattered sayings in reference to the senses
which we find attributed to him, do not help us much
towards the solution of such a problem. Sense-perception
was necessarily (according to his doctrine of νοῦς ἀμιγής)
effected by the relation of *unlike* to *unlike*, or rather of *contraries*, to one another. The sensory act implied, for
Anaxagoras, as for Aristotle, a change (ἀλλοίωσις) of some
sort in the organ of perception. This appeared possible
only if the organ and the object were dissimilar. Thus the
reflexion in the eye, on which seeing depends, is formed in
the part of the eye which is different in colour from the
object. We perceive heat and cold by touch only when
the object touched is hotter or colder than the organ.
So with the other senses. We perceive all qualities in the
object according to the excess or defect of them in the organ.
But all qualities exist in our organs [3], though in different
proportions ; so that the contrasts required for perception
of objects are always possible in experience. This doctrine,
however, of perception by contrast (of qualities within to
qualities without the organism), together with the other
doctrine of πᾶν ἐν παντί, does not go far to clear up the
distinctive and general features of sense-perception, or
furnish us with a point of view from which to contemplate

Marginal note: For Anaxagoras, who held that the soul is absolutely heterogeneous to the objects of the physical world, the interaction implied in perception is quite different from other kinds of interaction. He does not, however, show us what the difference is. We only know from him that perception takes place by the interaction of contraries. But these are physical, and the part played by soul in the relation of *percipiens* to *percipiendum*—in other words, the peculiarity involved—is left in obscurity.

---

[1] For the conception of αἰσθήσεις, as well as αἰσθητά, too small to be
noticeable, at least 'actually,' cf. Arist. *de Sens.* vi. 446ᵃ 7-15.

[2] Cf. Arist. 404ᵇ 1-3.

[3] Theophr. *de Sens.* §§ 27-8 ; Diels, *Dox.*, p. 507. 18 πάντα γὰρ
ἐνυπάρχειν ἐν ἡμῖν.

or pursue this subject apart from physical science. The contraries here referred to as required for perception are physical on both sides. Whence they derive their contrariety, or how the heterogeneity of the ψυχή, which is active in perception, takes effect we are not informed. The soul presides over the interacting contrary qualities of the perceiving sense and its object; that is all we know. True to his notion of perception by dissimilarity, Anaxagoras regards all exercise of the senses as accompanied by, or involving, discomfort or distress, consciously or unconsciously. In proof of this he points to the effects of time and age in dulling sense, and also to those of over stimulation, e. g. by too loud a sound, too brilliant a light, &c. He (as we have seen) held the view that in larger animals, with their larger sensory organs, sense-perception is more perfect than in others [1]. These vague observations constitute what we know of his theory of sensation in general. Needless to say, it is impossible to ascertain from them what settled views (if any) he entertained as to the common and peculiar characteristics of sensation.

Diogenes.

§ 8. Diogenes of Apollonia, holding as he did that air was the divine being, the *principium* of all things, the *fons et origo* of sense and thought and order in the world, the *deus in nobis*, endeavoured to give details respecting the sensory function of animals, and in connexion with the air within them—especially, or in the first instance, that around the brain, but ultimately that also in the region of the heart. As air was not only the *principium* of thought and sense, but also of things, for Diogenes, as for Empedocles and Democritus, it was axiomatic that *like* is perceived by *like*. We of course look as vainly to him, as to the others, for a distinctive and common account of the various kinds of sense-perception, such as Plato and Aristotle desire and attempt to supply. The internal air on which hearing, seeing, and smelling most immediately depend, is that in

Diogenes, who made air the supreme agency of sense and intellect, as also the substance of all that is real, could not hold that there is ultimately any peculiar feature in the inter-action of sense-organ and object to

---

[1] Theophr. *de Sens.* §§ 31-4.

distinguish
this from
other inter-
action. For
him psy-
chology
in the last
result mer-
ges itself
in physics.

or around the brain. Diogenes may, however, have held that sense involved a faculty of synthesis—a faculty of combining the data of sense. If so, then for him this faculty probably had its centre or seat in the thorax[1]. If this be so, his position would exhibit some approximation to that of Aristotle, making us curious to know more about it. It is not, however, hard for Theophrastus[2] to show that the psychology of Diogenes, like that of Empedocles, provides no ultimate discriminant between sensory and other processes, but tends rather to merge psychology in physics. When Diogenes, for example (after the manner of Empedocles to some extent), explains ὄσφρησις by the συμμετρία between the odour, wafted to the organ of sense, and the air around the brain, in consequence of which συμμετρία the odour and the said air are blended together; Theophrastus naturally asks: what then is there to distinguish this from all other kinds of κρᾶσις? Diogenes must either deny that there is anything to distinguish them, or acknowledge that he has omitted to state it, if there is. He would probably, if pressed to choose, have accepted the former alternative.

Plato.

Plato's
general de-
finition of
sensation:
a move-
ment com-
mon to
soul and
body, but
proceeding
through
the body
to the soul.
The dif-
fusion of
sensations
through

§ 9. Plato is the first writer who confronts the problem before us with a clear conception of its meaning. He defines sensation in general (αἴσθησις) as a 'communion of soul and body in relation to external objects. The *faculty* belongs to the soul; the *instrument* is the body. Both in common become by means of imagination apprehensive of external objects[3].' In the *Philebus* Plato himself says: 'Suppose that some of the affections which are in the body from moment to moment exhaust themselves in the body alone before—or without—reaching the soul, thus leaving the latter unaffected; while others pass through both, and

---

[1] According to the doubtful testimony of the *Placita*, Aët. iv. 5. 7, Diels, *Dox.*, p. 391, Diogenes placed τὸ ἡγεμονικόν in the ἀρτηριακὴ κοιλία τῆς καρδίας.　　　　　　　　　　[2] *De Sens.* § 46.

[3] Plut. *Epit.* iv. 8, Diels, *Dox.*, p. 394. 'By means of imagination'= διὰ φαντασίας. This gives to φαντασία the prominence which later psychologists attributed to it, but which it does not really, in this connexion, receive from Plato.

impress on both a sort of tremor of a quite peculiar kind, the body, in which both—body and soul—participate. ... When body owing to the mo- and soul in this way partake of this common affection and bility of its tissues. are moved by this common movement, if you should call The parts this movement sensation (αἴσθησις) you would speak quite formed of earth, and correctly[1].' In the *Timaeus* again Plato gives his general therefore conception of sensory affection. 'We have[2] yet to consider immobile, are without the most important point relating to the affections which sensation. Plato, in his concern the whole body in common, viz. the cause of the conception pleasurable and painful qualities in the affections which of αἴσθησις, fails to we have discussed, and also the processes which involve distinguish sensations produced through the bodily organs, and are the cogni- tive accompanied by pains and pleasures in themselves. This element from then is how we must conceive the causes in the case of feeling. every affection, sensible or insensible, recollecting how we defined above the source of mobility and immobility; for in this way we must seek the explanation we wish to find. When that which is naturally mobile is impressed by even a slight affection, it spreads abroad the motion, the par- ticles producing the same effect upon one another, until, coming to the centre of consciousness[3], it announces the property of the agent; but a substance that is immobile is too stable to spread the motion round about, and thus it merely receives the affection but does not stir any neighbouring part; so that, as the particles do not pass on one to another the original impulse which affected them, or transmit it to the entire creature, they leave the recipient of the affection without sensation[4]. This happens in the case of the bones, hair, and generally the parts formed of earth[5]; while the former conditions apply chiefly to sight

[1] *Phileb.* 33 D–34 A. From this passage, with the exception of the διὰ φαντασίας, an insertion borrowed from later psychology, that quoted above from the *Placita* seems derived.

[2] *Tim.* 64 A–C (Archer-Hind's version for the most part). In what follows αἴσθησις is confusedly treated as = feeling *plus* cognitive sensation.

[3] τὸ φρόνιμον: I cannot render it with Mr. Archer-Hind the 'sentient part': it includes more than this.   [4] ἀναίσθητον παρέσχε τὸ παθόν.

[5] Cf. Arist. *de An.* iii. 13. 435ᵃ 24 seqq.

P 2

and hearing, because these contain the greatest proportion of fire and air [1].' In another passage [2] he explains the cause of sensation, and its disturbing effects upon intelligence, as resulting from interaction between the elements which form the body and those external to it. 'For great as was the tide sweeping over them (sc. the bodies of newly created creatures) and flowing off—the tide which brought them sustenance—a yet greater tumult was caused by the effects of the bodies that struck against them; as when the body of any one came in contact with some alien fire that met it from without, or with solid earth, or with liquid glidings of water, or if he were caught in a tempest of winds, borne on the air; and so the motions from all these elements rushing through the body penetrated to the soul. This is in fact the reason [3] why these have all alike been called, and are still called, sensations (αἰσθήσεις). Then, too, did they produce the most wide and vehement agitation for the time being, joining with the perpetually streaming current in stirring and violently shaking the revolutions of the soul, so that they altogether hindered the circle of the Same by flowing contrary to it, and they stopped it from governing and going.' Plato does not in these passages distinguish sensation, as element in cognition, from feeling. The disturbing effects referred to by him are really due to the emotions connected with pleasure and pain. Aristotle also regards sensation as an affection common to body and soul, and beginning with the former [4].

Plato's description

§ 10. Further light is thrown upon Plato's conception of

---

[1] With this passage cf. that of Aristotle 459ᵃ 28–ᵇ 5, where the latter illustrates the transmission of sensation from point to point by the way in which heat is diffused through the body from the first point of contact to the ἀρχή. The ἕως τῆς ἀρχῆς of 459ᵇ 3 seems to correspond in a way to the μέχριπερ ἂν ἐπὶ τὸ φρόνιμον of Plato above: *Tim.* 64 A–C.

[2] *Tim.* 43 B–D (Archer-Hind). Here Plato, by his account of the agitation in the bodily tissues of newly created beings, seems to give or suggest the explanation adopted by Aristotle (*de Mem.* 450ᵇ 5) of the feebleness of the intelligence and memory of very young children.

[3] As if to connect αἴσθησις with ἀσθμαίνω, √ ἄϝ-η-μι.

[4] 436ᵇ 6 ἡ δ' αἴσθησις ὅτι διὰ τοῦ σώματος γίνεται τῇ ψυχῇ δῆλον καὶ διὰ τοῦ λόγου καὶ τοῦ λόγου χωρίς.

sensation by a passage in the *Theaetetus*[1]. He discusses the Protagoreo-Heraclitean doctrine that 'man is the measure of all things,' from the point of view of its effects upon objective knowledge. The doctrine is based upon the Heraclitean maxim πάντα ῥεῖ. This maxim applied to the subject of sensation or sensory perception results as follows. Protagoras held with Heraclitus that all physical things are in incessant motion. Motions are innumerable, but all fall into two classes, the *passive* and the *active*[2]. Things have their so-called qualities only by acting or being acted on. But activity and passivity are always relative : hence no quality belongs to anything *per se*. Only by interaction or relation of some sort are things determined in quality. We cannot say that they are *per se* anything in particular : or even that they *are*, at all. They only *become* : they are always *becoming*, not *being*. Our sensory presentations arise by the concurrence of the aforesaid kinds of motion— the active and the passive. The active belongs to what we call the αἰσθητόν or object of sense ; the passive belongs to the percipient or subjective organ[3]. When an object comes into contact with our sense-organ, so that the object acts on the organ, and the organ is acted upon by the object, a sensation, on the one hand, arises in the organ, while on the other hand, the object appears endowed with certain qualities. Thus arise in the organ sensations of *seeing, hearing, smelling, cooling, burning, pleasure, pain, desire, fear,* &c. ; while in the object arise *colours, tones,* &c. Some objects consist of slow motion, e. g. those which we call objects of touch. These produce their effects only on what is near them. Others are of quick motion, and

Margin note: of sensation as element in cognition from the point of view πάντα ῥεῖ. Sensation consists in mere becoming: in relations of a merely transitory kind. The αἴσθησις and αἰσθητόν, with ὁ αἰσθανόμενος, are thus lost in the flux of an ever changing process. Explanation from this point of view of what is meant by object and organ of sense: as well as by the sensible qualities commonly ascribed to things.

---

[1] The Protagoreo-Heraclitean scepticism, which stimulated Plato to epistemology, is also most fruitful for psychological speculation. That of Gorgias, on the other hand, is metaphysical in the main, and of little help for psychology. A perfect epistemology must have sounded the depths of sensational scepticism.

[2] *Theaetet.* 156 A τῆς δὲ κινήσεως δύο εἴδη, πλήθει μὲν ἄπειρον ἑκάτερον, δύναμιν δὲ τὸ μὲν ποιεῖν ἔχον, τὸ δὲ πάσχειν.

[3] It will be observed that Aristotle in the same way fixes the relation of object to organ as active to passive.

reach far; such are the objects of sight. The above results, however, viz. sensation in the organ and quality in the object, occur only in the said contact, and last only while it lasts. The eye does not see when not affected by colour; the object is without colour when not seen by an eye. Nothing therefore is or becomes what it is or becomes for itself and in itself, but only in relation to the subject perceiving; and the object presents itself differently to the subject according to the varying constitution of this subject. Things are for each man what they appear to him; and they necessarily appear to him according to his state or condition at the time. There is no objective truth. There are no universally valid propositions: no science, but only opinion [1].

§ 11. Thus Plato in the operations of sense *per se* finds, according to the above doctrine of Protagoras, nothing fixed or stable, which could form the basis of knowledge. Nor can we doubt that if he had stopped at the point of view of empirical psychology, as he conceived it, he would have been a devoted and enthusiastic follower of Heraclitus and Protagoras. He constructed, however, an epistemology by which he rescued the work of thought and belief from this disordered and chaotic condition. He was unable to discover in sense-perception *per se* any ποῦ στῶ—any fixed point to which the scattered data of sense could rally [2], and which could therefore constitute a starting-point for science. He asked himself the question how the interaction of subject and object in sense-perception *per se* differs from the physical interaction between things in nature, and was convinced that, for the school of Heraclitus and Protagoras at all events, there is *no* difference. One cannot read Plato's energetic and eloquent words without perceiving that up to the present stage of the argument he is with Protagoras heart and soul. Here then we discover a wide gulf separating him from his pupil, Aristotle. The latter did not think it necessary to go outside the province

Side note: It was Plato's purpose to construct a system of epistemology which should replace the despair of knowledge thus produced. For the school of Protagoras the interaction of *percipiens* and *percipiendum* does not differ from purely physical interaction. Sensation did not for Plato, as for Aristotle, contain in itself a principle of synthesis. For the basis of

---

[1] Plato, *Theaetet.* 156 A–157 C; Zeller, *Pre-Socratics*, (E. Tr.) ii. 449.
[2] Cf. Arist. *An. Post.* 100ª 11.

of perception itself to discover a germ of the synthetic power which should lay the foundation of *experience* ; an experience capable of being developed, under the presiding help of universal conceptions, into *science*. Having no conception of a κοινὴ αἴσθησις, or synthetic faculty of sense, Plato treated the subject of αἴσθησις with scant respect, being chiefly interested always, wherever he returns to it, in showing how untrustworthy it is as an element of knowledge. He did not find in it the characteristics which Aristotle found—critical and comparative power, proportionality, the quality of μεσότης. Aristotle brought downwards to sense the characteristics of intelligence. He could not assent to the theory of a complete breach between the lower and the higher faculties of mind. Plato denuded sense of all synthetic power, and, for the explanation of the possibility of scientific knowledge, which he as well as his pupil had at heart, had to fall back altogether upon the activity of the understanding. How the sensibility and the understanding, having in this way no principle of community between them, should be harmonized, was a question which Plato could hardly answer. Aristotle tried to solve it by endowing sense with synthetic faculty, which he ascribed, as we shall see, to that particular department which he calls the κοινὴ αἴσθησις. Thus he tried to fill the breach which Plato had made. He saw that a theory of mind, which ignores the activity and implicit generality of sense, is as false as one which disregards or denies the all-regulating power of reason. Plato's idealism had not succeeded in penetrating to the dark recesses of sense ; that of Aristotle, no less lofty but far more attentive to the details of concrete living experience, was at least a deliberate attempt to interpret sense in terms of reason.

objective knowledge Plato looked altogether to understanding and reason, which he sharply differentiated from sensation. Thus he was forced to admit, when brought face to face with the question of our chapter—what is the feature in sensation generally which distinguishes it from physical interaction? that there is no such feature.

Aristotle.

§ 12. It will be found that there is, according to Aristotle, a complete parallelism between at least the sentient soul, as a whole, and any one of its so-called parts ; also between the bodily organism which is the instrument of the former,

Parallelism of whole and part in sentient soul; also

in bodily organism as its instrument. Sentient soul to body as *form* to *matter*. Sensation generally: the faculty of apprehending the form of objects without their matter. This true of the sentient soul and body as whole; and also of each 'part' of soul and the organ of this. The distinction of form from matter has both a physical and a non-physical aspect, and so introduces us to a way of differentiating the relation involved in sensation from a merely physical, e. g. mechanical, relation between bodies.

and the particular portion of the body which forms the instrument of the latter. In consequence of this parallelism Aristotle can illustrate, as he does, his conception of soul as entelecheia of body by comparison with visual power, as entelecheia of the eye. In order, therefore, to ascertain what his conception was of the characteristic of sensation generally, in which, while all its forms agree, they all differ from merely physical operations, we shall not only consider what he says directly on the latter point, but also what he says of the sentient soul as a whole, so far as it bears upon our question. I say the *sentient* soul; because difficulties arise as to the intellectual functions and their connexion for Aristotle with the sensory functions, owing to which we can scarcely adduce his general account of ψυχή as a whole in order to illustrate his view of the meaning of sense. It is in developing his view of the relation of soul—especially the sentient—with body in general, that he expounds the idea of the soul being to the body as *form* is to *matter* ; on which idea his explanation of sensation in general rests also. For him the first essential characteristic of sensation in general is the power of sense to apprehend the form of objects without the matter[1]. In this all the senses, in all their manifestations, agree with one another; and in this essential characteristic they differ from inanimate things operating on one another according to merely physical laws. The distinction between form and matter, seeming the key to that between psychical and non-psychical, is fundamental in the philosophy of Aristotle; and although it connects itself properly with his metaphysics it is also of essential importance, if we are to understand his psychology of sense, that we should clearly conceive the way in which he applies this distinction, *first*, to the relation of soul and body, or of sense and sense-organ; and *secondly*, to the relation of sensory

---

[1] He agrees with Plato in the definition of αἴσθησις as a κίνησίς τις διὰ τοῦ σώματος τῆς ψυχῆς, but this definition, having served its purpose of connecting empirical psychology with the sphere of physics, is left behind, and a more characteristic and fruitful definition is sought for. Cf. 436ᵇ 6 with 424ᵃ 16; Zeller, *Arist.* (E. Tr.) ii. 58.

apprehension—sense-perception—wherein the knowing subject perceives by sense the qualities of an object. Soul *is*
form and *apprehends* form; and the same is true of each
sense-organ (*qua* animate) and its function. For we are
seeking, be it remembered, the respect in which the relation
of *percipiens* to *percipiendum* differs, according to Aristotle,
from a merely physical, e. g. a mechanical, relation.

§ 13. Aristotle [1] arrives at his most comprehensive view
of ψυχή as follows. There is a class of things called
substances (οὐσίαι), i. e. determinately existing things.
Any such thing may be viewed (*a*) as to its matter, (*b*) as
to its form, (*c*) as to the whole (οὐσία) which results from
the union of the two [2]. Matter is mere *potentiality*, form
actuality. The latter may have grades, e. g. a lower which
corresponds to ἐπιστήμη, and a higher which corresponds to
τὸ θεωρεῖν, or the *exercise* of ἐπιστήμη. Now the commonest
instances of substances are furnished by bodies, especially
natural bodies (φυσικὰ σώματα). Of the latter some have
life—by this being meant a process involving the maintenance of *nutrition, growth*, and *decay* in such bodies.
Every natural body having life is an οὐσία, with all the
implications above stated. Such living body cannot *per se*
(sc. *qua* body) constitute soul. The body *qua* matter is the
subiectum (τὸ ὑποκείμενον); while the soul, in virtue of which
the body is qualified as living, if a substance at all, is such
in only a formal sense—οὐσία ἡ κατὰ λόγον, or εἶδος. Such
substance as this—the οὐσία ἡ κατὰ λόγον or εἶδος—is the ἐντελέχεια, or *actualization*, as distinguished from the δύναμις,
or *potentiality*, of the living body. Bearing in mind that
ἐντελέχεια has the grades above illustrated, the one corresponding to ἐπιστήμη, the other to τὸ θεωρεῖν, we next observe

Marginal note: Definition of soul rests on two conceptions, (*a*) that of the analysis of οὐσία into *form* and *matter*, (*b*) that of *actuality* as distinct from *potentiality*.

---

[1] 412ᵃ 1–414ᵃ 28. οὐσία in Aristotle generally = anything subsisting for
itself, forming no inherent part or attribute of anything else, and not
requiring a substratum different from itself. πρῶται οὐσίαι are distinguished from δεύτεραι οὐσίαι as individuals from *genera* and *species*.
The use of the term οὐσία respecting ψυχή must be carefully watched at
the point where ψυχή comes to be spoken of as the οὐσία ἡ κατὰ λόγον
of the ζῷον.

[2] ὕλη is used first by Aristotle as the philosophical term for 'matter';
but such usage *might* have been suggested by Plato, *Tim.* 69 A.

that, as ἐντελέχεια of living body, ψυχή answers to the former of these. For the possession of soul, by a living body, is consistent with the non-exercise of its faculties, for instance, during sleep. The capacity for such exercise is chronologically prior, in the individual, to the actual exercise. Hence we call soul the *first* ἐντελέχεια of a living body, or of a natural body capable of living. Such potency or capacity belongs to bodies which possess organs, and therefore to vegetable as well as to animal bodies. Thus we formulate a definition sufficiently general to apply to all kinds of soul, if we state that it is *the first ἐντελέχεια of a natural organic body*. With this definition as expressing the nature of the *sentient* soul only we shall here have to do.

The terms form and matter derived from objects of sense. Form and matter even in these are only notionally distinguishable. But this notional distinction imparts the character of idealism to all experience from its very inception onwards and upwards.

§ 14. Without clearly understanding Aristotle's distinction of matter and form, we could not understand his theory of sensation. There is one fixed word for matter, viz. ὕλη, but form is expressed by several : σχῆμα, μορφή, εἶδος. From the frequent use of the two first, it would appear that the philosophical distinction was imported from the ordinary or vulgar use of μορφή and ὕλη, to distinguish the material of an object from its shape, by which, therefore, this distinction in its primary form is best illustrated. A lump of wax has always and must have some shape. The shape and the wax are inseparable except by abstraction—an act of thinking. The shape must have a matter or material, the material a shape. The shape and material are different indeed, but do not differ as, e. g., two lumps of wax would differ from one another. These are locally and really separable ; not so the shape and material of one lump. The shape of one lump of wax cannot perish without the material sharing its fate ; nor can the material perish— it cannot even be thought away—without the shape also vanishing. If the lump ceases to have any form it ceases to exist ; and so, too, if it ceases to have any matter. We may name the shape and the material separately, and by different names, but we cannot even imagine a material substance without some shape, or a shape without material. Matter and form are thus correlative terms notionally (λόγῳ) distinct, i. e.

distinguishable by an effort of mental abstraction, and by this only. Such distinction borrowed from objects in space was transferred by Aristotle to every concrete individual; not merely those possessing physical properties, but all others, including the entities with which metaphysical speculation undertakes to deal. In regard to every individual thing (τόδε τι) of any kind, therefore, Aristotle distinguishes (1) its matter, (2) its form, (3) the composite consisting of both. Neither matter nor form by itself constitutes the individual—the τόδε τι. It is constituted or consists of both together. This distinction of form and matter is, as made by reason or thought, the first step towards the idealizing of experience, and the introduction, or discernment, of the characteristic which distinguishes sensation generally from purely mechanical or other kinds of physical interaction. In virtue of it, or our power to make it, experience and all that it can contain is from the first endowed with a character derived from mind.

§ 15. To form Aristotle gives precedence in rank and importance. The reason of this for him is, no doubt, that form, though itself unknowable in nature apart from matter, is what renders things capable of being known. All the determinate qualities of things, all the predicates by which they can be the subject of conversation or reasoning, come under the head of form. The determination of the 'form' of a thing is a progress in the complete knowledge of that thing. The reverse process, by which knowledge of form is obliterated, would ultimately leave our minds a blank. For *mere* matter is a mere negative. It has *per se* no predicates, and nothing real could be known about it. As, therefore, scientific and all knowledge advances *pari passu* with further determination of the form of a subject—and as science confined to mere matter would be impossible—indeed inconceivable—it was natural for Aristotle to give the higher place in dignity to form as compared with matter. Form is on the side of clearness and knowledge; matter, on that of confusion and ignorance. But for a single *res completa*, or for a real world, we, in Aristotle's opinion, require both.

The progress of knowledge is a progressive 'information' of matter. Of mere matter, i.e. matter without form, we have no apprehension. Hence form is for Aristotle the 'higher' of the two. By form we know matter: not vice versa.

Affinity be-
tween the
two dis-
tinctions
of (a)
actuality
and poten-
tiality, (b)
form and
matter.
The fact
of soul
being
actualiza-
tion of a
body with
definite po-
tentialities
renders
such ideas
as that of
transmigra-
tion of
souls
absurd.

§ 16. The distinction between matter and form is allied to the distinction between potentiality (δύναμις) and actuality (ἐνέργεια, ἐντελέχεια [1]), also of capital importance in Aristotle. It is not hard to see the affinity between the two distinctions. Matter is that which exists only potentially; before anything can be a τόδε τι—can exist at a particular place in a particular time—it must have *form*. Unformed matter is something which can only be conceived as possibility: something which is conceived as nothing yet, but which is capable of *becoming* anything, we do not yet know what, according to the form it may assume. Nature exhibits no instances of such potentiality, such unformed matter, in the absolute sense; but relatively speaking, many natural things illustrate it. It is seen especially in the processes of organic life, such as that of growth from seed to tree. The seed is the tree in potency, or formed imperfectly; the tree is the seed in actuality, or perfectly formed. The process is one from matter less formed to matter more formed; but even at the lowest steps we can find no matter that has not already some form. When the potentiality of some particular matter has been completely actualized, it has, in Aristotle's phrase, reached its ἐντελέχεια—its final consummation. In the successive steps of the process, however, each higher stage is ἐνέργεια compared to the lower; δύναμις as compared to those above it. The idea of the soul entering into, or passing by transmigration through, a variety of different bodies is absurd. It is not with every casual body that a given form of soul will unite itself. To suppose otherwise is as erroneous as to suppose that a carpenter could do his work with a flute as well as with hammer or saw [2].

The σῶμα
has an
existence
of its own;
it is a τόδε
τι—an
οὐσία

§ 17. Accordingly we may see what Aristotle meant by speaking of the animate body as οὐσία of which the σῶμα *per se* is the ὕλη, while the soul *per se* is εἶδος. For the σῶμα to have life is to have realized in it certain antecedent potentialities, which belonged to the ὕλη from which the living

---

[1] The difference of these may be neglected here.
[2] Cf. 407b 14-25. This is directed against the Pythagoreans and Plato's *Phaedo*.

body has sprung. Ψυχή is the realization of such potentialities. The ζῷον is the τόδε τι. Its ψυχή is that in virtue of which it lives—that which is the seal and mark of the potentialities of its σῶμα *qua* ὕλη. The soul is not a τόδε τι, neither is it something joined to, and capable of separation from, the σῶμα [1], any more than form generally from matter. It is ψυχή, however, that gives meaning or intelligibility to the organic body whose functions are adapted to its maintenance, and employed for its sake. Thus the ἐντελέχεια and the τέλος are identical. While, however, the ψυχή is no τόδε τι—no concrete individual thing—we cannot say this of σῶμα. The latter indeed taken *per se*, and without soul as a dead body might be, is no longer what it was when animated or fit for the habitation of soul; it is no more an animal body than an ὀφθαλμός deprived or incapable of vision (ὄψις), such as an eye of stone, would be an eye in the same sense as one with its native power. It could now have the name it formerly bore only in an ambiguous or homonymous way. Yet, though not the same as what it was, it is a concrete individual thing; which could not be said of its ἐντελέχεια, the ψυχή *per se*, out of relation to the σῶμα. The body when lifeless is still a substance, a τόδε τι, though no longer ἔμψυχόν τι. Therefore body cannot be said to be itself the εἶδος or form of soul. In other words soul cannot be explained materially— as consisting of any form of material body however fine. Body is always of the nature of a *subiectum*: the subject of attributes and predicates: not itself an attribute or predicate. We can no more say that body *is* the soul of an animal, than we could say that the wax *is* the shape or form of the cube of wax before us. Its cubicalness is a predicate of the wax as a subject, and this relation is irreversible. Thus, and for the analogous reason, we could not say that in a given living ζῷον the body is the soul, or in other words, that the soul is material. The cubicalness is a quality predicable of the wax, and now belonging to it

(marginal note:) having not only matter, but a *form* too. Such is the case with even a lump of matter of any sort. The soul is not an οὐσία of this kind: but only an οὐσία κατὰ λόγον: a *notional* entity. The body is to it only τὸ ὑποκείμενον, hence body cannot be the form of soul, and soul cannot be explained on a purely materialistic hypothesis. To get an explanation of the attributes of the body, we must look to its form—its οὐσία κατὰ λόγον— soul.

---

[1] In this Aristotle seems to attack the very basis of the main argument of Plato's *Phaedo*.

as the result of a process of change. Just so in the living body, its soul—its being alive—is the quality which informs and determines it to its intelligible character.

Condition of a body's having life is that it should have organs: and in the case of the animal soul, organs of sense.

§ 18. The soul, then, is the actualization of the potentiality of life, and this it is in virtue of its being the form of the living body. But it is only a stage—the first stage —in a process of actualization. With it ends the process upwards from lifeless ὕλη to ὕλη which now lives ; and with it again begins another process upwards from *mere* life, as in vegetables, to the life which has intelligence (νοῦς) in its sublimest energy. That the body should live, organs are necessary. That further determination or development of soul should take place—that, for example, it should rise from its lowest grade such as plants exhibit to the next above it—that of sentiency which all animals exhibit— further organs are necessary. These are the instruments of its activity or functionality: the organs of sense.

Parallel between whole animated body and each of its sentient organs.

Each sense is the πρώτη ἐντελέχεια of each sensory organ: the whole ψυχὴ αἰσθητική is that of the whole animated organism.

Just as the soul is the first entelechy of living body, so each sense is the first entelechy of the organ adapted to its function. Each sense is the form, while its organ (a portion of the body) is the matter. The senses all postulate the living body as their substratum or ground of possibility ; in their manifestations of function, and in their development, they each offer the closest parallel to the sentient soul as a whole in its relationship to the body as a whole. This parallelism is stated by Aristotle himself. As each sensory organ is organic to that sense, so the whole σῶμα is organic to ψυχή, and is qualified as such an ὄργανον[1]. The soul, not being material, is not a magnitude. Again, we must not ask whether soul and body are one, any more than whether the wax and the figure it bears are one, or generally whether any material and that of which it is the material are one. Soul is called an οὐσία—a sub-

Relation of soul and body. We cannot say that they are *one* and *the same thing*: neither can we say that they are *two things*.

---

[1] Cf. 645ᵇ 14 ἐπεὶ δὲ τὸ μὲν ὄργανον πᾶν ἕνεκά του, τῶν δὲ τοῦ σώματος μορίων ἕκαστον ἕνεκά του, τὸ δὲ οὗ ἕνεκα πρᾶξίς τις, φανερὸν ὅτι καὶ τὸ σύνολον σῶμα συνέστηκε πράξεώς τινος ἕνεκα πλήρους. This is confined by Aristotle to the lower part of ψυχή, and does not apply to the distinctively noëtic part, which is *possibly* χωριστόν, and which belongs to the subject of πρώτη φιλοσοφία, not of 'physics.'

stance or essence—but this must not be taken to mean that
it is a τόδε τι. It is an οὐσία ἡ κατὰ τὸν λόγον—an ideal or
formal substance—the actualization of the idea underlying
the potentiality of body to live. Without it the living body
would no longer live: its structure and organs would have
lost their meaning, or would not fulfil the idea˙ which
informs them. '. . . We can see this[1] by comparison with
certain particular organs and their functions. If the eye
(ὀφθαλμός) were an animal (ζῷον), then, by analogy, its soul
would be its visual faculty (ἡ ὄψις). This (ὄψις) is the
form or ideal substance of the eye (οὐσία ὀφθαλμοῦ ἡ κατὰ
τὸν λόγον). So the eye is the matter (ὕλη) of the visual
faculty (ὄψεως), lacking which it would be an eye no longer
in the same meaning of the term as before, but only in
some other, just as an eye carved in stone or painted
in a picture might bear this name. We must conceive
what is true, in this manner, of the part as true also of the
living body as a whole. For as each sensory function is
to its sensory organ, so is the whole sentient soul (ἡ ὅλη
αἴσθησις) to the whole sentient body as such. . . . As
seeing (ὅρασις) is the full consummation (ἐντελέχεια) of the
potentiality of the eye, so waking[2] is that of the potentiality of the whole living body. The soul is the realization
of the potentiality of the organic body, in the way in which
vision as a power is that of the organ of vision. Considered *per se*, the body is that which has only the potency
of living. As the " pupil " and its visive function (ὄψις)
together make up the eye (ὀφθαλμός), so the soul and the
body together make up the animal (τὸ ζῷον).'

§ 19. The foregoing has been needful to prepare us in
some measure to understand the comparatively brief section[3] in which Aristotle, having previously given a detailed
account of the special senses, recurs to the theme of
sensation generally, in order to state the characteristics
which distinguish it from all material interaction. Αἴσθησις

Marginal notes: As pupil and vision the living eye, so body and soul make up the ζῷον.

Marginal notes: Thus in sense-per-ception form apprehends form: the soul (which is the form of body)

---

[1] 412ᵇ 6–28.
[2] ἐγρήγορσις, what we might call complete consciousness.
[3] 424ᵃ 16–ᵇ 3.

through its parts (which are the form of their respective bodily organs) apprehends the form (i. e. the qualities) of the objects of sense-perception. But sense only apprehends form in the individual, not in universals. Implicit universality of sense. The particular sensory organ (as distinct from the function) is the part of the living animal in which appears this faculty of apprehending form apart from matter. Relation of particular organ to its faculty like that of body to

is, he says[1], a form of γνῶσις. We have to conceive αἴσθησις in general as the power which animals possess, in virtue of their ψυχή and αἰσθητήρια, of apprehending sensible objects in their forms without their matter[2], as wax takes the mark (σημεῖον) of the seal ring, without taking the iron or the gold of which the latter may be composed, but quite indifferently as to this material element. In the same, or in an analogous, way, sense-perception is related to its objects. It apprehends the colour or taste, or other sensible quality of things, being affected by each thing not in so far as such thing is a τόδε τι or substance, but in so far as it is a τοιονδί, i. e. possesses particular *quality*[3]. For form apprehends form. The soul, which is the οὐσία ἡ κατὰ λόγον of the whole animate body, *informs* the sensory organ ; and the latter by its form becomes apprehensive of the forms of objects. Though sense thus grasps the form in objects, it differs from intelligence in not grasping the universal as such. It only seizes the form in the individual τόδε τι, i. e. in a given thing at a given time and place. Yet even so, we can observe the implicit universality of knowledge from its commencement in sensible experience. For even in the individual, however limited as to place and time, the form is implicitly universal ; and αἴσθησις, being not τοῦδέ τινος, but τοῦ τοιοῦδε[4], has the implicitly universal as its object. So much for the general character of αἴσθησις or sense-perception.

A sensory organ, on the other hand, in its primary[5] conception, is that part of a living animal in which the faculty of apprehending form apart from matter appears. This faculty depends on the constitution of the organ : no part can be such an organ unless it occupies the position of a *mean* between the qualities which are extremes in the scale of sense to which it refers[6]. The sense (αἴσθησις) and

---

[1] 731ᵃ 33 γνῶσίς τις, cf. 458ᵇ 2, 432ᵃ 16.

[2] 424ᵃ 17 τὸ δεκτικὸν τῶν αἰσθητῶν εἰδῶν ἄνευ τῆς ὕλης, cf. 425ᵇ 23, 434ᵃ 29. εἰδῶν in 424ᵃ 17, required on general grounds, and supported by its use in 434ᵃ 29, is certainly sound.

[3] οὐχ ᾗ ἕκαστον ἐκείνων λέγεται ἀλλ' ᾗ τοιονδὶ καὶ κατὰ τὸν λόγον.

[4] *Vide* 87ᵇ 28, 100ᵃ 16. [5] 424ᵃ 24.

[6] For this thought that the organ must be a mean between the

its organ (αἰσθητήριον) are in a way the same and yet not the same [1]. They are different in conception (λόγῳ) or in their way of manifesting themselves (τῷ εἶναι). That which perceives is, *qua* part of σῶμα, a μέγεθος or magnitude; but the essential idea or function of perception is no magnitude or material, but a ratio or power of some kind inherent in the perceiving organ [2]. From these considerations (viz. that the faculty of a sense-organ depends on its occupying a due mean or proportion between any two different objects in its scale) it is plain why excessive impressions from sensible objects of any sense injure or destroy the organ. If the motion set up by the object is too strong for the organ, the essential mean or proportion is disturbed; and this being disturbed, sensory power is lost; just as the musical quality of a lyre is lost if it be struck so violently as to break the strings [3].

§ 20. The fact that there are three kinds of soul—the *nutrient* (and generative), the *sentient* (and motor), and the *intellectual*—is consistent with the unity of soul as a whole. Aristotle illustrates this by reference to the unity of higher geometrical figures, which still implicitly contain the lower. Thus the quadrilateral is one, though it contains the trilateral. The nutrient is contained in the sentient soul; the nutrient and sentient in the intellectual; yet the

Margin notes: sentient soul as whole. The organ, like the whole σῶμα, is a magnitude: the faculty is not, but rather a *ratio* or *proportion*.

Margin notes: Unity of soul consistent with plurality and diversity of its faculties. Illustrated by geometrical figures like

extremes—or any two different qualities—in the scale of αἰσθητά to which it refers, and hence must not itself have any of the qualities in a determinate degree, but only in such a way as to be relatively, e. g., cold as compared with a hot object, hot as compared with a cold, cf. Plato, *Tim.* 50 D–E; also Arist. 429ᵃ 15 seqq., and § 24 *infra*.

[1] Just as are ψυχή and σῶμα.

[2] ἀλλὰ λόγος τις καὶ δύναμις ἐκείνου. Editors make ἐκείνου = τοῦ αἰσθητοῦ; Bonitz (*Ind.* 437ᵃ 48) takes it as = μεγέθους, and (*Ind.* 206ᵇ 17) as = τοῦ αἰσθητοῦ. It appears to me to be a subjective genitive, referring to τὸ αἰσθανόμενον in ᵃ 26, i.e. the subject-organ, whose perceiving power consists in this λόγος. The mistake which Aristotle here aims at correcting is like that of the one who should regard the musical function of a lyre as a magnitude, and identify this function with the strings, pegs, and material framework of the lyre, omitting to take account of, e. g., the ratios of the strings on which the musical function depends.

[3] 424ᵃ 31.

the quadri-sentient and intellectual are each actually one, though
lateral, potentially several ; just as the quadrilateral is actually one
which is
one though though capable of division into two trilaterals. Plants, as
made up
of two well as animals, have life, and therefore soul. Aristotle
trilaterals. denies them, however, even the rudiments of sensation,
Plants have
soul, but pointing out the reason (as he regards it) why they cannot
not sense. possibly possess this. No doubt they are (he says) affected,
Reason of
this. Thus e. g., by the cold and hot, i. e. they are cooled and
Aristotle
answers the heated. Hence one might overhastily assume that they
question as have a perception of cold and hot. This would be a mis-
to the com-
mon and take. Their mode of affection is not that of animals. The
peculiar plant lacks the primary requisites of sense. Plants have no
feature in
sensation organs possessing the essential μεσότης, which would give
generally. discrimination of the degrees of heat ; and therefore they
are incapable of apprehending the form of heat apart from
the matter of the hot thing. When plants come into relation
with external objects, to be affected by these they must
receive the matter with the form [1]. Thus a plant's touching
is but physical contact. As sense apprehends material
objects in their form, and as intellect apprehends immaterial
objects, so plants apprehend the material object only in its
matter. Thus it is that Aristotle answers the question : what
is the feature common and peculiar to sensation generally—
the feature in which all sensory functions agree, and in which
all differ from purely physical interaction ? Thanks to the
fact of the sensory organ being (or having in its constitution)
a λόγος of all the differences possible in its sensible province,
so that it can present itself, as a mean, to any two such
differences and discriminate them, it is capable of apprehending the form, i. e. the qualities, of objects apart from
their matter. Thus the ἀλλοίωσις involved in sensation is
no purely physical change. It is a process in which the first
ἐντελέχεια of the organ—its potentiality of such apprehension
—is converted into the second ἐντελέχεια or actualization of
its potentiality.

Sensation   § 21. For all αἴσθησις involves ἀλλοίωσις [2] of the organ by
involves a
change in the object. When the hand is plunged into water of exactly

---

[1] De An. ii. 12. 424ª 16–ᵇ 3.   [2] For §§ 21–22, cf. 416ᵇ 32–418ª 4.

its own temperature, it feels the water neither hot nor cold[1]. the percipient. In determining the nature of this ἀλλοίωσις or qualitative change of the *percipiens*, Aristotle also settles (to his satisfaction) the old question, whether perception is effected by a relation of like to like or of unlike to unlike. This he does in such a way as to reconcile the apparently inconsistent theories of, e. g., Empedocles and Anaxagoras on this point. A similar question is, he says, possible respecting the relation between the body nourished and the food which nourishes it. Is nutrition effected by the agency of like on like or of unlike on unlike? Aristotle replies : there is a previous question as to what exactly nutriment *is*. Is it the digested or undigested food? Manifestly it is the former. The question, therefore, may be answered in two ways. If by nutriment we mean food not yet digested, then nutrition is effected by the agency of unlike upon unlike; but if by nutriment we mean digested food, nutrition is effected by the agency of like upon like. A process of ἀλλοίωσις has intervened between the taking of the food and its thorough digestion, in which process the food which was at first unlike the body has become assimilated to it: the unlike has become like[2]. Thus he introduces his settlement of the analogous question respecting perception. The object sets up a change in the percipient. The former is in this relation active, the latter passive. The perception for which the subject is naturally fitted is developed into actuality by the object perceived, the form of the object being impressed upon the percipient, i. e. the qualities of the object which the percipient is adapted to perceive being apprehended by it. This relationship between the two is the kind of qualitative change—ἀλλοίωσις—in which perception is developed. At the moment when this qualitative change, produced in the percipient by the object, begins—i. e. when the former commences to be affected— then the object is unlike the percipient; when, however, the ἀλλοίωσις has completed itself and the *percipiendum* has become a *perceptum*, in the moment of actualized per-

Marginal notes: the percipient. Nature of this change. Perception not simply relation of like to like, or of unlike to unlike. It is a relation in which what was unlike becomes like. Illustration from nutrition and 'assimilation' of food.

---

[1] 424ᵃ 2 seqq.　　　　[2] Cf. *de An.* ii. 4. 416ᵇ 3–10.

ception, the percipient has become like the object. The latter has assimilated the former to itself. Both are now qualitatively alike. The question, therefore, whether perception results from an affection of unlike by unlike (as Anaxagoras held), or of like by like (as Empedocles believed), admits of being answered either way according as one regards the initial or the final stage in the process of ἀλλοίωσις in which perception consists. If the former is thought of, Anaxagoras' answer would be correct; if the latter, the correct answer would be that of Empedocles[1]. A process has intervened in this case as in that of nutrition between the incipiency and the termination of the relation between agent and patient. The organ therefore is qualitatively changed.

The sensory faculty is (like the sentient ψυχή as whole) a πρώτη ἐντελέχεια, prior to the moment of perception, in which its relative potentiality is actualized. The 'object in general' is a thing per se; it exists with qualities capable of being perceived, even when not perceived. Thought can supply its own objects—universals.

§ 22. This change will be understood only if we remember that the sensory faculty is nothing but a faculty until confronted by its object. It is something which exists only potentially, until the object stimulates it. By this stimulation it acquires actuality. It must wait for an object, i.e. something different from itself, in order to be actualized, i. e. to perceive. Were this not so, the sensory organs would perceive themselves; which, however, they can no more do than an axe or saw can cut itself. The process of ἀλλοίωσις, which we have been describing here, is a process from the sense δυνάμει to the sense ἐνεργείᾳ. The ἐνέργεια or ἐντελέχεια, with which a sense-organ is primarily endowed, is that which it derives from, or has in virtue of, the whole ψυχή, of which it is a particular organ. Such ἐνέργεια is, however, only the πρώτη ἐνέργεια (or ἐντελέχεια) of the organ, as capable of functioning, i. e. as αἰσθητικόν. This first grade of actuality is itself potentiality as compared with higher grades. The case is (in reference to the particular part of soul engaged in one sense, as well as in reference to the whole sentient soul) like that of ἐπιστήμη and θεωρία, to use Aristotle's illustration. If a person is a scholar or man of science, he is in virtue of this able to exhibit or apply knowledge in a certain way; given certain conditions,

---

[1] 418ᵃ 4 πάσχει μὲν οὐχ ὅμοιον ὄν, πεπονθὸς δ' ὡμοίωται καὶ ἔστιν οἶον ἐκεῖνο. Galen, De Placit. Hipp. et Plat., § 636, remarks that sense-perception is not, as some say, an ἀλλοίωσις, but rather a διάγνωσις ἀλλοιώσεως.

he does so. This potentiality of his corresponds to the grade which every sensory faculty occupies in the absence of an object to stimulate its organ. On the other hand, when such a person is exercising his knowledge in some particular concrete case [1], he furnishes the parallel for the actually percipient organ of sense after it has been affected, and while yet affected, by its object. A change has passed over the organ of sense, but not one which impairs it. There are two kinds of change which a thing may undergo; one in a direction depriving it of its qualities or functions; the other in the way of developing or realizing its powers [2]. The change which the percipient undergoes, when affected by the *percipiendum*, is a change of the latter sort, one which brings the faculty from potentiality to actual realization, like the change from ἐπιστήμη to θεωρία which fulfils the potency of the ἐπιστήμων.

Sense must wait to be affected by its objects —individuals; the universals are within the soul. The particulars or individuals are outside the soul, and outside the body. Only the form of them is inside the soul, and this first at moment of perception.

The object which causes the change has its own actual existence in the world, apart from the relation of sense. It would exist even if no one perceived it. It actually exists, and is potentially perceptible. So, conceived in relation to an absent object, the sensory organ is perceptive, or capable of perceiving it. The object has its own actual qualities [3]—its form, which sense finds in it at the moment of perception. Thus, for Aristotle, the object is what Kant would call a *Ding an sich*.

Between sense and thought, however, though paralleled for the above illustration, there is the great difference that thought can discover its own objects within itself, for it deals with universals (τὰ καθόλου). Sense-perception must await stimulation from without, as it can only deal with particulars (τὰ καθ' ἕκαστον) [4]. Universals are in a manner within the soul itself [5]. Hence it follows that thinking is in one's

---

[1] 417ᵃ 29 ὁ ἤδη θεωρῶν ἐντελεχείᾳ ὤν, καὶ κυρίως ἐπιστάμενος τόδε τὸ Α.

[2] δύο τρόπους εἶναι τῆς ἀλλοιώσεως, τήν τε ἐπὶ τὰς στερητικὰς διαθέσεις μεταβολὴν καὶ τὴν ἐπὶ τὰς ἕξεις καὶ τὴν φύσιν 417ᵇ 14–16.

[3] Cf. 426ᵃ 20–25, 7ᵇ 35 seqq., and 1010ᵇ 36.

[4] τοῦ μὲν τὰ ποιητικὰ τῆς ἐνεργείας ἔξωθεν, τὸ ὁρατὸν καὶ ἀκουστόν, ὁμοίως δὲ καὶ τὰ λοιπὰ τῶν αἰσθητῶν.

[5] ἡ δ' ἐπιστήμη τῶν κοθόλου, ταῦτα δ' ἐν αὐτῇ πώς ἐστι τῇ ψυχῇ.

own power when one wishes to make the effort; but it is not in one's power to perceive always when he wishes to do so. There must be present a particular object of perception before this faculty of sense can be realized [1].

§ 23. We have seen that, as the nutrient soul can exist without the sentient, but the latter cannot exist without the former, so the sense of touch can exist without the other senses, while without it these cannot exist [2]. And we may assume that as the nutrient soul is present with and accompanies—or is the foundation of—every exercise of the sentient, so the sense of touch is implied as at least accompanying every exercise of the other senses. What then is its exact relation to each of them in actual exercise? or has it any? Are we to suppose that it *merely* accompanies, and has no assignable office? Such was not the opinion of Democritus, as we have already observed. Can it have really been the opinion of Aristotle himself? He allows that taste is a modification of touch. When we come to deal with the common sense—that central bureau which receives and elaborates the reports of the several senses—we shall have reason to think that on this point the two philosophers agreed. At all events, Aristotle's theory of the evolution of soul requires a close relation between touch and the other senses of which it is the pre-supposition (see p. 248, n. 1). The ascending forms of soul are like the ascending figures. As the triangle is implicit in the tetragon, so the faculty of nutrition—or the nutrient soul—is implicit in the sentient soul. We seem to be led up by him to the parallel thought of an ascending scale within the sentient soul—a scale which reaches from ἁφή at its lower to ὄψις at its higher extremity. We have an involution of the sense of touching in every other sense, however highly developed [3]. But Aristotle does no more than bring us to the threshold of this conception. He nowhere (except in the case of γεῦσις, which is ἁφή τις) explicitly defines the relationship between the other senses successively and that of touch. Yet we may, with much

Sense of touch can exist apart from the other senses : not these without it. It is implied throughout the operation of all the higher senses, as nutrition is implied throughout all sensory life. Democritus held all the other senses to be differentiated from it. How far did Aristotle really (despite verbal protests) agree with Democritus here? Suggested order of senses in ascending scale according to the meaning of Aristotle : *touching, tasting, smelling, hearing, seeing.* A sense is higher the more

---

[1] *De An.* ii. 5. 417ᵇ 24.   [2] 415ᵃ 3-5.   [3] Cf. 435ᵃ 18.

probability, infer his view of their respective relationship to it, by simply reversing the order in which he arranges the senses for discussion. When he states [1] that ὄψις is the sense *par excellence,* he doubtless means that this sense, in a greater degree than any other, exhibits the power of apprehending form apart from matter. Touch possesses this power, but in the lowest degree. Taste comes—or would seem to come— next above touch, for sensations of taste proper are impossible without contact of the tongue with the sapid substance, and γεῦσις is ἀφή τις. It, however, superadds a determination of form foreign to mere touch *qua* touch : the sapid qualities of body are known through it alone, as they could not be by mere touch. Next in order as we go up comes smelling, which is allied on the one hand to tasting and touching— being subservient directly in its most important use to the purpose of tasting—and on the other hand to hearing and seeing, in virtue of its operating through a medium (τὸ ὑγρόν) with which the media of hearing and seeing are in a certain way identical. For the medium of hearing, viz. air, is ὑγρόν, and the ὑγρόν and the διαφανές, as we learn from the constitution of the κόρη, have much in common. Next above smelling comes hearing, and the scale culminates in the sense of seeing. Hearing apprehends less of the matter, more of the form of its object than smelling does : and the same can be said of seeing as compared with hearing. Seeing is the most pure—touching, the least pure—form of sense. Thus the progress in the ascending scale of sense is at the same time a progress towards the scale of intelligence, from the threshold of which again (if we can determine a threshold), we should proceed still upwards step by step guided by the same clue, the higher step being always that which leads towards the purer form—towards the universal. Finally, though νοῦς apprehends its objects only under conditions determined by perception, yet it endeavours to free them more and more from all such conditions.

§ 24. Each sense is capable of perceiving objects which

Side note: the form of the object without the matter is apprehended by it. This ascent brings us to the threshold of intelligence (as distinct from sense) i. e. of a νοῦς which strives to apprehend pure form.

Side note: Each sense is a μεσότης,

---

[1] 429ᵃ 2 ἡ ὄψις μάλιστα αἴσθησις.

and *therefore can discriminate contraries and differences in its modality. More detailed explanation of the μεσότης and the λόγος involved in each sensory faculty. Each αἴσθησις a formal dynamic unity. Each province, or modality, a generic unity. Basis of formal unity of each sense, the λόγος or μεσότης.*

are contraries—opposites in the same genus [1]. This power it owes to its involving what Aristotle calls a μεσότης between the opposite extremes in the scale to which its object belongs. To this its discriminative power is due [2]. For Aristotle this doctrine of μεσότης is of cardinal importance in the theory of sense-perception. Without understanding it we must fail to grasp his explanation of how αἴσθησις apprehends form without matter. Each αἴσθησις or sensory faculty is for him a unity [3], ruling as it were over its own province which is also one and consists of its αἰσθητά. The unity is, of course, qualitative or formal, not quantitative. That of the faculty is an unity δυνάμει; that of its province, an unity γένει. The *sensibilia* which constitute the province are all homogeneous *inter se*, and heterogeneous with those of every other sense. Thus seeing presides over or discerns (κρίνει) the province including colour [4]. Colour is a province lying between and bounded by the opposites white and black. These are one in kind, or genus, though opposite as species. Between these opposites come other species which mediate between them, and which Aristotle endeavoured to arrange in a scale of succession reaching continuously from the one opposite to the other. Seeing presides over all these species alike, comparing and distinguishing them. This power, he tells us, it possesses in virtue of its being a μεσότης or λόγος. It is a μεσότης *qua* standing in a middle character between both extremes—white and black—or between any other pair of different species or different colours in the scale, so that it can relate itself to either at the same time as to the other. It is a λόγος or ratio in the sense that it involves in its organ a λόγος τῆς μείξεως of the physical elements which constitute its αἰσθητά, and therefore is capable of taking the 'form' of

---

[1] 424ᵃ 10 ἔτι δ' ὥσπερ ὁρατοῦ καὶ ἀοράτου ἦν πως ἡ ὄψις, ὁμοίως δὲ καὶ αἱ λοιπαὶ τῶν ἀντικειμένων.

[2] 432ᵃ 16 τῷ κριτικῷ ὃ διανοίας ἔργον ἐστὶ καὶ αἰσθήσεως.

[3] For the difficulty which Aristotle finds in applying this to the sense of touch, see TOUCHING, §§ 9-10 *supra*.

[4] Besides colour there are other objects of seeing, viz. fire and the phosphorescents. These, though not possessing colour in the ordinary sense, have it in the same sense in which light has colour.

any of them indifferently [1]. So a lyre in tune is a μεσότης or λόγος to the variety of chords or airs which may be played upon it. It is capable of sounding high or low notes indifferently; and has in its tension, or in the relative tensions of its strings and of the frame on which they are strung, the due harmonic ratio to all the sound solicitations to which it may be called upon to respond. But until struck, the lyre is silent. That which entitles each sense [2] to be called *one*, and also constitutes the condition of its sensory power, is this form—this λόγος or μεσότης which characterizes it. Thus it is that Aristotle transforms the doctrine of Empedocles and others of his predecessors, viz. that each sense requires for its exercise a συμμετρία between the object and the organ; and that each is affected by the object either as its like or its unlike. Instead of a material συμμετρία, such as that between ἀπόρροιαι and πόροι—the mechanical conception of Empedocles—Aristotle substituted a rational or formal symmetry; while instead of the ἀλλοίωσις, which was a purely physical effect, he substituted the conception of an ἐπίδοσις εἰς αὐτό. Thus by the application of his peculiar notions of matter and form on the one hand, and of δύναμις and ἐνέργεια (or ἐντελέχεια)

[margin note] Aristotle's transformation of the doctrine of Empedocles and others as to the necessity of συμμετρία between object and organ of sense. For their conception of a physical ἀλλοίωσις he substitutes that of an ἐπίδοσις εἰς αὐτό.

---

[1] ὡς τῆς αἰσθήσεως οἷον μεσότητός τινος οὔσης τῆς ἐν τοῖς αἰσθητοῖς ἐναντιώσεως· καὶ διὰ τοῦτο κρίνει τὰ αἰσθητά. τὸ γὰρ μέσον κριτικόν, 424ᵃ 4.

[2] This power, which Aristotle seems again and again to ascribe to each sense *per se*, more properly belongs to the *sensus communis*. In ordinary experience the several senses are not divorced from the *sensus communis*, but normally act in communication with it; whence it is that Aristotle allows himself to demit its powers to them, in the passages in which he is not *contrasting* its functions with theirs. Each of the special senses seems at times, according to Aristotle, to be a rudimentary *sensus communis* in regard to the specific differences which fall under its ken. As the whole sentient soul, or *sensus communis*, divides itself, so to speak, into the so-called five senses, so each of these again sub-divides itself, consistently with its dynamic unity, into a multitude of particular activities, not only distinct in time, but also in kind, from one another. The actual object of a single energy of the same sense is *numerically* one; the possible object of all its activities is *generically* one; while between these falls the *specifically* one possible object of each of its separate kinds of activity. Cf. 447ᵇ 9 seqq.

on the other, he revolutionized the conception of the relation between sense-organ and object which had been accepted by his predecessors up to and including Plato.

Qualitative unity of percipiens and percep-tum at moment of actual per-ception, i.e. when the αἰσθητόν has assimilated the αἰσθη-τικόν to itself. No converse operation of the percipiens on the per-ceptum in perception. The per-cipiens and percipien-dum are necessary correlates, yet the latter has its own proper ex-istence with qualities potentially percep-tible, in which it is prepared to reveal itself when the moment of its being

§ 25. Aristotle (as we have repeatedly observed) conceives the relation between a sense-organ and its object as one between patient and agent. In the *de Sensu* [1] he speaks of having in the *de Anima* explained how the αἰσθητόν in general is related to αἴσθησις ἢ κατ' ἐνέργειαν. In perception the object transforms the subject-sense from potentiality to actuality. This is a perfecting of the sense—an ἐπίδοσις εἰς αὐτὸ καὶ εἰς ἐντελέχειαν [2]. When the transformation or ἀλλοίωσις is complete, i.e. when the particular sense is actually perceiving its object, then the *percipiens* and *perceptum* are qualitatively one. When the *percipiendum* has become *perceptum*, the unlike have become like. This proposition is only another way of stating that the sense has received or apprehended the form of the object [3]. There is no *reciprocal* relation, in Aristotle's opinion, between the object and the organ [4]. There is a participation between the two, related as patient to agent, in a common fact, the resultant of which is the perception. Here we are reminded of the Protagoreo-Heraclitean theory, already stated [5] above, which Plato sets forth in the *Theaetetus*. But Aristotle holds with the unquestioning fidelity of a 'natural Realist' that the 'common fact' is one in which the object is revealed in its true, i.e. independent, qualities. The object exists independently, as well as being an αἰσθητόν, or a 'possibility of perception.' The relation between τὰ αἰσθητά and αἱ κατ' ἐνέργειαν αἰσθήσεις is sometimes described as one of unity; at other times as one of similarity [6]. The meaning in

---

[1] 439[a] 13.   [2] 417[b] 6 εἰς αὐτό—not αὐτό. Cf. [b]16, ἐπὶ τὴν φύσιν.

[3] μία μέν ἐστιν ἡ ἐνέργεια ἡ τοῦ αἰσθητοῦ καὶ ἡ τοῦ αἰσθητικοῦ, τὸ δ' εἶναι ἕτερον, 426[a] 15.

[4] The passage in which alone such relation is asserted, 459[b] 23 seqq., is certainly spurious.   [5] Cf. VISION, § 32, and Plato, *supra*, § 10.

[6] The unity becomes absolute in the case of the objects of thought or νοῦς. In the case of those of sense-perception it does not go beyond the stage of *similarity* ; but this is *unity of form*.

either case is the same : that τὸ αἰσθητικόν has taken the form of τὸ αἰσθητόν. When the eye actually perceives, it has apprehended the colour—which as quality belongs to the form—of its object. How far Aristotle carries this doctrine appears from the passage in which he states that there is a real meaning in saying that the organ or subject of seeing, when regarded as its own object, is coloured [1]. The κόρη is *per se* of no particular colour, but holds the mean between any two colours as well as between the extremes of black and white. In virtue of this its quality of μεσότης—which again involves its bearing a λόγος or proportionality to its object—it is capable of apprehending all colours, i. e. of taking any given colour, as form.

perceived comes. The relation between ἡ αἴσθησις ἡ κατ' ἐνέργειαν and τὸ αἰσθητόν is one of unity of form.

§ 26. The objects of sensation in general are classified by Aristotle [2] as τὰ ἴδια, τὰ κοινά, and τὰ κατὰ συμβεβηκός. The two former are said to be properly and in themselves perceptible [3]. The ἴδια are illustrated by the examples of colour, sound, taste. They are defined by two marks, (a) that they are perceptible by one and only one sense, (b) that it is not possible to be mistaken respecting them [4], or at all events that error respecting them is at its minimum. One cannot be mistaken in thinking that what he sees is colour or what he hears is sound, though he may easily be so as to what the coloured or sonant thing is.

Classification of objects of sensation in general. (a) τὰ ἴδια, (b) τὰ κοινά, (c) τὰ κατὰ συμβεβηκός.

The κοινά are illustrated by κίνησις and ἠρεμία, ἀριθμός, σχῆμα, μέγεθος [5]. These are said to be κοινά, because they are ἴδια to no one sense but common to all ; for—the writer goes on—κίνησις is perceptible by both touch and sight [6].

---

[1] 425ᵇ 22 ἔτι δὲ καὶ τὸ ὁρῶν ἔστιν ὡς κεχρωμάτισται· τὸ γὰρ αἰσθητήριον δεκτικὸν τοῦ αἰσθητοῦ ἄνευ τῆς ὕλης ἕκαστον.

[2] For § 24 cf. *De An.* ii. 6. 418ᵃ 7-25.   [3] καθ' αὑτὸ φαμὲν αἰσθάνεσθαι.

[4] περὶ ὃ μὴ ἐνδέχεται ἀπατηθῆναι ; qualified, however, 428ᵇ 18 ἡ αἴσθησις τῶν ἰδίων ἀληθής ἐστιν ἢ ὅτι ὀλίγιστον ἔχουσα τὸ ψεῦδος.

[5] In *de Sens.* i. 437ᵃ 9 some MSS. give στάσις instead of ἠρεμία, some omit this altogether. In 442ᵇ 5, we have τὸ τραχὺ καὶ τὸ λεῖον, τὸ ὀξὺ καὶ τὸ ἀμβλὺ τὸ ἐν τοῖς ὄγκοις, added.

[6] 418ᵃ 18. That the word πάσαις is hardly meant to be pressed appears not only from this illustration, but also from 442ᵇ 6 κοινὰ τῶν αἰσθήσεων εἰ δὲ μὴ πασῶν, ἀλλ' ὄψεώς γε καὶ ἀφῆς. A wholly different reason for this application of the term κοινά to the objects so strangely confined in

Τὰ κατὰ συμβεβηκὸς αἰσθητά are not directly perceived objects of sense, but rather inferences from direct perceptions. One sees a white object, but says or thinks that he sees, e. g., 'the son of Diares.' That this is not a direct perception is obvious from the mere fact that the organ of vision is nowise affected by the object in its incidental character [1]. The colour affects the κόρη; the magnitude is also, as stated above, καθ' αὐτὸ αἰσθητόν [2]; but the fact that the white object is the son of Diares does not at all impress the organ of sense: this fact is merely associated incidentally—κατὰ συμβεβηκός—with the colour [3]. Aristotle observes that, of the objects καθ' αὐτὰ αἰσθητά, τὰ ἴδια are κυρίως αἰσθητά, and are those to which the essential nature of the special senses is properly adapted [4]. The physical natures of τὰ ἴδια—or of three of them— discussed by Aristotle, de Sensu, iii–v, have been already referred to in their proper places.

The medium of sensation in general : the notion on which the theory of it was based. The medium has a common nature with the αἰσθητόν and the αἰσθητήριον.

§ 27. The nature of the medium and its relation to the organ of perception was for the Greek psychologists of primary importance. Their epistemology was rooted in physiology, and this in physics. In the connexion between 'external' things and the organism, through the medium, they seemed to find a sufficient account of the possibility of the cognition of the external things. The theory of Empedocles for the explanation of our faculty of objective cognition was that the organs of sense and of cognition in general are composed of the very same elements as the things outside the organism, and that *therefore* knowledge of the latter is accessible through these

these illustrations appears in 425ᵃ 27 τῶν δὲ κοινῶν ἤδη ἔχομεν αἴσθησιν κοινήν : the κοινά are the direct objects of the κοινὴ αἴσθησις. But if this be the reason, what are we to think of the places in which the other reason is given and almost contradicted straightway by the illustrations? See *infra*, pp. 282–4.

[1] οὐδὲν πάσχει ᾗ τοιοῦτον ὑπὸ τοῦ αἰσθητοῦ.

[2] An ambiguity lurks here : it is, as appears, e. g., from 450ᵃ 9, καθ' αὑτό αἰσθητόν only to the κοινὴ αἴσθησις, being κατὰ συμβεβηκός to ἡ ἰδία.

[3] ὅτι τῷ λευκῷ συμβέβηκε τοῦτο οὗ αἰσθάνεται.

[4] In this distinction the way is prepared for the doctrine referred to in the above notes, that the κοινά are directly perceptible only to ἡ κοινὴ αἴσθησις.

organs. There are, accordingly, in the organs the primordial air, fire, earth, water, of which all things whatever consist. By like we know like. By the fire within us we see fire, by the water we see water, by the earth, earth, and by the air, air. This notion of identity of elements in objects and organs, with its implied explanation of knowledge, was adopted even by those who asserted the heterogeneity of ψυχή and the objects of knowledge. The difference arising from such heterogeneity for them was that instead of knowing like by like we know each thing by its contrary: hot by cold, white by black, &c. So Anaxagoras, who (with Alcmaeon and Heraclitus) held the theory of cognition by contraries, required for explanation of knowledge the assumption within the organism of all the elements which constitute external objects, though only in order that each external *percipiendum* might thus have in the organism its necessary opposite. We have seen already how Aristotle endeavoured to reconcile these opposing views of cognition. He held that perception is not simply an affection of like by like or of unlike by unlike, but of unlike by an unlike which, however, becomes like, having assimilated the percipient to itself in that process of ἀλλοίωσις which every perception involves. With Empedocles and Plato he held the doctrine of the above four elements, to which he ascribed four fundamental contrary attributes hot, cold, dry (solid), moist (fluid). Of these the bodily tissues are formed[1]; and of the tissues again the organs are constituted. At the basis of his whole theory of perception there is for him, as for his predecessors, the thought that the fundamental community of elementary constitution in αἰσθητά and αἰσθητήρια is the cause of our being able to perceive objects. The ἀλλοίωσις (by which he reconciles these different views) implies in every case a medium by, as well as through, which αἰσθητά and αἰσθητήρια are brought into correlation. For this medium has a common nature with the αἰσθητόν

---

[1] Cf. 389ᵇ 27 ἐκ μὲν γὰρ τῶν στοιχείων τὰ ὁμοιομερῆ, ἐκ τούτων δ' ὡς ὕλης τὰ ὅλα ἔργα τῆς φύσεως. The ὁμοιομερῆ in the body are composed of homogeneous parts. Thus all the parts of flesh are flesh, all those of bone are bone, and so on.

and the αἰσθητήριον. Thus the required conditions of perception are established (see further, §§ 31–34 *infra*).

Aristotle's realism as distinguished from the materialism of Empedocles and Democritus, and from the sensational idealism of Protagoras. Physical basis of the μεσότης of each sensory organ: physical constitution of the organ. The fundamental contrarieties inherent in the four elements, the physical basis of the possibility of perception.

§ 28. Aristotle rejected the naïve materialism of Empedocles and Democritus[1]. He also rejected the sensational scepticism of Protagoras. He took a middle course, holding that things potentially perceptible exist in themselves, while faculties or potentialities of perception 'exist' in our organs. It is not true, he says[2], that nothing would exist if it were not perceived. Yet when perceived it is by virtue of its form, not of its matter, that it is so; and for us its form is due to the act of mental apprehension which perception involves. At the actual moment of perception the thing *qua* perceived and the organ *qua* perceiving, are so related as to be, in form, an unity. He did not, with the early physiologists, regard the sense-organs as mere channels by which the elements of things outside are conducted into the organism, and so the things are known[3]. We do not take in the matter but only the form of things. As the noëtic soul is the τόπος or εἶδος εἰδῶν, i.e. the place or form of forms, so each faculty of perception in the sentient soul is an εἶδος αἰσθητῶν, a form of objects of sense[4]. But each sensory organ by its elementary constitution is or exhibits a μεσότης, i.e. it can present itself as a discriminant (κρίνειν) between any two διαφοραί within its province. Thus the faculty of touch, in virtue of the constitution of its organ, distinguishes[5] between any two degrees of heat, or, as Aristotle says, between hot and cold. This μεσότης, however, is, on its physical side, derived from the proportion in which the στοιχεῖα are combined in the organ. In every organ the four elements, earth, air, fire, water, are combined. These elements are endowed with the fundamental contrary qualities of heat, coldness, fluidity, solidity,

---

[1] Notwithstanding that Empedocles (cf. § 30 *infra*) admitted that the λόγος τῆς μείξεως constituted the true φύσις of things, his position was to all intents and purposes materialistic; he did not distinguish form from matter.   [2] See note 3, p. 229, *supra*.

[3] 431ᵇ 29 οὐ γὰρ ὁ λίθος ἐν τῇ ψυχῇ ἀλλὰ τὸ εἶδος. Cf. 429ᵃ 28.

[4] 432ᵃ 2 ὁ νοῦς εἶδος εἰδῶν καὶ ἡ αἴσθησις εἶδος αἰσθητῶν.

[5] τὸ γὰρ μέσον κριτικόν.

which are so related as to produce in the elements a fundamental community of nature, whereby their μεῖξις is possible[1]. In virtue of this community they are capable of affecting, and being affected by, one another. The same qualities and elements form αἰσθητά as form αἰσθητικά. When, therefore, a given αἰσθητόν, e.g. a certain temperature, affects its αἰσθητικόν, e.g. when a warm object affects the sense of touch, what happens is this: the θερμόν of the object works upon the organ, producing in the latter an ἀλλοίωσις, by which the temperature of the organ gradually becomes assimilated to that of the object. This physical ἀλλοίωσις is the *sine qua non* of perception; when it is complete, then τὸ αἰσθητήριον ἐνεργεῖ: then we perceive the object as hot. But it is not *qua* fire internal (in the organ) and external (in the αἰσθητόν) that organ and object come into the relation of patient and agent; it is rather *qua* containing contrariety. The organ is relatively cold, the object relatively hot, and this contrariety flows from the common constitution of organ and object[2]. The four elements have affinity with one another, and are capable of μεῖξις, just because of the contrary qualities which they each possess. Earth is cold and dry; water is cold and moist; air is hot and moist; fire is hot and dry. Thus each of them has one quality contrary to one of each other. But contraries, though opposites, are opposites in the same genus. Hence the fundamental community. Thus for Aristotle, as for Empedocles, but in a different way, the fact of the organs being composed of the same elements as the objects is the ground of the ἀλλοίωσις in which perception consists.

§ 29. The sensory organs then, like the organism in general, are composed of the four elements. We are told[3] Sensory organs consist of the

---

[1] 331ᵃ 12 seqq. ὅτι ἅπαντα πέφυκεν εἰς ἄλληλα μεταβάλλειν, φανερόν· ἡ γὰρ γένεσις εἰς ἐναντία ἐξ ἐναντίων, τὰ δὲ στοιχεῖα πάντα ἔχει ἐναντίωσιν πρὸς ἄλληλα διὰ τὸ τὰς διαφορὰς ἐναντίας εἶναι.

[2] 441ᵇ 8–15 πάσχειν γὰρ πέφυκεν τὸ ὑγρὸν ὥσπερ καὶ τᾶλλα ὑπὸ τοῦ ἐναντίου ... ᾗ μὲν οὖν πῦρ καὶ ᾗ γῆ οὐδὲν πέφυκε ποιεῖν καὶ πάσχειν οὐδ᾽ ἄλλο οὐδέν, ᾗ δ᾽ ὑπάρχει ἐναντιότης ἐν ἑκάστῳ, ταύτῃ πάντα καὶ ποιοῦσι καὶ πάσχουσι.

[3] 302ᵃ 21–3.

four ele-
ments in
various
propor-
tions. The
ὁμοιομερῆ
of which
organs con-
sist are
themselves
composite.

that σάρξ (which, *plus* τὸ ἐντός, is the organ-medium of touching) contains potentially both earth and fire. Again [1], it is not enough when defining σάρξ to state that it is a σύνθεσις of fire, earth, and air; we should also determine the proportion in which the elements are combined in it. Moreover [2] all mixed bodies, such as exist in this world, contain in their composition all the simple bodies: earth, water, air, and fire. This is proved by the process of nutrition in the case of animal bodies; for all such bodies are nourished by food, which consists of the same elements of which they are composed. The tissues (ὁμοιομερῆ), of which the organs are built [3], are formed of water and air by the agency of the hot and cold, which are the active principles, the dry and moist being the passive, in elemental compounds [4]. The nutrient process in animals has as συναί-τιον the activity of the fire in their organisms [5]. There are in the αἰσθήσεις [6] fire, earth, and the other στοιχεῖα. For the sense of touch not only earth but fire is indispensable [7], since by this sense we discern the hot and cold, as well as the other opposites of which σάρξ is a λόγος [8].

True φύσις
of a body is
the ratio in
which the
elements
are com-
bined in it.
Origin of
this ratio,
something
outside and
beyond
each body.

§ 30. The λόγος of the mixture of elements in a body is that which constitutes its true nature. Empedocles was led by the constraining power of truth itself [9] to declare that the οὐσία or φύσις of compounds like ὀστοῦν consists in the λόγος τῆς μείξεως αὐτῶν, not merely in some one, or two, or three, or even all, of the elements of which it is com-posed. This λόγος has an origin altogether outside the mere ingredient elements. The hot and cold operating on the dry and moist could produce in these the qualities (πάθη) of hard, soft, and so on, but not the proportion which is the distinctive feature of a natural body. This pro-portion or λόγος is, in individual living bodies, derived from ὁ γεννήσας ὁ ἐντελεχείᾳ ὤν, which (or who) is its efficient cause [10]. Discussing the sense of touch [11], Aristotle says that

[1] 642ᵃ 23, Plat. *Tim.* 82 C.   [2] 334ᵇ 31–335ᵃ 12.
[3] Cf. 647ᵃ 2 seqq.   [4] 384ᵇ 30, 378ᵇ 10.   [5] 416ᵃ 12 seqq.
[6] 417ᵃ 4–5 where αἰσθήσεις = αἰσθητήρια.
[7] Cf. Plat. *Tim.* 31 B–C; Arist. 435ᵃ 11–24.   [8] 429ᵇ 14.
[9] 642ᵃ 17–24.   [10] 734ᵇ 28–36.   [11] 423ᵃ 12–424ᵃ 15.

the animate body cannot consist of air and water alone. It must also contain something solid (στερεόν τι). Hence earth, too, must be an ingredient in it. Such is the case with σάρξ and its analogue. As we perceive objects of sight and smell through their proper media, air and water, so we perceive the objects of touch through the medium of the flesh, with this difference between the cases, that we perceive the former at long distances from the organism, the latter only close by it. The σάρξ then is, by virtue of the γῆ contained in it, the organ and medium (or organ-medium) of touch, *qua* discerning hard and soft; and by virtue of the πῦρ, it is the organ and medium *qua* discerning differences of temperature. The objects of touch are the διαφοραί of body *qua* body; those, that is, by which the elements themselves are distinguished, viz. hot, cold, solid, fluid. The organ (says Aristotle) which perceives these is that of touch. To perceive is to be passively affected in a certain way. The organ is potentially such as the object is actually. In touching, therefore, the organ is potentially, while the object is actually, e.g. hot or solid. If the organ or its medium (e. g. the flesh of the hand) be qualitatively like in temperature with the object, the latter cannot produce the requisite ἀλλοίωσις, and we perceive the object neither as hot nor as cold; and so it is moreover with the perception of solidity. In touching, as well as in exercising the other senses, the percepts, to begin with, present themselves as 'extremes' (ὑπερβολαί), between which the αἰσθητικόν comes as a mean. This capacity of the αἰσθητικόν to present itself as a mean, so becoming a δύναμις[1] κριτική—a faculty of 'discerning' between the contrary poles of quality involved in the αἰσθητά, is, as we have already said, rooted in the λόγος of the elements which constitute the organ. The organ of touch is not absolutely, or *per se*, hot or cold, or hard or soft, but a mean between all pairs of differences coming under either category.

§ 31. The media of the organs of touch and taste are altogether internal to the body. That of touch is the Media in-
ternal and
external

---

[1] Cf. 99ᵇ 35, 432ᵃ 16.

σάρξ (with the skin), which covers or forms the periphery of the body; that of taste is the 'potentially moist' σάρξ of the tongue. The organs of seeing, hearing, smelling, have media external to the body; but though external, these media have a peculiarly close relationship not only with the objects[1] but also with their respective organs, so that they have their internal lodgment or representation in every case within the bodily organ. Thus the organ of hearing has air as external medium, but a portion of air is also lodged in, or built into, the organ itself[2]. The organ of seeing has the diaphanous for its medium. Externally this is the air: but internal to the organ there is a cell full of water[3]. This water as internal medium co-operates with the air as external, for both act visually in virtue of their common property τὸ διαφανές. It is not easy to gather a definite idea respecting the internal and external media of smelling from the various statement of Aristotle respecting this sense. In the case of animals which respire he regards the medium of smell as air. This externally is affected by the odorous object and transfers the affection continuously to the olfactory organ, by which it is then inhaled and conducted to the 'point of sense.' Thus for such animals air internal and external to the organ constitutes the medium of smell. But for the class of animals which do not respire some different medium must be assumed. Fish can smell, as can other subaqueous creatures. Consequently Aristotle infers that the common medium of smelling in the case of all creatures which possess this power is τὸ διαφανές—not, however, as such, but qua capable of absorbing or contracting the effect of ἔγχυμος ὑγρότης[4]. At all events, the medium of smell and the essential constituent of the organ of smell consist either of air or water[5], i.e. of common elements.

---

[1] e.g. the colour of objects is the διαφανές in them.

[2] 420ᵃ 9.

[3] Anatomy had not taught Aristotle to distinguish *two* cells.

[4] 443ᵃ 1.

[5] ἡ μὲν γὰρ κόρη ὕδατος, ἡ δ' ἀκοὴ ἀέρος, ἡ δ' ὄσφρησις θατέρου τούτων, 425ᵃ 4.

§ 32. There is one passage [1], however, in which Aristotle speaks with apparent decision, and in a very different way, of the constitution of the olfactory organ and of its object. Summing up at the end of a long polemic against Empedocles and Plato, who regarded the essential part of the visual organ as consisting of fire, Aristotle, having corrected what he thought amiss in their views of the eye, as well as in those of Democritus, proceeds as follows: *Aristotle's inconsistency (real or apparent) as regards the essential constituent element in the organ of smelling.*

'If the facts be as here stated, and if we must refer the essential part of each of the sensory organs to some one of the elements, we must suppose that in the visual organ this consists of water; in the organ of hearing it consists of air; while in that of ὄσφρησις it consists of fire [2]; for what ὄσφρησις is actually this τὸ ὀσφραντικόν [3] is potentially. Since it is the object (αἰσθητόν) that causes the faculty (αἴσθησις) to actualize itself, the faculty or its organ must possess, to begin with, the corresponding potentiality [4]. Now odour, the object of ὄσφρησις, is fumid evaporation, which arises from fire.' Thus the organ of smelling is potentially hot, i.e. potentially it possesses the quality of fire. Hence this organ has its proper place near the brain. . . . The essential organ of touch (τὸ ἁπτικόν) consists of earth; and that of taste is a form of touch. Hence the organ of these two lies near the heart, which is a counterpoise to the brain, being as it is the hottest, while the brain is the coldest, of the bodily parts [5].

---

[1] 438^b 16–439^a 5.

[2] Bonitz, *Ind. Arist.* 538^a 30, appears right in his suggestion that in πυρὸς δὲ τὴν ὄσφρησιν, 438^b 20, the last word = *organ* of ὄσφρησις. The course of the argument which follows requires this; though it is awkward that in the same line ὄσφρησις is also used to mean the realized perception.

[3] =τὴν ὄσφρησιν, 438^b 21.

[4] If when actualized in ὄσφρησις it is actually hot, it must prior to such ὄσφρησις be potentially so.

[5] There are involved in this passage several difficulties for readers who expect or wish to find Aristotle in his writings perfectly consistent with himself. *First*, the assertion that ὀσμή is 'fumid evaporation' is vehemently contradicted, 443^a 21 seqq. *Next*, the assertion that ὄσφρησις is essentially fire is opposed to 425^a 5 ἡ δ' ὄσφρησις θατέρου τούτων (sc. ἀέρος ἢ ὕδατος). *Finally*, in this latter passage also we read τὸ δὲ πῦρ ἢ οὐθενὸς ἢ κοινὸν πάντων, which denies that πῦρ is the

§ 33. Since the organs of touching and tasting have, according to the various standpoints from which Aristotle regards them—the current or popular, and that which he approved of—either no medium or no external medium ; and since moreover the organ of touch is either (according to the popular view) distributed all over the periphery of

essential constituent of any particular organ of perception, while here it represented as potentially constituting ἡ ὄσφρησις. The argument of Bäumker (*op. cit.*, pp. 47–8), assented to by Neuhäuser (*Arist. Lehre von dem sinnlichen Erkenntnissvermögen*, p. 21), Zeller (Arist. ii, p. 63 n. E. Tr.) and others, that the particle εἰ being read, as it probably should be, before δεῖ in 438ᵇ 17, we may regard the whole passage as written by Aristotle from an alien standpoint, does not carry conviction. Nowhere does Aristotle object to the *principle* which connects the separate organs of sense, respectively, with certain elements as essential constituents. On the contrary he accepts it, and makes it the basis of his argument, e. g., in 647ᵃ 9–14. The main objection urged in *de Sens.* ii. is to the fact that Empedocles, Plato, and probably others (including e. g. Alcmaeon), regarded the eye as constituted of fire ; for that they found a difficulty in making the five organs square with the four elements 437ᵃ 21, does not contain an objection against this general principle ; nor does Aristotle explicitly recur to the latter point, on which his difficulty was as great as theirs. But his dogmatic assertions here that τὸ ἁπτικόν consists of earth and τὸ ὀσφραντικόν, or ἡ ὄσφρησις, of fire, are scarcely to be reconciled with the statements of the *de Anima* (425ᵃ 5–6, 435ᵃ 11 seqq.). And besides this, the explanations of ὀσμή here and later in the *de Sensu* (443ᵃ 21 seqq.) are irreconcilable with one another. The best way of getting over the difficulty is to suppose that he does not mean to say that the ἁπτικόν consists of earth *alone*, but only predominantly ; which is certainly what he means in other places. But with regard to ὄσφρησις or τὸ ὀσφραντικόν this is not effectual as a solution. Such discrepancies as remain, however, may be explained either on the hypothesis of interpolation, or on that of a change of views on the part of Aristotle. The *de Sensu* seems to contain preliminary essays on certain subjects of the larger work *de Anima*, which may therefore (notwithstanding many references, e. g. 436ᵃ 1 seqq.) be regarded as possibly later. It is not to be supposed that Aristotle in his earlier works held the same views as in his later ; any more than that Spinoza, while still a follower of Descartes, held the views of the author of the *Ethica*. He doubtless passed through a long process of mental development, and the many works connected with his name, even when they are, like the *de Sensu* and *de Anima*, of unquestionable authenticity as a whole, could not be expected to be everywhere in agreement with one another. As well might one expect to find in Kant's early essays the 'Copernican thought' of the *Critique of Pure Reason*. See *infra*, pp. 245 n. 3, 248 nn. 1 and 2.

the body, or (according to his own view) vaguely regarded as ἐντός τι; there are several passages in which these organs of non-mediated perception, or rather of perception by contact [or quasi-contact; *vide* TOUCHING, § 13], are set in contradistinction to the others, and the name αἰσθητήρια seems almost appropriated, for the time being, to the latter. Thus [1], at the beginning of the third book of the *de Anima*, having declared that we perceive by touch all the tangible qualities of body, and that, when we perceive the other qualities, we do so by organs which act through media composed of the elements, Aristotle proceeds to treat these mediated organs as if they alone were called αἰσθητήρια. He expressly asserts that αἰσθητήρια are composed only of air and water—as if the organs of taste and touch were not αἰσθητήρια at all, or as if, being αἰσθητήρια, they could be regarded (in defiance of the fairly consistent teaching of other places) as composed solely of air and water [2]. But in this place we must remember that the organ or organs which act by contact have been already sufficiently dealt with in the opening lines; and that the αἰσθητήρια referred to in the sequel are only those which perceive διὰ τῶν μεταξύ, i. e. by external media: viz. those of seeing, hearing, and smelling. These of course may be declared to consist essentially of air or water; for the contrary qualities of fire and earth (the remaining elements) are only perceptible by τὸ ἀπτικόν, and cannot be essential constituents in organs destined to act through external media, and not by contact with their objects [3]. The moisture in which the object of

---

[1] 424[b] 21 seqq.   [2] 425[a] 7-9.

[3] It seems inexplicable how one who is so well acquainted with Aristotle as Bäumker should in his otherwise excellent work *Des Aristoteles Lehre von den äussern und innern Sinnesvermögen*, pp. 47–8, where he endeavours to rescue Aristotle from inconsistencies, assert that the only media are air and water. 'Luft und Wasser sind und bleiben die bevorzugten Stoffe, welche einzig and allein, wie als Medien, so als Grundmaterie der Organe auftreten.' This statement is based upon a contracted view of the matter, in which Bäumker overlooks the fact of σάρξ being a medium, and omits to look beyond what is contained in *de An.* iii. 1. 424[b] 30-425[a] 9. Moreover, he does not see that even there, τὸ ἀπτικόν being disposed of, the

taste must be contained, if it is to affect the organ and so be perceived, is not an external medium. For tasting contact is always necessary[1], and this moisture is ἁπτόν τι. Taste, therefore, has no external media, but only the same medium which touch, of which it is a form, possesses. Taste is a kind of touch, but with a certain distinctive power of its own.

No sense exists beyond the

§ 34. There exists no sense beyond those known to us as 'the five senses[2].' The argument by which Aristotle tries αἰσθητήρια whose essentials are air and water are only those of seeing, hearing, and smelling. He also overlooks the argument of *de An.* iii. 13 (435ᵃ 11–ᵇ4) in which, while showing that τὸ τοῦ ζῴου σῶμα cannot be ἁπλοῦν, or composed solely of any one element, Aristotle proves that earth and fire are elements in the organ of touch, whose medium is σάρξ. As regards the question whether the *only media* are air and water, we have above said more than enough to show that whereas, indeed, air and water are the sole *external* (i. e. extra-organic) media, they are not the *sole* media, earth and fire being essential constituents of σάρξ, the intra-organic medium of touch and taste. Further untenable assertions of Bäumker here are (*a*) 'that it is in the medium not in the organ that the perceived affection which is potential in the αἰσθητόν *per se* is first actualized' ('Erst in jenem Medium tritt die wahrgenommene Affektion, die in dem Gegenstande an sich nur potentiell angelegt ist, aktuell auf'). (*b*) That according to Aristotle (differing in this from the ancients) 'the organs are not brought into relation with the objects as such, but the qualities of the objects must correspond to their respective media' ('dürfen die Organe nicht zu den Objekten als solchen in Beziehung gebracht werden, sondern ihre Beschaffenheit muss den zu ihnen gehörigen Medien entsprechen'). With regard to (*a*) we may remark simply that a πάθος in the external medium, as such, is as yet no percept at all ; not having affected the organ, it produces no αἴσθημα. To do this, it must have affected the *internal* medium, and so the organ, of sense. With regard to (*b*) ; if the organ is not to be brought into relation with the object as such, what, we may ask, is the purpose of *de An.* ii. 5, 416ᵇ 35–418ᵃ 4, which is devoted to the discussion of the question whether *like* is perceived by *like* or *unlike* by *unlike*, and concludes thus: τὸ δ' αἰσθητικὸν δυνάμει ἐστὶν οἷον τὸ αἰσθητὸν ἤδη ἐντελεχείᾳ, καθάπερ εἴρηται· πάσχει μὲν οὖν οὐχ ὅμοιον ὄν (sc. τὸ αἰσθητικόν), πεπονθὸς δ' ὡμοίωται καὶ ἔστιν οἷον ἐκεῖνο? The passages quoted by Bäumker to justify his views on the above points are far from adequate to their purpose. But we cannot here go into the details of a full discussion.

[1] 422ᵃ 10–14.

[2] 424ᵇ 21–425ᵃ 13. Though Aristotle here names them 'the five,' he was, as we have already seen, perfectly aware that touch is differentiable

to prove this most difficult proposition is obscure, but may so-called 'five senses.'
be outlined thus. Assuming [1] that there exists no body
or affection of body other than those known to us in this Aristotle's argument for this conclusion.
world [2], our present five senses make all the bodies in this
sphere accessible. Hence if we assumed any further sense,
it would either have no object, or would merely duplicate
some existing sensation ; either of which suppositions would
be intolerable. Therefore no further sense beyond the five
is to be assumed.

The stress of the argument is laid by Aristotle on the
second proposition, viz. that our present senses give us
the perception of all known bodies; which is thus proved.
The four elements are the basis of all existing σώματα and
their πάθη. In our bodily organs of perception, and the
media through which they act, all the elements are
functionally employed; hence by their elementary constitution our present organs bring us into acquaintance
with all the bodies and affections of bodies in the world.
If a particular αἴσθησις were lacking, this could be only
because its fitting αἰσθητήριον was so. But no αἰσθητήριον
which would be of service for actual perception is lacking.
Hence we possess all the αἰσθήσεις, and there is none beyond
'the five.' The proposition that our present organs by
their elementary constitution make us acquainted with all
σώματα and their πάθη is shown to be true as follows. All
possible qualities of body are exhausted in two classes,
those perceived through external media and those not

---

into several senses; especially into those of temperature (the perception of the 'hot and cold') and of pressure and resistance (the
perception of the 'hard' and 'soft,' 'solid' and 'fluid'). Thus Reid
was not, as Lord Kelvin (*Popular Lectures and Addresses*, 'The Six
Gateways of Knowledge,' p. 262) says, the 'first to point out the
broad distinction between the sense of roughness or resistance and the
sense of heat.'

[1] This assumption, of course, involves a *petitio principii* : for if
there were other bodies with other πάθη there would have to be
other αἰσθήσεις.

[2] 425ᵃ 11–13 εἰ μή τι ἕτερον ἔστι (=*exists*) σῶμα καὶ πάθος ὃ μηθενός
ἐστι τῶν ἐνταῦθα σωμάτων. This assumption, although not mentioned
till the end, is the major of the whole deduction.

so perceived. Touch and taste give us knowledge of (or the faculty of knowing) all possible tangible qualities, i. e. all those which do not require an external medium. The remainder are perceived by the remaining senses ; for their organs consist of the elements which constitute external media, viz. air and water. All the externally non-mediated αἰσθητά are ἁπτά: and ἀφή *per se* is capable of perceiving all these. Touch has its organ and medium framed essentially of earth and fire, which, through their πάθη, represent to us the διαφοραί of σῶμα *qua* σῶμα. Thus, so far as these two elements go, nothing that exists in our world is unprovided for by touch [1]. The externally mediated αἰσθητήρια, on the other hand, provide for the perception of the non-tangible properties of things ; and this they do by their being essentially constituted of air and water, which are the only elements capable of serving as external media. But they are sufficient, for they mediate for all αἰσθητά not already provided for through touch. Thus either mediately or immediately (or rather by media external *and* internal, or media internal *only*) access is given us, by our organs of perception, to knowledge of all the bodies and properties of body which exist in our world, of which we can form any conception. Hence no other αἴσθησις is to be assumed [2]. The higher animals possess already

[1] In 425ᵃ 5–7 we read that fire ‘either belongs to no one of the three externally mediated organs, or else it belongs to all alike,’ since it lies at the root of life and sensation. Earth, too, has no special connexion with any of these three sense-organs, though it lies with fire at the basis of touch. Thus earth and fire are related to the three externally mediated organs just so far as these are related to the organ of touch (see § 23 and §§ 28–9 *supra*).

[2] We must suppose that Aristotle regards τὸ ἁπτικόν throughout this passage as including both taste (of which nothing is expressly said) and touch. We must further bear in mind that (for reasons already given), when an organ is said to be composed of water or of air, this only means that in its composition the water or the air is the ingredient essential for its function, the latter depending on the λόγος or ratio which either bears to the other elements in the organ. To imagine Aristotle saying that one single element could constitute any sensory organ, or, indeed, any other part of the body, would be to imagine him throwing overboard the teaching of his Physiology and Physics.

all the αἰσθητήρια that are either (a) possible in point of constitution from the four elements, or (b) requisite for the perception of existing σώματα and their πάθη. To restate the points of Aristotle's argument more briefly. Our faculty of perception in general (τὸ αἰσθητικόν) is equipped with the needful means of perceiving all αἰσθητά. It has, by ἀφή, the means of perceiving all which do not need an external medium, i. e. all whose διαφοραί belong to body *qua* body, and characterize the two στοιχεῖα, fire and earth. It has, by organs constituted of air and water, the means required for perceiving all the αἰσθητά which do need an external medium : i. e. those whose διαφοραί do not depend on fire and earth. No αἰσθητόν, therefore, remains inaccessible to perception with our present senses [1].

---

[1] In the parenthetic words 424ᵇ 30 ἔχει δ' οὕτως to 425ᵃ 2 δι' ἀμφοῖν Aristotle shows how it is conceivable that there should be a *reduction* in the number of αἰσθητήρια, or a *duplication* of αἰσθήσεις or (what comes to the same thing) of αἰσθητά; but leaves it plain that in no such case could we imagine the list of our αἰσθήσεις to be usefully increased. For (a) we can conceive one αἰσθητήριον so constituted as to perceive two heterogeneous αἰσθητά; as, for example, if air is medium for both ψόφος and χρόα, and if it be necessary that an αἰσθητήριον essentially of air should perceive both of these. Again (b) we can also conceive two αἰσθητήρια so constituted that either might perceive the same αἰσθητόν as the other ; as, for example, if air and water are each a competent medium of χρόα, a person with two organs essentially consisting the one of water, the other of air, should with either perceive χρόα. But neither (a) nor (b) would point the way towards an increase in the list of useful αἰσθήσεις. The former would give us the same two αἰσθήσεις and αἰσθητά as we have, only by one organ instead of two. The latter only brings us to the conception of two different organs employed in giving us one and the same αἴσθησις or αἰσθητόν.

PART III. SENSUS COMMUNIS

§ 1. WE now come to one of the most interesting portions of the ancient Greek psychology—the theory of the faculty of synthesis at its earliest stage. The name which heads the chapter is a translation of the term κοινὴ αἴσθησις[1], which was used first by Aristotle for this faculty. It is necessary here, as before, to consider how much of what he had to say regarding it was to be found in the speculations of his predecessors. As, however, these did not, at least until Plato's time, undertake the discussion of the faculty of synthesis as such, we must content ourselves with stating the functions ascribed by Aristotle to the κοινὴ αἴσθησις, and seeing how these functions were dealt with by preceding psychologists. To this department of ψυχή, then, variously named by him ἡ κοινὴ αἴσθησις, τὸ κρῖνον, τὸ πρῶτον αἰσθητικόν, he assigned (*a*) the power of discriminating and comparing the data of the special senses, all of which are in communication with it ; (*b*) the perception of the 'common sensibles,' τὰ κοινά, of which principal are κίνησις σχῆμα ἀριθμὸς μέγεθος and χρόνος ; (*c*) the consciousness of our sensory experiences, i. e. the power by which we not only perceive, but perceive *that* we do so ; (*d*) the faculty of imagination, i. e. reproductive imagination—τὸ φανταστικόν ; (*e*) the faculty of memory and reminiscence, μνήμη καὶ ἀνάμνησις ; and (*f*) the affections of sleeping and dreaming. To ascertain, therefore, how much of Aristotle's theory respecting this had been anticipated, we must survey the works of his predecessors. As they do not (until we reach Plato) distinctly formulate the idea of a synthetic faculty, we can only examine what they may have done to explain the various phenomena of mind abovementioned as attributed by Aristotle to the agency of the

---

[1] Though Aristotle uses this actual term but seldom (cf. 425ᵃ 27, 450ᵃ 10, 686ᵃ 31), often employing equivalents like πρῶτον αἰσθητικόν, &c., yet as a convenient name for an important conception it was generally adopted by his followers, and in its Latin form continued to play a great part throughout the psychology of the Middle Ages.

κοινὴ αἴσθησις. We shall find before Plato very little in the remains of the old psychologists on this important subject of synthesis. We have already recounted what they had to say of the special senses and sensation generally ; and from this it is clear that they did not neglect the presentative department of psychology. As regards the representative, however, they do not seem to have taken nearly the same pains. They referred the above-named functions to ψυχή, or νοῦς, in a vague and general fashion ; feeling perhaps that these functions were too complicated and obscure for treatment in detail with any prospect of success. Before Plato, moreover, we find no record of any serious psychological treatment of memory or imagination.

§ 2. Owing to the parallelism in Aristotle's theory between psychical wholes and parts, the consideration of the *sensus communis* will divide itself into sections corresponding to the divisions adopted with reference to each of the special senses. This, their common centre, has its function and organ, its objects, and its medium, and will have to be investigated with reference to each of these. As we have premised that none of the pre-Platonic psychologists distinctly conceived such a subject as this, our treatment must (following such records as we possess) be of a piecemeal character, according as we find reason to suppose that each, or any, of the writers with whom we have to do, took or would naturally take a particular view of any of the functions of the common sense, or ascribed any of them to some particular organ.

Sensus communis must be studied as to its function and organ, its objects, and its medium, just like each particular sense.

Alcmaeon.

§ 3. Of the function of a *sensus communis*, or of synthetic function in general, Alcmaeon had no distinct idea, as far as his remains and the testimony respecting him can be trusted for information. We know, indeed, that he is said to have distinguished sensibility or sense-perception (αἰσθάνεσθαι) from intelligence (τὸ ξυνιέναι), and to have confined the possession of the latter to human beings. But he has left no evidence to show where he regarded αἴσθησις as ending or ξύνεσις as beginning, or how he would

Alcmaeon. No treatment by him of synthetic function, either intellectual or sensuous. Perhaps an implication of it in the word

ξυνιέναι =
intelli-
gence ; so
seeming *ex
vi termini,*
to ascribe
synthetic
function
(as Plato
did) to
under-
standing.
Brain
would for
him (as
also for
Plato)
have been
organ of
synthetic
faculty.
Sleeping—
a pheno-
menon
which
depends on
the blood.

distinguish these. Except, then, for the form of this word ξύνεσις, which *implies* synthesis in its notion, and seems to ascribe it (as Plato did) to understanding, we have no hint that Alcmaeon paid attention to it. Its importance remained submerged under a familiar name, and it eluded discussion. As little do we know of any classification of objects of sense-perception by him in which he would distinguish the data of special from those of ' common ' sense. If, however, he had had a conception of this sense, he would probably have assigned the brain as its organ. There can be no doubt that he silently included the functions of the common sense under those of ξύνεσις, and we have abundant evidence that for him the brain was the organ of intelligence, and that, moreover, all the several αἰσθήσεις are connected with it and cannot discharge their functions if their connexion with it is disturbed [1]. Sleeping (which according to Aristotle is an affection of the *sensus communis*) results, according to Alcmaeon (as well as to his successors, including Aristotle), from the retirement of the blood into the larger blood vessels, while ' waking ' (i. e. full consciousness) returns after its rediffusion [2]. This might seem to imply that for Alcmaeon the blood would have been the chief organ of consciousness. But we know that sensation was for him impossible without the co-opera-tion of the ἐγκέφαλος with each sense ; and therefore, most pro-bably, as Siebeck[3] remarks, it is to this organ that he would have assigned the *consciousness* of sensation, which Aristotle ascribes to the organ of the *sensus communis*, viz. the heart.

---

[1] Theophr. *de Sens.* § 26 ἁπάσας δὲ τὰς αἰσθήσεις συνηρτῆσθαί πως πρὸς τὸν ἐγκέφαλον, διὸ καὶ πηροῦσθαι κινουμένου καὶ μεταλλάττοντος τὴν χώραν· ἐπιλαμβάνειν γὰρ τοὺς πόρους, δι' ὧν αἱ αἰσθήσεις. Cf. also Plut. *Epit.* iv. 17, 1, Diels, *Dox.*, p. 407, where, however, the term τὸ ἡγεμονικόν shows how far we are from the text of Alcmaeon. This Stoic term is pro-bably derived from the Aristotelean τὸ ἡγούμενον, 1113ᵃ 6. Plato, no doubt, refers to Alcmaeon in *Phaedo* 96 B : ὁ τὰς αἰσθήσεις παρέχων τοῦ ἀκούειν καὶ ὁρᾶν καὶ ὀσφραίνεσθαι. It is to Alcmaeon and Plato that Aristotle probably alludes, 469ᵃ 22 : διὸ καὶ δοκεῖ τισὶν αἰσθάνεσθαι τὰ ζῷα διὰ τὸν ἐγκέφαλον.

[2] εἰς τὰς αἱμόρρους φλέβας, Plut. *Epit.* v. 24, Diels, *Dox.*, p. 435.

[3] *Geschichte der Psychol.*, p. 103.

Empedocles.

§ 4. We miss, in the information which we have respecting Empedocles, anything which would show that he had a conception of the synthetic faculty as something which it was the duty of a philosopher—or even a psychologist— to discuss; for to reason from his metaphysical conceptions of φιλία and νεῖκος to psychological analogues of synthesis and analysis would be merely fanciful. He gives no psychological classification of the objects of sense, and whatever is to be known respecting his attitude towards the *sensus communis* must be altogether, as in the case of Alcmaeon, due to inferences more or less doubtful. We know that for him the blood—more especially that in the region of the heart—was the seat or organ of intelligence. As he did not really distinguish sense from reason or intelligence[1], this must show that the blood would have been for him the organ of a central faculty of sense had he distinctly formed a conception of this. But we have no information as to how he regarded the ἀπορροαί, which entered the pores of each sense, as co-ordinated and marshalled into the service of a systematic experience. He does not exhibit a feeling of the need of any such process; but the blood (in which the elements are most perfectly mixed) would, no doubt, have, for him, supplied the organic means towards it. In his theory of 'temperaments[2],' by which men possess talents according to the perfection of the κρᾶσις of the elements in various parts of the body, he seems to betray a singular absence of any perception of the need of systematization of sensory data under some controlling central power. Aristotle notices this fault in the psychology of Empedocles, and complains that he does not provide any central force to combine or keep together and co-ordinate either the various energies or the elemental parts

[Margin note: Empedocles—lack of a conception, on his part, of the necessity of a synthetic faculty of any sort. 'Each element in us perceives its like outside us.' Whatever synthesis was possibly contemplated by him must have had its instrument in the mixture of the elements contained in the blood, especially that round the heart or in the heart. His theory of 'temperaments,' adverse to the conception of a central synthetic faculty. Aristotle criticizes the neglect of synthetic function as a defect]

---

[1] E. Rohde, *Psyche*, § 464, note 2, holds that Empedocles did draw this distinction, though admitting that for him τὸ νοεῖν was only σωματικόν τι. Cf. Arist. 427ᵃ 22.

[2] Cf. Theophr. *de Sens.* § 11. The man who has the elements most perfectly mixed in the tongue is the orator; he who has the mixture perfect in the hand is the artist, and so on.

in the psychology of Empedocles. of the soul[1]. The supposition that the blood, especially that around the heart[2], would, as central organ of perception, have taken, for him, the place of the heart itself as conceived by Aristotle, might seem to be confirmed by his theory of sleeping. This affection is produced by a 'symmetrical cooling of the blood[3].' The organ immediately affected in sleeping is, one would think, the organ of consciousness. But this theory of sleeping, as dependent on the blood, is common to him with Alcmaeon and Plato, for whom, however, the brain was the central organ of sense-perception.

Democritus.

Democritus did not discuss the faculty of synthesis; nor distinguish sensibility from intelligence, as psychical entities or functions. He allocated certain faculties of soul to certain parts of the body. He is credited with § 5. Democritus did not put to himself the question—what is the faculty by which the data of sense are combined and distinguished, by which we are conscious of our mental acts, by which we imagine, remember, &c.? He drew no dividing line between αἴσθησις and νοῦς as psychical[4] entities. For him all knowledge, sensory and other, is effected by mechanical interaction between the atoms of bodies and those of the soul[5]. It results from εἴδωλα (or δείκελα, to use the more general expression) ἔξωθεν προσιόντα. The soul atoms were divided or distributed all over the body. Notwithstanding this he seems (so far as we can trust our authorities) to have located certain mental faculties in particular parts of the body[6], and even to have anticipated the tripartite division of Plato who assigned the intelligence, the faculty of energy, and the faculty of desire, to the brain, the heart or thorax, and

[1] De An. i. 5. 410ᵇ 10–13 ἀπορήσειε δ' ἄν τις καὶ τί ποτ' ἐστὶ τὸ ἑνοποιοῦν αὐτά (sc. τὰ στοιχεία), and 411ᵃ 26–ᵇ 7 πότερον πάσῃ νοοῦμεν . . . τί οὖν δήποτε συνέχει τὴν ψυχήν;

[2] αἷμα γὰρ ἀνθρώποις περικάρδιόν ἐστι νόημα, Frag. 109, Diels, Vors.p. 212.

[3] Plut. Epit. v. 24, Diels, Dox., p. 435 κατάψυξιν τοῦ ἐν τῷ αἵματι θερμοῦ σύμμετρον.

[4] He distinguished, however, between the evidential value of αἴσθησις and νοῦς, between σκοτίη and γνησίη γνῶσις, Sext. Math. vii. § 138.

[5] ἐκεῖνος μὲν γὰρ ἁπλῶς ταὐτὸν ψυχὴν καὶ νοῦν· τὸ γὰρ ἀληθὲς εἶναι τὸ φαινόμενον, Arist. de An. i. 2, 404ᵃ 27.

[6] Cf. pseudo-Hippocr. Epistulae ix. 392 L περὶ φύσιος ἀνθρ., Diels, Vors., p. 470, where Democritus is said to have called the brain φύλαξ διανοίης; the heart (καρδίη) βασιλίς, ὀργῆς τιθηνός; the liver (ἧπαρ) ἐπιθυμίης αἴτιον.

the liver or abdomen, respectively. He is also credited[1] having made both with a bipartite division of the soul, placing τὸ λογικόν in a tripartite the thorax, while distributing τὸ ἄλογον all over the body. and a bipartite In fact, however, we can depend very little on information division of coming from a pseudo-Hippocratean writer of the second the soul. century, or from the *Placita*, respecting points like this.

According to the physical principles of Democritus, sense and thought result from emanations coming to us from things and entering the pores of our bodies, but especially the pores of the proper organs, penetrating to the atoms of the soul, and so in some way bringing to our minds the ideas of the things from which they have come. Thus it is with the perceptions of our waking life; and thus it is also that we dream when asleep. For in sleep, too, εἴδωλα of things and persons stream into our bodies, or, being already lodged in them, then become active, and visions of the persons or things from which they originate arise in our minds[2]. Sleeping, according to Democritus, is a cooling Sleeping, of the heat-atoms of the body, or rather the expulsion, the expulsion of under the pressure of the environment, of a certain number a certain number of of them[3]. This cooling affects the outer parts chiefly, and heat-atoms the vital heat retires to the interior, sc. to the neighbourhood and soul-atoms, of the heart. Amid these vague and indefinite notions we with concannot discover any inkling of a synthetic faculty by which centration of the vital the effects of ἀπορροαί in the way of sensation were collected heat round the heart. and arranged for the purposes of systematic experience.

§ 6. We might, at first sight, expect to discover, in His connexion with what Democritus says of φαντασία, some references to φαντασία clue to his attitude respecting the central sense. But we give no clue to a find at once that by φαντασία he does not mean the repro- doctrine

---

[1] Plut. *Epit.* iv. 6, Diels, *Dox.*, p. 390.

[2] Arist. *de Div. per Somn.* ii. 464ᵃ 5 ὥσπερ λέγει Δημόκριτος εἴδωλα καὶ ἀπορροὰς αἰτιώμενος. Cf. Lucret. iv. 747–66 (Giussani), and Plut. *Sympos.* viii. 10, § 2 ὅ φησι Δημόκριτος, ἐγκαταβυσσοῦσθαι τὰ εἴδωλα διὰ τῶν πόρων εἰς τὰ σώματα καὶ ποιεῖν τὰς κατὰ τὸν ὕπνον ὄψεις ἐπαναφερόμενα: from which it would appear that the εἴδωλα, which are ever coming when we are awake, sink deeply into our bodies, destined in sleep to arise, as it were, 'from the depths' and present themselves to consciousness.

[3] Cf. Arist. 472ᵃ 2–15, 404ᵃ 5–16.

of central
sense, or of
synthesis,
on his
part: for
it means
only pre-
sentation.
He formed
no theory of
representa-
tion, of
memory,
or remi-
niscence.

ductive imagination, but merely the presentative faculty:
that faculty whereby things appear, or present themselves,
to us in ordinary perception. He taught that the 'secondary
qualities' (as they were called by Locke) have no objective
existence: they are only affections of our sensibility
according as it is qualitatively altered [1]. The same thing
that appears (φαίνεσθαι) to us sweet may appear to others
bitter, &c. As regards the function of reproductive imagination, therefore, which Aristotle ascribed to the κοινὴ αἴσθησις,
we cannot ascertain that Democritus held definite views,
any more than as to the κοινὴ αἴσθησις itself. To complete
our discomfiture we are unable to discover that he formu-
lated a theory of memory or recollection. In no way,
therefore, can we find a point of contact between his
doctrines and that of the κοινὴ αἴσθησις of Aristotle. He
seems to have been too much immersed in the details of
physics and physiology to spare time or thought for the
more abstract and higher aspects of psychology.

Anaxagoras.

Anaxa-
goras could
not con-
sistently
have held
a theory
of *sensus
communis,*
or synthetic
faculty of
sense. For
he could
not, except
by a
miracle,
make soul
and body
communi-
cate with

§ 7. If there is any proposition which may be implicitly
believed respecting the teaching of Anaxagoras, it is that
for him νοῦς[2] was ἀμιγής, i. e. absolutely free from all
admixture of the elements[3] of the μεῖγμα. This being
so, it is impossible to understand how any principle of
community could connect it with the material body; or
how there could be a κοινὴ αἴσθησις with an αἰσθητήριον to
correspond, in which the soul and the infinitude of elements
— ὁμοιομερῆ — should be really related to one another.
Only a 'miracle[4]' could bring about such communion
for Anaxagoras. Accordingly, sleeping—for Aristotle a
function of ἡ κοινὴ αἴσθησις—is for Anaxagoras an affection

---

[1] πάντα πάθη τῆς αἰσθήσεως ἀλλοιουμένης, ἐξ ἧς γίνεσθαι τὴν φαντασίαν,
Theophr. *de Sens.* §§ 63-4.

[2] He refers to νοῦς also as ψυχή: cf. Arist. 404[b] 1-3, Schaubach,
Anax. p. 113. This he did probably when descending from the
teleological to the mechanical standpoint: the ground of Socrates'
complaint against him.

[3] Cf. Arist. 405[a] 16, 429[a] 18.

[4] Cf. Eurip. *Frag.* 1007 (Nauck) ὁ νοῦς γὰρ ἡμῶν ἐστιν ἐν ἑκάστῳ θεός.

of the body only, not of the soul[1], an opinion to which he was probably led *a posteriori* by the activity of the mind in dreams as well as *a priori* by this theory of νοῦς (or ψυχή) ἀμιγής. Yet, despite this theory, Anaxagoras appears to have held an exoteric form of his doctrine of ψυχή, in which, as his 'final' causes were displaced by mechanical causes, so his views of soul approached somewhat nearer to those of ordinary psychology. His teaching respecting the special senses shows traces of this. Can we, even from this standpoint, discover in him any evidence of a doctrine of synthesis—of the faculty by which the data of the several senses are combined and distinguished? If so, what would for him have been its organ? We saw that, in explaining the faculty of hearing (ἀκοή), he regarded ψόφος as making its way ἄχρι τοῦ ἐγκεφάλου. Censorinus tells us that Anaxagoras held the brain to be the source of all the senses[2]. It seems at all events certain that for him, in general, νοῦς or (its equivalent in his psychology) ψυχή would have fulfilled the functions of κοινὴ αἴσθησις— have supplied consciousness, memory, &c., as well as distinguishing and comparing the phenomena of sense. As to the particulars of the manner of its doing so, we can say nothing. We can only rest on hypotheses respecting the matter. Theophrastus[3], distinguishing the teaching of Clidemus from that of Anaxagoras, says: 'Clidemus taught that, while the senses of seeing, smelling, tasting, and touching, independently perceive their objects, the senses —or rather the organs—of hearing merely convey their report to νοῦς, which is that which properly and directly hears[4]; though he does not, as Anaxagoras did, make νοῦς

[margin note:] one another. Sleeping is, for Anaxagoras, an affection of body only. Censorinus seems to say, for it is very doubtful) that he connected all the senses with the brain. The only principle of synthesis must for him have been νοῦς, or ψυχή taking the place of νοῦς in popular terminology. How it co-ordinated the data of sense we have nothing to inform us.

---

[1] Plut. *Epit.* v. 25, Diels, *Dox.*, p. 427 ... σωματικὸν γὰρ εἶναι τὸ πάθος, οὐ ψυχικόν.

[2] Cens. *de die Natali*, vi. 1 'Anaxagoras cerebrum, unde omnes sunt sensus (sc. ante omnia iudicavit increscere)' : unless here the clause 'unde ... sensus' be inserted by Censorinus *de suo*, as the indicative suggests.　　　　　[3] *De Sens.* § 38; Diels, *Dox.*, p. 510.

[4] μόνον δὲ τὰς ἀκοὰς αὐτὰς μὲν οὐδὲν κρίνειν, εἰς δὲ τὸν νοῦν διαπέμπειν, οὐχ ὥσπερ Ἀναξαγόρας ἀρχὴν ποιεῖ πάντων τὸν νοῦν : where Diels observes on μόνον 'nam qui praecedunt sensus ipsi iudicium ferunt.'

the ἀρχὴ τῶν πάντων [1].' Though Clidemus did not, like Anaxagoras, make νοῦς the explanatory principle of all things in general, he regarded it as the true percipient subject in the case of hearing. The implication by contrast here would certainly seem to be that the subject in the case of every sense was for Anaxagoras νοῦς itself, while the sensory organ was but a mere instrument or channel. But it is almost idle to speculate as to how Anaxagoras would have conceived a theory of synthesis, when of this faculty itself he does not appear to have felt the necessity.

Diogenes of Apollonia.

Diogenes discussed memory and reminiscence. His antici- pations of the theory of Aris- totle. The central organ of intelli- gence for Diogenes: the air round the brain in connexion with the air in the thorax, or round the heart.

§ 8. Diogenes, who (notwithstanding his revival of the theory of Anaximenes which made *air* the *principium* of all things) is one of the most interesting of the pre-Platonic psychologists with whom we have undertaken to deal [2], stands alone among the latter in having discussed, even though indirectly, the subject of memory and reminiscence. He seems to have held a theory of the psychical function of the air in (or around) the brain in its relation with that in (or around) the heart in the thorax ; which reminds one of Aristotle's doctrine of the connexion of three of the senses with the brain, or rather with the membrane surround- ing this, and then with the heart, to which the brain or its membrane was only an intermediate station. We have already seen how he connected the several special senses with the air in the brain : how the eye, when images fall on the pupil, conveys its message by means of the air in this organ to the inner air, and so on [3]. The air animates the whole body, being conducted through it with the blood in the veins. Thinking is due, he says, to the activity

---

[1] Zeller (*Pre-Socratics*, ii. 369, E. Tr.) infers that Anaxagoras made Νοῦς the true subject of perception in the case of each and all of the αἰσθήσεις : this would seem to require πασῶν instead of πάντων.

[2] Parmenides also seems to have formed a theory of μνήμη, making it to depend (like διάνοια in general) on a due κρᾶσις of cold and hot in the body. Cf. Theophr. *de Sens.* §§ 3–4.

[3] Theophr. *de Sens.* §§ 39–42.

of pure and dry air, for moisture impedes intelligence. Cause of forgetful-
Hence infants are of weak intelligence: they have too ness, and of
much moisture[1], hence the air is not able to circulate weakness of memory
freely through their bodies but is confined within the in children.
breast. For lack of ducts—the necessary means of such
circulation of air—plants are destitute of intelligence. The
cause of the passionate and fickle disposition of infants
is the same. Hence, too, the tendency of young children
to *forgetfulness*. As the air does not penetrate freely to
all parts of their body they are lacking in intelligence[2].
A proof of the proposition that the obstruction of the Conditions
air in the breast causes mental difficulties is found in of memory and of
the distress which persons feel who endeavour to recollect. remi-
This feeling they have in the breast[3]. When they have niscence.
recovered the idea for which they have sought in this
effort, the obstructed air is set free, and they experience
a feeling of relief[4]. The air being the primary agent
of mind, if it becomes obstructed in its chief seat—the
breast, into which it passes in respiration—mental power is
impaired, and mental efforts are thwarted, until the air again
secures free passage for itself. We notice here how closely
Diogenes approaches to Aristotle, who made the organ of
central sense, of which ἀνάμνησις is a function, the heart
or the region of the heart[5]. A further partial coincidence Theory of
between Aristotle and Diogenes appears in their treatment sleeping.
of the affection of sleeping. According to Diogenes[6], sleep
comes on when the blood has forced the air that is in the
veins back into the breast. Sleep is, according to Aristotle
also, an affection of this same region of the breast, which
was the seat of the κοινὴ αἴσθησις. In the *Placita* we read[7]
that Diogenes placed τὸ ἡγεμονικόν (which term, however,
raises suspicion of the authenticity of the statement) ἐν τῇ
ἀρτηριακῇ κοιλίᾳ τῆς καρδίας, ἥτις ἐστὶ πνευματική. If this

---

[1] Theophr. *de Sens.* §§ 44–5.   [2] ξύνεσις.   [3] περὶ τὰ στήθη.
[4] With the above cf. Arist. *de Mem.* 453ᵃ 14–31 and 453ᵇ 3–10.
[5] Cf. Panzerbieter, *Diogenes Apoll.* pp. 90–3.
[6] Plut. *Epit.* v. 24; Panzerb. p. 90; Arist. *de Somno, passim.*
[7] Aët. iv. 5. 7; Diels, *Dox.*, p. 391; Panzerb., pp. 87 seqq.

statement has a basis of truth, we must regard those of the passages in which the air around the brain is said to be the percipient subject as only provisionally true: this air has to convey the messages of sense to the air of the thorax before consciousness of sensation arises. It may be that Diogenes, like Aristotle, made the environment of the brain only an intermediate stage in the process of sensation as regards three senses—hearing, seeing, and smelling; while touching and tasting, of which he says nothing definite, were regarded by him, as by Aristotle, as having direct communication with the central seat of sense-perception [1]. On the whole it appears that Diogenes possessed in a marked degree a perception, which Alcmaeon had in a slight measure, but which Democritus and Empedocles did not possess at all, of the necessity for a central organizing faculty, whether of sense or intelligence, on which consciousness and memory depend; and that he regarded this as seated chiefly in the air in the region of the heart—whether in the lungs [2] or, as the compiler of the *Placita* tells us, in 'the arteriac cavity' of the heart.

Diogenes had a conception (which Empedocles and Democritus lacked) of the necessity of a synthetic faculty.

Plato.

§ 9. Plato of course does not even name a κοινὴ αἴσθησις, but he investigated carefully the function of synthesis whose importance was paramount in his psychology. He ascribed it not to sense, as Aristotle did, but to thought. Yet there is reason for regarding this difference—from the psychologist's point of view, not from that of the metaphysician or epistemologist—as one of method more than anything else. No psychologist has ever been able to answer satisfactorily the question where sense-perception ends and thinking commences. In order, therefore, to be in a position to compare Aristotle's doctrine of κοινὴ αἴσθησις with Plato's doctrine of

Plato denied synthesis to sense, and ascribed it to thought or intelligence. Yet he in many ways paves the way for Aristotle's theory of sensus communis. We may, therefore,

[1] Cf. Arist. 469ᵃ 12 δύο αἰσθήσεις φανερῶς ἐνταῦθα (sc. εἰς τὴν καρδίαν) συντεινούσας ὁρῶμεν, τήν τε γεῦσιν καὶ τὴν ἀφήν, ὥστε καὶ τὰς ἄλλας ἀναγκαῖον.

[2] Diogenes probably held that the κοιλίαι of the heart communicated directly with the lungs. Cf. Arist. 496ᵃ 22 καὶ εἰσὶν [sc. αἱ κοιλίαι] εἰς τὸν πνεύμονα τετρημέναι πᾶσαι.

the synthetic faculty so far as these may coincide, we shall here consider what information the latter has left us respecting the faculty whereby the ᵃdata of sense are combined or distinguished; also respecting imagination, memory, reminiscence, and the other functions claimed for the κοινὴ αἴσθησις by his great pupil.

[margin: compare Plato's and Aristotle's theories of synthesis, distinct though they were in kind.]

§ 10. In the *Theaetetus* it is that Plato most emphatically exhibits his appreciation of the importance of the synthetic faculty. 'With the eyes one discerns black and white objects; with the ears one perceives grave and acute tones; at least so people say. This account of the matter is not, however, scientifically accurate. We do not see *with* the eyes; rather we see *through* them. We do not hear *with* the ears, but *through* them also. It would surely be strange if we had placed within us, like so many warriors in Trojan horses [1], a multitude of sensory faculties (αἰσθήσεις) which did not tend to unite in some one form— call it soul or some other name—*with* which we truly perceive all that we do perceive through these senses as through instruments [2].' The organs through which one perceives things *hot, hard, light,* or *sweet*, are parts of the body. When we perceive such an object through some one faculty (δυνάμεως), it is not possible for us to perceive the same through any other faculty. We cannot by sight perceive the objects of hearing, nor can we by hearing perceive the objects of sight. But if you think something concerning *both* of these objects *in common*, it cannot be through either organ singly that you do so [3]. Sound and colour are two different objects, unlike one another. In thus thinking of them as distinct from each other, as together

[margin: The soul itself (not the organs of special sense) the true faculty of perception. These are but instruments or channels of the soul's activity. It is not with the eyes but through them that one sees. We have not a multitude of different sensory faculties within us like the warriors ensconced within the Trojan horse. To think something common to several sensory faculties]

---

[1] Cf. Galen. *de Placit. Hipp. et Plat.* §§ 631-3.

[2] ἐς μίαν τινὰ ἰδέαν, εἴτε ψυχὴν εἴτε ὅ τι δεῖ καλεῖν, πάντα ταῦτα ξυντείνει, ᾗ διὰ τούτων οἷον ὀργάνων αἰσθανόμεθα ὅσα αἰσθητά, *Theaet.* 184 D.

[3] *Ibid.* 185 A εἴ τι ἄρα περὶ ἀμφοτέρων διανοεῖ, οὐκ ἂν διά γε τοῦ ἑτέρου ὀργάνου, οὐδ' αὖ διὰ τοῦ ἑτέρου περὶ ἀμφοτέρων αἰσθάνοι' ἄν. Notice the choice of verbs employed in each clause, by which Plato would seem to desire to fence off the action of the synthetic faculty altogether from that of sense-perception. He has used αἰσθάνεσθαι just above (see last note) to denote the action of ψυχή operating *through* the αἰσθήσεις.

two, while each is *one*, it cannot be by the agency of either sight or hearing singly that one forms a conception which thus embraces both [1]. Common characteristics of diverse sense-percepts are not themselves perceived by the special organs of sense. The soul itself, independently of sense, 'inspects' the attributes common to objects of the different senses—their several unity, their difference *inter se*, &c. [2] There is no special organ at all, formed of a bodily part, instrumental to the soul's action in perceiving these common attributes [3]. Here Plato recognizes the function of synthesis as necessary for the co-ordination and systematization of the data of sense, but denies that it belongs to sense, or has a bodily part, analogous to the eyes or ears, connected with it as its instrument.   In 184 D, however, by the very terms he employs ($\hat{\eta}$ . . . αἰσθανόμεθα) he shows how closely his thought approximates to that of Aristotle. He did not speak, it is true, of a πρῶτον αἰσθητικόν or of a κοινὴ αἴσθησις, yet by this passage the thought of such a faculty might have been suggested to Aristotle. This is confirmed by the use of the word κοινά in the same connexion. Plato does not employ the term τὰ κοινά here, as Aristotle did, to signify 'common sensibles,' i. e. objects capable of being perceived by all the senses in common ; such e. g. as κίνησις. According to Aristotle, κίνησις is perceptible by *any* sense, being a common object to all, or at least to sight and touch. According to Plato no one sense can perceive the κοινά. Even here, however, the difference between Aristotle and Plato is not so great : for, after all, the κοινά were for Aristotle only αἰσθητὰ κατὰ συμβεβηκός in relation to any one sense, while they were directly αἰσθητά to the κοινὴ αἴσθησις, fulfilling as this did the function here ascribed by Plato to ψυχή. With this, and the use of αἰσθανόμεθα as referred to above, the thought of ἡ κοινὴ

Marginal notes (left column):
we require a faculty different from any one of these : whether we call it ψυχή or by any other name. 'The soul has not need of any bodily instrument' in thinking of the *common* features of various sensibles. Different use of the term τὰ κοινά in Plato and Aristotle. Yet this difference is not absolute. For τὰ κοινά, if = objects of the κοινὴ αἴσθησις, in Aristotle are exactly parallel to Plato's κοινά, the objects perceived by the soul itself as common to the data of several senses.

---

[1] *Theaet.* 185 B οὔτε γὰρ δι' ἀκοῆς οὔτε δι' ὄψεως οἷόν τε τὸ κοινὸν λαμβάνειν περὶ αὐτῶν.

[2] ἀλλ' αὐτὴ δι' αὑτῆς ἡ ψυχὴ τὰ κοινά μοι φαίνεται περὶ πάντων ἐπισκοπεῖν, *Theaet.* 185 D.

[3] *Ibid.* δοκεῖ τὴν ἀρχὴν οὐδ' εἶναι τοιοῦτον οὐδὲν τούτοις ὄργανον ἴδιον ὥσπερ ἐκείνοις (sc. as the *proper* sensibles have).

αἴσθησις lies obvious to the reader's mind. As κοινά in his sense of the word, i. e. as objects of the ψυχή so acting through the αἰσθήσεις, Plato names (a) οὐσία καὶ τὸ μὴ εἶναι, (b) τὸ ὅμοιον καὶ τὸ ἀνόμοιον, (c) ἓν καὶ πολλά, (d) τὸ καλὸν καὶ τὸ αἰσχρόν, (e) τὸ ἀγαθὸν καὶ τὸ κακόν (*Theaet.* 186 A).

§ 11. The *presentative* faculty—φαντασία. The same wind which to one man is cold is to another warm: and it *is* so because it *appears* (φαίνεται) so. This 'appearing' is the work of sense: φαντασία and αἴσθησις are of essentially the same nature, and possess similar evidential value throughout the various provinces of sensation[1]. So Plato observes, tracing the character of subjective or Protagorean idealism—or rather *sensationism*. In this 'appearing,' however, which Plato treats with such scant courtesy, lies the foundation of experience, since the presentative is the foundation of the re-presentative element[2].

Out of such 'appearing' arises *memory*, by which we have knowledge of past time, or by which there *is* for us a past. The soul, says Plato, is like a book[3]. Memory and perceptions meet at the moment when such perceptions occur, and thereupon memory as it were inscribes a record of the perceptions in our souls. When this record is true, true opinion arises in our souls; when the 'secretary of records' within us[4] inscribes what is not true, the resulting opinion is false. But there is another artist at work within us at the same time as memory. This other is the painter (ζωγράφος)—*Imagination*. He, succeeding the recording secretary, paints in the soul likenesses (εἰκόνες) of the things perceived—transferring from the eye or other organ of sense the sensible data which are to be matter of

Marginal note: Φαντασία in Plato = (a) presentation, (b) re-presentation. But in this its second function (b) Plato only refers to it by figurative terms. The word is generally used by Plato in the first sense (a). Memory (the γραμματεύς within us), imagination (the ζωγράφος within us). The records of memory refer to the past. The pictures of imagination may refer to past or future. On

---

[1] *Theaet.* 152 B-C. Here φαντασία is clearly a different thing from the faculty of reproductive imagination as defined by Aristotle (429ᵃ 1) κίνησις ὑπὸ τῆς αἰσθήσεως τῆς κατ᾽ ἐνέργειαν γινομένη. Cf. *Theaet.* 152 B τὸ δέ γε φαίνεται αἰσθάνεσθαί ἐστιν; ἔστι γάρ.

[2] The synthesis involved in φαντασία at this its first stage (wherein ideas of objects are *presented* to the mind) is what psychology should most earnestly examine. Needless to say Plato did not pay much attention to it; nor did Aristotle.

[3] δοκεῖ μοι ἡμῶν ἡ ψυχὴ βιβλίῳ τινι προσεοικέναι, *Phileb.* 38 E.

[4] *Phileb.* 39 A ὁ τοιοῦτος παρ᾽ ἡμῖν γραμματεύς, sc. μνήμη.

them are
built ex-
pectations
when they
have this
latter
reference
(ἐλπίδες εἰς
τὸν ἔπειτα
χρόνον).
Definitions
of *memory*
and *remi-
niscence.*
In remi-
niscence
the soul
acts with-
out the
body.

opinion or discourse. Thus a person sees images of those data somehow painted within him. The likenesses of true opinions and words are true, those of the false are false[1]. But it is not to the past and present alone that these writings and paintings have reference ; they refer also to the future[2]. Thus arise *expectations* (ἐλπίδες εἰς τὸν ἔπειτα χρόνον) as to the future, such as we are filled with our whole lives through. Memory is a conservation of percep-tion[3]. Reminiscence is, however, different from memory[4]. Whenever the soul by itself within itself as far as possible[5] retraces and retrieves a lost piece of perception or learning, we say that it recollects (ἀναμιμνῄσκεσθαι). Reminiscence, or recollection, is the power which the soul by itself, and, as far as possible, without the body, has of recovering experiences which it had before in common with the body.

Forgetting. *Forgetting*, on the other hand, is simply the exit of memory[6], which, again, is to be distinguished from unconsciousness, the negative state expressed by the word ἀναισθησία. Of course if we are completely unconscious we are thereby without all our former αἰσθήσεις and μαθήματα. This, how-ever, is not what happens when we simply forget. We are conscious enough in all respects, save in that of the par-ticular αἴσθησις or μάθημα which has left our minds[7].

Illustra-
tion of the
formation
of memory:
the ' wax
tablet '

§ 12. The operation of memory in the first instance—the way in which the scribe or secretary takes his records—is further described by the following simile. There is as it were in the mind of man a block of wax for receiving

[1] *Phileb.* 39 B–C. Here we find Plato raising the subject of the reproductive imagination, the psychical faculty described or defined by Aristotle in the preceding note.

[2] *Phileb.* 39 D.   [3] *Phileb.* 34 A σωτηρίαν αἰσθήσεως.

[4] In what follows I neglect as irrelevant all reference to the dis-tinctively Platonic theory of ἀνάμνησις, suggesting pre-existence and the doctrine of Ideas.

[5] ὅταν ἀπολέσασα μνήμην εἴτ' αἰσθήσεως εἴτ' αὖ μαθήματος αὖθις ταύτην ἀναπολήσῃ πάλιν αὐτὴ ἐν ἑαυτῇ, καὶ ταῦτα ξύμπαντα ἀναμνήσεις καὶ μνήμας που λέγομεν, *Phileb.* 34 B–C with *Phaed.* 75 E. This passage of the *Philebus* (34 B–C) forms the original of much that is in Arist. *de Mem.* ii *ad init.*, 451ᵃ 18 seqq.

[6] μνήμης ἔξοδος.   [7] *Phileb.* 33 E.

impressions [1]. In different persons it is of different sizes *within us.* and different qualities also, being in some harder, moister, *On this percep-* or purer than in others. It is the gift of Mnemosyne, the *tions or* mother of the Muses, to men. When we wish to remember *thoughts* *are in-* aught that we see, or hear, or think, within ourselves, we *scribed.* hold the wax to the perceptions or thoughts, and take *On the qualities of* impressions of these in it as if stamped there by a seal ring. *this wax* *and its* We remember and know what is printed there as long *fitness for* as the impression lasts; but when it is effaced, or when no *receiving* *and retain-* impression has been taken, we forget, and do not know. *ing distinct* *and clear* Now when the wax in the soul of any one is deep and *impres-* abundant, and smooth and well-tempered, the impressions *sions* *depends* which pass through the senses and sink into the heart *the good-* of the soul (as Homer says in a certain passage in which he *ness or* *badness of* indicates the likeness of the soul to wax [2]), being pure and *memory.* *Good* clear and finding a sufficient depth of wax, are lasting. *and bad* Minds such as these easily learn, and easily retain what *memory* *explained* they learn, nor are they liable to confusion. They have in *and illus-* them plenty of room, and having clear impressions of things, *trated.* they quickly distribute these in their proper places on the block. Such are called wise or clever men. When, on the contrary, the heart of any one is ' shaggy [3],' a quality which the all-wise poet commends, or muddy, or of impure wax, or very soft, or very hard, there is in the mind a corresponding defect. The soft are good at learning, but apt to forget; the hard are the reverse; the ' shaggy,' or rugged, or gritty, or those who have an admixture of earth or dung in their composition, have the impressions indistinct; so have also the hard, for there is no depth in them. The soft, too, are indistinct, for their impressions are easily confused and effaced. Still greater is the indistinctness when all are jostled together in a little soul which has no room. Such are the natures which have false opinion; for when they see or hear or think of anything, they are slow in assigning the right objects to the right impressions—in their stupidity they confuse them, and are

---

[1] κήρινον ἐκμαγεῖον.   [2] κῆρ (=κέαρ), κηρός.
[3] ὅταν λάσιόν του τὸ κέαρ ᾖ. The heart, or the region round the heart, is for Aristotle the organ of central sense.

apt to see and hear and think amiss—and such men are said to be deceived in their knowledge of objects and ignorant[1]. In this famous simile, Plato, in his picturesque way, portrays the functions of sensation, memory, and imagination. The *stamping* of the impressions is the presentative φαντασία—sense-perception. The memory or *retention* of them, when the objects which stamped them are gone, is due to the representative φαντασία—the reproductive imagination.

§ 13. But here, too, Plato proceeds to develop the difference between mere retention of impressions and the power of recalling them to mind at need: the difference between memory and reminiscence[2]. To do this he introduces another, and equally famous, simile. Suppose a person to have caught a great many wild doves, or other birds, and to keep them in an aviary at home. In one way we may say of him that he always *has* them, because he is the possessor of them; but, in another way, he may have none of them the while. They are merely in his power, in his enclosure, so that he can catch any of them when he wants, and let it go again, and do this as often as he likes. Now to apply this. Suppose that there is in each one's mind an aviary of all sorts of birds, some in great flocks apart, some in small groups, others solitary, flying anywhere and everywhere. Suppose further that the birds are kinds of knowledge; that when we were children the aviary was empty; but that whenever a person has gotten and confined in the enclosure a kind of knowledge he may be said to have learned or discovered the thing which is the subject of the knowledge: and that, therefore, he *knows* it. . . . When the various forms of knowledge are flying about in the aviary, and he, wishing to capture a certain sort of knowledge out of the general store, takes the wrong one by mistake, getting hold of the ring-dove when he wants the pigeon: in this way we may

---

[1] *Theaet.* 191 D–195 A, from Jowett's Translation.

[2] No one can fail to be struck with the fundamental resemblances between Plato here and Aristotle in the *de Memoria*.

suppose false opinion to arise. When he catches the one
he wants, his opinion is true [1].

In the former of these two sensuous images—the block of
wax and the *columbarium*—we have an exact, though fanciful,
parallel for Aristotle's κύριον αἰσθητήριον, at least on its passive
side. Nowhere else does Plato so closely approach the
Aristotelean conception [2]. Even here he does not seem to
treat it quite seriously, but leaves it before us rather as a
piece of fancy work than a serious product of psychological
analysis. The block of wax represents the mere *retention*
of ideas—memory: the dovecote represents their active
recall—reminiscence. He does not go to the length of
saying that there is any one particular organ or bodily
part analogous to the wax or pigeon-house; he does not
assign its function to the heart or brain. Had he done
so, it would have been more natural for him to choose
the former, the brain being the instrument of *reason*,
according to the *Timaeus*. He has thus, however, skilfully
enough delineated the functions of sensation, memory, and
imagination.

§ 14. To return to his conception of Reminiscence: we
shall find that in the *Phaedo* in connexion with this subject
he has as genuine, if not as highly developed, notions
respecting the 'Association of Ideas' as his pupil Aristotle
exhibits. He there observes that if a person recalls
anything by reminiscence, he must at a former period
have known that thing. Now if a person sees or hears
something or perceives it by some other sense, and thereby
gets the idea not of it alone, but also of something else
the knowledge of which is different, a person is properly
said to recollect (ἀναμιμνήσκεσθαι) the latter—the thing of
which he thus gets the idea. Thus a person on seeing
a lyre, or cloak, which a friend was wont to use or wear,
gets into his mind at once the idea of the friend, and this

Association of ideas in reminiscence. Anticipations of Aristotle.

---

[1] *Theaet.* 197 D seqq., Jowett's Trans.

[2] It will be noticed that it is to the heart, not to the brain, that the
similes, however obscurely, point as the organ of such a faculty of
sensus communis.

is reminiscence. The process of association is especially noticeable for the way in which it recalls to mind things which, through lapse of time or for some other reason, one had quite forgotten. The reminiscence may take place either (*a*) from the similarity of the idea, which recalls the other, to this other, as when the picture of Simmias recalls the idea of Simmias; or (*b*) without any such similarity, as in the case of the lyre, the sight of which recalls the idea of the friend who used to play upon it [1].

Formation and nature of δόξα— the faculty of judgment at its lowest grade.

§ 15. It is germane to the subject to adduce here Plato's account of opinion (δόξα)—the faculty of judgment at its lowest grade. Opinion results from memory and sense. What happens is like this: A person sees an object at a distance, not quite distinctly. His curiosity leads him to discern it clearly and pronounce what it is that he sees. 'What is it that I see?' he would say to himself: 'What is the object that presents (φανταζόμενον) itself as standing beside the cliff yonder beneath the tree?' Next he might make answer to himself and say: 'it is a human being,' thereby guessing correctly, or he might mistake and say: 'What I see is something made by shepherds— a figure of a human being.' If in company with some one, he would give audible utterance to these attempts to pronounce; his efforts at opinion (δόξα) would take the form of discourse (λόγος). But if he is alone he proceeds to discuss (διανοούμενος) the matter with himself, keeping it to himself for a good while [2]. Thus αἴσθησις, φαντασία, μνήμη, δόξα, διάνοια, and λόγος are brought into relation with one another; the object of presentation is compared with that of memory or thought, and a judgment or opinion, true or false, is formed of the relation between them [3].

(*Phaedo*) Plato's specula-

§ 16. Notwithstanding that in the *Theaetetus* Plato speaks of the soul as being, by itself, without the use of

---

[1] *Phaedo* 73 C–E. For association of *interests* superadded to and reinforcing association of *ideas*, cf. *Lysis* 219-20.

[2] *Phileb.* 38 C seqq.

[3] Here, it may be observed, we have to do with what Aristotle calls the perception of τὰ κατὰ συμβεβηκός.

any bodily organ, able to recover by reminiscence its temporarily lost impressions, he in various places speaks of it, and even of its highest functions, as having a bodily seat or organ. 'I speculated,' says Socrates [1], 'as to whether the blood is the part of us with which we think and perceive [2], or else the air, or the fire, within us; or whether it is none of these, but the brain is that which supplies the sensations (ὁ παρέχων τὰς αἰσθήσεις) of hearing, seeing, and smelling [3]; and whether from these arise memory (μνήμη) and opinion (δόξα), while from memory and opinion, when fixed and stable (λαβούσης τὸ ἠρεμεῖν κατὰ ταὐτά), arises scientific knowledge (ἐπιστήμη).' Here the organ ᾧ φρονοῦμεν is evidently made to include reference to the processes of sense-perception, and also to those which immediately follow—memory and the other processes referred by Aristotle to the κοινὴ αἴσθησις. Thus the Platonic Socrates enumerates all or most of the suggestions made by former writers to explain the 'seat' of perception and thinking—by Empedocles and Kritias (αἷμα), Diogenes of Apollonia (ἀήρ), Heraclitus (πῦρ), and Alcmaeon (ὁ ἐγκέφαλος). In the *Timaeus* Plato himself adopts the last of these suggestions, making the brain the seat of the intellectual functions of soul. Hippocrates, as well as Alcmaeon, had already held the brain to be the essential organ of sense and thought. 'This is that which interprets for us the impressions derived from the air (ἡμῖν τῶν ἀπὸ τοῦ ἠέρος γενομένων ἑρμηνεύς) if it is in a healthy condition; but it is the air that supplies it with intelligence (τὴν δὲ φρόνησιν αὐτῷ ὁ ἀὴρ παρέχεται) [4].'

§ 17. 'In it (the spinal marrow) the Demiourgos implanted and fastened the several kinds of souls; and according to the number and fashion of the shapes that Soul should have, corresponding to her kinds, into so many similar forms did he divide the marrow at the outset of his distribution. That which should be as it were a field,

[side note: tions as to the organ of the faculty of synthesis.*]*

[side note: (*Timaeus*) Tripartite division of soul and allocation of its parts to bodily organs. The soul*]*

---

[1] *Phaedo* 96 C, with Archer-Hind's notes.
[2] ᾧ φρονοῦμεν: cf. ἐπὶ τὸ φρόνιμον, *Tim.* 64 B, which also evidently includes sense-perception.
[3] He does not mention touching and tasting here.
[4] Hippocr. *de Morbo Sacro,* 17.

of plants. The αἴσθησις of plants is not perception but feeling.

to contain in it the *Divine* seed, he moulded in a spherical form, and this part of the marrow he called the brain (ἐγκέφαλος), with the view that, when each animal was completed, the vessel containing it should be the head. That which was to have the *mortal* part of the soul he distributed into moulds at once round and elongated [i. e. the vertebral column]. All these forms he named marrow, and from them, as from anchors, he put forth the bonds to fasten all the soul; and then he wrought the entire body round about it; first building, to fence it, a covering of bone[1].' Thus for Plato the cerebro-spinal marrow was the organic seat of *intelligence* (νοῦς), *courage* (θυμός, or τὸ θυμοειδές), and *appetite* (τὸ ἐπιθυμητικόν). The cerebral portion was given to νοῦς; the thoracic portion to θυμός; the abdominal, to ἐπιθυμία. We learn further in the *Timaeus*[2] that the third part of soul, which plants as well as man possess, is in man seated between the midriff and the navel (μεταξὺ φρενῶν ὀμφαλοῦ τε ἱδρῦσθαι); that in virtue of it plants have—not, indeed, the 'sense' which is an element of cognition, but only—feeling, pleasant or painful, with the accompanying appetites or impulses[3].

The three parts of soul in the *Timaeus*.

§ 18. The three souls or parts of soul were connected through the cerebro-spinal marrow on which they were all 'strung' together. The head was the separate abode of the immortal[4] soul; the mortal soul was planted apart

---

[1] *Tim.* 73 C–D (Archer-Hind).   [2] 77 B.

[3] ᾧ δόξης μὲν λογισμοῦ τε καὶ νοῦ μέτεστι τὸ μηδέν, αἰσθήσεως δὲ ἡδείας καὶ ἀλγεινῆς μετὰ ἐπιθυμῶν. In this sentence αἰσθήσεως means not the sensory factor, or element, of knowledge, but what is generally known to modern psychologists as 'feeling': the pleasurable or painful element in consciousness. It is in this sense that Plato here ascribes αἴσθησις to plants (φυτά). Aristotle denies it of plants in this as well as in the sense of perception, making it the attribute of ζῷα exclusively. As for the Greeks the term αἴσθησις had to express the sense of pleasure or pain as well as the factor of cognition, so with us till lately the word 'feeling' did duty for both, and is commonly used in this ambiguous way in the works of English writers of the last century. Plato distinguishes cognitive αἴσθησις from ἡδονῇ καὶ λύπῃ μεμειγμένος ἔρως, *Tim.* 42 A. In *Philebus* also (e. g. 32 D) ἡδονή and λύπη together='feeling,' cf. § 19 *infra*.

[4] For what follows see Grote, *Plato*, iii. 272–5. In the *Phaedrus* 246 B θυμός and ἐπιθυμία seem reckoned in with the immortal soul, the body only being mortal.

from it in the trunk, with the neck as an isthmus of
separation between the two. 'Again, the mortal soul was
itself not single but double : including two divisions, a better
and a worse. The gods kept the two parts separate;
placing the better portion in the thoracic cavity nearer
to the head, and the worse portion lower down, in the
abdominal cavity : the two being divided from each other
by the diaphragm, built across the body as a wall of
partition.' 'Above the diaphragm, and near to the neck,
was planted the energetic, courageous, contentious, soul ; so
placed as to receive orders easily from the head, and to
aid the rational soul in keeping under constraint the
mutinous soul of appetite, which was planted below the
diaphragm. The immortal soul was fastened or anchored
in the brain, the two mortal souls in the line of the spinal
marrow continuous with the brain ; which line thus formed
the thread of connexion between the three. The heart
was established as an outer fortress for the exercise of
influence by the immortal soul over the other two. It
was at the same time made the initial point of the veins—
the fountain from whence the current of blood proceeded
to pass forcibly through the veins round to all parts of
the body. The purpose of this arrangement is, that when
the rational soul denounces some proceeding as wrong
(either on the part of others without, or in the appetitive
soul within), it may stimulate an ebullition of anger in the
heart, and may transmit from thence its exhortations and
threats through the many small blood-channels [1] to all the
sensitive parts of the body; which may thus be rendered
obedient everywhere to the orders of our better nature.
. . . The third or lowest soul, of appetite and nutrition,
was placed between the diaphragm and the navel. This
region of the body was set apart like a manger for containing necessary food : and the appetitive soul was tied
up to it like a wild beast; indispensable, indeed, for the

---

[1] For Plato, as for Aristotle, the blood-vessels take the place of nerves,
conveying sensations through the body; cf. *Tim.* 65 C, 67 B, 70 A seqq.,
77 E.

continuance of the race, yet a troublesome adjunct, and therefore placed afar off, in order that its bellowings might disturb as little as possible the deliberations of the rational soul in the cranium, for the good of the whole. The gods knew that this appetitive soul would never listen to reason, and that it must be kept under subjection altogether by the influence of phantoms and imagery. They provided an agency for this purpose in the liver, which they placed close upon the abode of the appetitive soul. They made the liver compact, smooth, and brilliant, like a mirror reflecting images ;—moreover, both sweet and bitter on occasions. The thoughts of the rational soul were thus brought within view of the appetitive soul, in the form of phantoms or images exhibited on the mirror of the liver [1]. When the rational soul is displeased, not only images corresponding to this feeling are impressed, but the bitter properties of the liver are all called forth. . . . When the rational soul is satisfied, so as to send forth mild and complacent inspirations,—all this bitterness of the liver is tranquillized, and all its native sweetness called forth. . . . It is thus through the liver, and by means of these images, that the rational soul maintains its ascendancy over the appetitive soul ; either to terrify and subdue, or to comfort and encourage it.'

'Moreover, the liver was made to serve another purpose. It was selected as the seat of the prophetic agency ; which the gods considered to be indispensable, as a refuge and aid for the irrational department of man. Though this portion of the soul had no concern with sense or reason, they would not shut it out altogether from some glimpse of truth. The revelations of prophecy were accordingly signified on the liver, for the instruction and within the easy view of the appetitive soul; and chiefly at periods when the functions of the rational soul are suspended— either during sleep, or diseases, or fits of temporary ecstasy.

---

[1] Plato rejects vaticination from victims. Tim. 72 B στερηθὲν δὲ τοῦ ζῆν [sc. τὸ ἧπαρ] γέγονε τυφλὸν καὶ τὰ μαντεῖα ἀμυδρότερα ἔσχε τοῦ τι σαφὲς σημαίνειν.

For no man in his perfect senses comes under the influence
of a genuine prophetic inspiration. Sense and intelligence
are often required to interpret prophecies, and to determine
what is meant by dreams, or signs, or prognostics of other
kinds: but such revelations are received by men destitute
of sense[1]. To receive them is the business of one class of
men ; to interpret them, that of another. . . . Such was the
distribution of the one immortal and the two mortal souls,
and such the purposes by which it was dictated. We cannot
indeed (says Plato) proclaim this with full assurance, as
truth, unless the gods would confirm our declarations.
We must take the risk of affirming what appears to us
probable[2].' In these three 'parts of soul' we have the
foundation laid by Plato of the future analogous division
of mental elements into those of cognition, feeling, and (will
or) desire.

§ 19. It may help us to understand Plato's distribution
better if, distinguishing αἴσθησις as we have done into two
elements, the element of feeling and the element of cognition, we refer the latter element of αἴσθησις uniformly to the
intellectual soul which has its seat in the *cranium*[3]. The
distinction is strongly marked for Plato, though he has
not the proper terms for expressing it. Plants have no
share in the cognitive αἴσθησις. This, therefore, we must
regard as coming under the part of soul ᾧ μανθάνει
ἄνθρωπος[4]. In the *Laws*[5] Plato implicitly confirms this
classification in the words ξυλλήβδην δὲ νοῦς μετὰ τῶν
καλλίστων αἰσθήσεων (sc. τῆς ὄψεως καὶ τῆς ἀκοῆς) κραθείς.

*Αἴσθησις
as element
of cogni-
tion to
be kept
separate
from
αἴσθησις
as element
of feeling.
Plato dis-
tinguishes
them, but
for want
of appro-
priate
terms for
each, this
distinction*

---

[1] There is another species of divination, that depending on divinely
inspired excitement or 'enthusiasm,' which also requires to be inter-
preted by calm reason. *Phaedr.* 244 A seqq., 265 A seqq.

[2] Grote, *Plato*, iii. pp. 272-5 ; Plato, *Timaeus* 69-73 ; cf. also
Phaedrus 246 A seqq. ; *Rep.* iv. 438 D seqq. ; *Laws* xii. 961 D, E.

[3] Plato himself aims at the above distinction, so important for
psychology, when in *Tim.* 69 D and 79 B, he divides αἴσθησις into
αἴσθησις ἄλογος, or αἴσθησις ἡδεῖα καὶ ἀλγεινὴ μετὰ ἐπιθυμῶν, on the one
hand, and, on the other, the αἴσθησις which is subservient to cognition.
The former is part of the lower or vegetative soul, that which φυτά
possess and which has no self-consciousness (*Tim.* 77 B). Cf. Zeller,
Plato 432 n., E. Tr.　　　　[4] *Repub.* 436 A.　　　　[5] 961 D.

is forgotten by readers. The cognitive αἴσθησις (or the αἴσθησις subservient to cognition) probably was conceived by Plato as belonging to the cranial part of soul.

In *Timaeus* 65 A, 71 A, we learn that ἔρως, αἴσθησις ἄλογος, ἡδονή, λύπη, θάρρος, φόβος, θυμός, ἐλπίς are seated in the thoracic and abdominal parts of soul ; whence it is obvious to infer that the other αἴσθησις—that conducive to cognition—belongs to the cranial part. Sight and hearing are ministers of reason[1]. Against this it might seem as if Plato attributes cognitive power to the lower or abdominal soul, when he says that images are presented on the mirroring surface of the liver for the purpose of warning or encouragement. But on examination of the passage (*Tim.* 71 B) we find that the effects conveyed to this organ from the brain only impress the appetitive part with *feelings* or *emotions*, without necessarily implying that it has any *cognitive* function[2].

Tasting referred by Plato to the heart. Touching proceeds, through the σάρξ, ἐπὶ τὸ φρόνιμον.

§ 20. It is at first somewhat surprising, after this, to find that Plato in explaining the physiology of tasting[3] refers its sensations to the heart. 'When earthy particles enter in by the small veins which are like test-tubes on the tongue extending from it to the heart[4], these give rise to astringent tastes.' Does the heart then, for Plato, as for Aristotle, take a direct share in the mechanism of sense? The sense of touching is for Aristotle that most obviously and directly traceable to the heart as its organ ; we cannot discover from Plato whether he connected it with this, as he contents himself with referring the consciousness of the sensations of touch to a movement propagated by the σάρξ onwards until it reaches the

---

[1] *Tim.* 47 B–C.

[2] ἵνα ... ἡ ἐκ τοῦ νοῦ φερομένη δύναμις, οἷον ἐν κατόπτρῳ δεχομένῳ τύπους καὶ κατιδεῖν εἴδωλα παρέχοντι, φοβοῖ μὲν αὐτό (sc. τὸ ἐπιθυμητικόν) ; also just before (71 A) εἰδότες δὲ αὐτό, ὡς λόγου μὲν οὔτε ξυνήσειν ἔμελλεν, εἴ τέ πῃ καὶ μεταλαμβάνοι τινὸς αὐτῶν αἰσθήσεως, οὐκ ἔμφυτον αὐτῷ τὸ μέλειν τινῶν ἔσοιτο λόγων, ὑπὸ δὲ εἰδώλων ... ψυχαγωγήσοιτο : from which we can see that the appetitive soul is only susceptible to non-rational effects in the way of feeling or emotion.

[3] Perhaps the fact that this sense belongs rather to *feeling* than to *cognition*, may serve to explain the reference of it to a non-cognitive part of soul ; but why then was it not directed towards the liver?

[4] περὶ τὰ φλέβια οἷόνπερ δοκιμεῖα τῆς γλώττης τεταμένα ἐπὶ τὴν καρδίαν, *Tim.* 65 C.

'centre of consciousness[1].' He does not speak of odours *Smelling* as affecting the brain; when they are disagreeable, in *affects all the part of* certain cases, they irritate all the cavity of the body lying *the cavity of the body* between the head and the navel[2]. Sound is, as we know, *betwixt the head and a stroke caused by the air, transmitted through the ears, *the navel.* affecting the brain and blood, and propagated 'to the soul'; *Hearing* and the motion produced by it, beginning in the head and *involves a motion be-* ending in the liver, is *hearing*[3]. He uses only vague terms *ginning* to designate the sensoria concerned in dreaming. Pungent *with the head and* tastes are caused by substances which affect the tongue *ending* and fly up towards the 'senses of the head[4].' From all *in the liver.* this we can see how difficult it is to gather what Plato *Did Plato* regarded as the common seat or organ of the αἰσθήσεις as *any one* elements of cognition, or, indeed, whether he held that there *part as organic* was any one such seat. The brain at one time (in accord- *to the* ance with the view that the function of synthesis is *senses in common?* intellectual) seems to be the organ to which the senses should refer their messages; while, soon after, the heart or the liver is found in possession of similar prerogatives.

§ 21. Plato suffers from the consequences of what Galen *Perplexi-* ascribes to his merit—the adoption of three ἀρχαί[5]. To *ties arising from his* this initial want of centralization are traceable the per- *tripartite* plexities into which he leads us, and which he must himself *division of* have felt, respecting the various sensory functions, and the *soul.* bodily parts concerned in each. This initial subdivision of the soul into 'parts,' located in three different portions of the body, makes it impossible for him to give a consistent or systematic account of the psychical facts. We cannot, therefore, elicit from his writings any evidence as to views of his own respecting a κοινὸν αἰσθητήριον. On several occasions, especially in the similes of the waxen block and the dovecote, he comes very near the thought of it; but he always employs images and metaphors from which we

[1] 64 B μέχριπερ ἂν ἐπὶ τὸ φρόνιμον ἐλθόντα.

[2] *Tim.* 66 D–67 A.   [3] *Tim.* 67 B.

[4] 65 E ὑπὸ κουφότητος ἄνω πρὸς τὰς τῆς κεφαλῆς αἰσθήσεις.

[5] Cf. Galen. *de Placit. Hipp. et Plat.* §§ 505 and 519, ὅτι μὲν οὖν εὐλόγως ὁ Πλάτων εἴδη τε καὶ μέρη ψυχῆς ὀνομάζει ταῦτα, μακροτέρων οὐ δέομαι λόγων.

cannot extract a clear or simple meaning. With regard, however, to the synthetic faculty which arranges the data of sense in memory, &c., we find that he has treated most of its functions in a way which closely anticipates much of what Aristotle afterwards taught. Not, however, attributing it, as Aristotle did, to sense, he ascribes to it functions which far transcend those ascribed to it by Aristotle. He lays what may have been the foundation of Aristotle's theory of it as the faculty which distinguishes and compares the data of sense, and of the theory of imagination, memory, and reminiscence. Indeed, the terms in which he expressed himself respecting these, and the similes he employed for the purpose of elucidating them, have remained part of, and have deeply influenced the language of, psychology, to the present day. In fullness of detail on such points Aristotle surpasses him ; but all the main or cardinal psychological ideas respecting the functions of synthesis are already, at least in outline, to be found in Plato. The difference between him and Aristotle on this point was mainly a difference of method. He chose to classify all functions of synthesis as parts of the activity of the understanding. This, indeed, as an epistemologist or metaphysician, he was wise in doing ; but for the purposes of empirical psychology Aristotle's attribution of synthesis to the faculty of sense is unquestionably sound.

Aristotle.

I. *Sensus communis in presentative consciousness.* Each sense is within its own province a faculty of comparing

§ 22. According to Aristotle each sense, regarded as subservient to cognition, is, as regards its proper αἰσθητόν, a δύναμις σύμφυτος κριτική[1], with the faculty of distinguishing and comparing all διαφοραί belonging to that αἰσθητόν. Thus ὄψις discerns black and white and all the colours between these. Such a measure of synthetic power Aristotle grants to each individual sense[2]. It must be

[1] 99[b] 35, 428[a] 4, 432[a] 16.
[2] Each αἴσθησις is a δύναμις, and a δύναμις is the possibility of contraries. The αἴσθησις occupies a middle position between the contrary properties in each sensory province, and hence is able to discern—τὸ γὰρ μέσον κριτικόν, 424[a] 6.

admitted that there is a confusion, or ambiguity[1], in Aristotle's statements respecting the individual senses and the *sensus communis*, which sometimes amounts to or involves contradiction. We find him occasionally referring to αἴσθησις as if each sense were *per se* an analogue of the *sensus communis*, with all its power of comparison and distinction, only in a narrower province. Again, from a changed point of view—as when he is urging the case against simultaneous perception of two objects by one sense[2]—the sensory function of each particular αἴσθησις becomes narrowed to such slender proportions that we cannot conceive how it is, even within its own province, a δύναμις κριτική, according to its definition. Something must be allowed for looseness in the use of the term αἴσθησις, by which at times the writer tacitly includes, at other times excludes, reference to the κοινὴ αἴσθησις. When, however, (*a*) the data of *different* senses are to be presented together to the mind and compared or distinguished, this cannot be done by any single special sense, and we must have recourse to the assumption of a κοινὴ αἴσθησις. Again, (*b*) when we perceive either the κοινά or the incidental objects of perception (τὰ κατὰ συμβεβηκός), we exceed the powers of any individual sense. The κοινά, which are at times said to be perceptible by each and every sense together with its proper αἰσθητόν, are really proper objects of no single sense, but are objects of ἡ κοινὴ αἴσθησις; and so, too, are the incidental perceptions, such as we have when, e.g. seeing a white object, we say, or think, that we see 'the son of Diares.' Thirdly, (*c*) when the question is asked how we perceive *that* we perceive— how we are conscious of perceiving, the answer (for Aristotle) is: through the agency of the *sensus communis*.

§ 23. *The distinguishing and comparing faculty of sense.* By what, asks Aristotle[3], do we perceive (αἰσθανόμεθα) that *white* differs from *sweet*? By sense-perception (αἰσθήσει) of course, for these objects are both αἰσθητά. But it cannot be the work of any single sense, even of the most

Marginal notes: and distinguishing. Thus each sense at times seems to have for Aristotle some of the powers of the *sensus communis*. Confusion arising from this in his exposition. (*a*) For comparing and distinguishing the data of *different special senses*, the agency of a common faculty is conspicuously necessary. So, too, (*b*) for perceiving τὰ κοινά and τὰ κατὰ συμβηκός. So (*c*) finally for perceiving *that* we perceive, i.e. for the consciousness of perception. A. The *sensus communis* as the distinguishing and comparing

---

[1] Cf. *infra.*, pp. 283, 325-8. [2] Cf. *de Sens.* vii. 447ᵇ 9-21.

[3] 426ᵇ 12-427ᵃ 16.

faculty of sense. Even the sense of touch, though so fundamental, cannot discharge this function, which is not confined to tactual perceptions. Nor can touch in concert with any other sense suffice. The act of comparison requires that the things compared be brought before a *single* judging function *at the same time.*

comprehensive of all—that of touching. It cannot at all events be done by the instrumentality of σάρξ. For σάρξ, to perceive *sweet*, has to come into contact with the object; though sight does not need to do so in order to perceive *white*. If, therefore, the organ which perceives both be that on which touching depends, this organ cannot be σάρξ[1]. Nor can the comparison be effected by the two senses, touching and seeing, acting together[2]. It is impossible for separate entities (κεχωρισμένοις) to pronounce that white is different from sweet. Both objects must be present to the judgment of one self-identical agency, not each to a different agency from the other, as if for instance *I* were to perceive the one and *you* the other[3]; for such would really be the case if two senses took part in the comparative judgment. That which pronounces white and sweet to be different αἰσθητά must be not two agents, but one and the same. And not only must it be one and the same agent, but its agency at the moment of comparison must likewise be one. It must act at one and the same instant of time with reference to both the things compared. The two must be perceived co-instantaneously in one single instant[4]. When the comparing faculty pronounces one of the things compared to be different from the other, then, too, it pronounces the other to be different from the one. The very relation of difference into which the objects are brought thus involves identity in the judging subject. Hence (*a*) this is self-identical, and (*b*) its judgment respecting the one thing takes place at the same instant[5] as its judgment respecting the other. In short it is but one comparative judgment.

---

[1] In 455ᵃ 20-25 we see how closely allied, for Aristotle, are the κοινὴ αἴσθησις and the sense of touching—τὸ ἁπτικόν. It occurs to him here (426ᵇ 15), therefore, that the sense of touching may to some seem to be the one which discerns *sweet* and *white*, for tasting which perceives sweet is a mode of touching. But—while he does not utterly discard this assumption, and indeed the organ of touch proper and that of the *sensus communis* are, at bottom, one—he is careful to show that the flesh—the medium of touching, cannot be the organ of such comparing and distinguishing sense.

[2] 426ᵇ 17.   [3] 426ᵇ 19.   [4] 426ᵇ 23.
[5] 426ᵇ 29 ἐν ἀχωρίστῳ χρόνῳ.

When I judge white to be different from sweet, at that same time I judge sweet to be different from white ; and I who judge am the same in both relations.

§ 24. There is need of explanation, however, if we are to understand how one and the same sensory faculty can thus act at one and the same time with reference to objects like white and sweet, which as perceived affect sense differently. The same subject cannot, so far as it is undivided (ἀδιαίρετον), and so far as it acts in an undivided time (ἐν ἀδιαιρέτῳ χρόνῳ), be affected at once with opposite movements (κινήσεις). In whatever way sweet moves the sense, bitter moves it in the opposite way ; and white moves it in a way different from either. Yet if, as experience teaches us, such comparison is a fact, the above simultaneous action must be possible somehow. Perhaps the solution is that the faculty which pronounces (τὸ κρῖνον) on the difference of such qualities (whether homogeneous or not) is *in itself* when it so acts, numerically one, undivided and indivisible[1]; yet, *in its relations*[2], not self-identical, but divided (κεχωρισμένον)[3]. If this be so, one and the same percipient subject would, in virtue of its partibility of relationship, apprehend the several objects, while in virtue of its local and numerical identity it would grasp them together, and bring them into one relation with one another[4].

§ 25. Yet is this explanation really admissible? The same numerically and locally (τόπῳ καὶ ἀριθμῷ) one thing may

How one and the same faculty can apply itself co-instan-taneously to different objects in the act of compari-son or distinc-tion. In one respect the sensory faculty is single : in another it is divisible and not single. This sug-gests the answer.

This answer not wholly

---

[1] ἀριθμῷ ἀδιαίρετον καὶ ἀχώριστον.

[2] τῷ εἶναι = in its relations *to the objects perceived*. Cf. 449ᵃ 10–20 where (ᵃ 20) τῷ λόγῳ = in relation *to the faculty of conception*.

[3] The difficulty with which Aristotle here contends is put sharply in *de Sens.* vii. 447ᵇ 17 seqq. It is there shown that so far as a sense is a *single* faculty (δύναμις) and the time of its action indivisible, so far its ἐνέργεια is and must be single. There is but one 'movement'—once for all—possible, in a single time-instant, for such a faculty. That such a faculty should perceive white and sweet, or any other two objects co-instantaneously, in order to compare or distinguish them could not be admitted. In the same chapter it is afterwards shown that there is a way of regarding sense in which it is *not* such a simple, single, faculty as this, but endowed with the breadth and comprehensiveness of the *sensus communis*.   [4] 427ᵃ 3.

280 SENSUS COMMUNIS

satisfactory without further explanation: for though the agent of comparison and discrimination may be *potentially* several as regards different objects, yet how can it be *actually* so? Illustration from the way in which the στιγμή or τὸ νῦν is actually both *one* and *two*.

in its *potential* relationships be (or exhibit) contraries, but not in its *realized* relationships, while remaining one and the same. As, for instance, the same surface cannot at once be white and black, so (it might be argued) the same one sensory faculty cannot at once receive the forms[1] of *white* and *black*. This difficulty is real, Aristotle admits; yet it may, he thinks, be met. In a passage of the *Physics*[2], arguing that ὁ χρόνος is ἀριθμὸς κινήσεως κατὰ τὸ πρότερον καὶ ὕστερον, the geometrical point, ἡ στιγμή, and the unit of Time, τὸ νῦν, are compared. Each has two aspects, in one of which it is a πέρας or limit. In this aspect the στιγμή is not a μόριον μήκους, and the νῦν is not a χρόνος. As in the space-line, so in the time-line, the 'now,' which some call a point, is at once the beginning and the end, according to the aspect in which we view it. It is the end of the past, the beginning of the future. Thus it would fittingly illustrate the position of the percipient subject in relation to different things and focussing them all at the same time. As the νῦν can be at once both beginning and termination, while numerically one and the same, so this subject, while preserving its self-identity, may be related at once to different, and even opposite, objects, such as black and white, or sweet and white[2]. The κοινὴ αἴσθησις, like each special αἴσθησις, is

---

[1] τὰ εἴδη: the distinctive function of sense is the reception of forms without matter.

[2] 220ᵃ 5–26 συνεχής τε δὴ ὁ χρόνος τῷ νῦν, καὶ διῄρηται κατὰ τὸ νῦν ... ἀκολουθεῖ δὲ καὶ τοῦτό πως τῇ στιγμῇ· καὶ γὰρ ἡ στιγμὴ καὶ συνέχει τὸ μῆκος καὶ ὁρίζει· ἔστι γὰρ τοῦ μὲν ἀρχὴ τοῦ δὲ τελευτή. Ἀλλ' ὅταν μὲν οὕτω λαμβάνῃ τις ὡς δυσὶ χρώμενος τῇ μιᾷ, ἀνάγκη ἵστασθαι, εἰ ἔσται ἀρχὴ καὶ τελευτὴ ἡ αὐτὴ στιγμή. By making στιγμή = τὸ νῦν here (427ᵃ 10, cf. 426ᵇ 28), with Brentano, we not only explain the phraseology, but we get a more appropriate simile. The point in the time-line at which the relationship between the different objects is realized is just that which could best illustrate Aristotle's attempt at explanation. A difference of time between the perception of one object and that of the other would be fatal to his explanation of comparison: and this difference is just what he smooths over by his ingenious simile. Time is the 'form of internal sense.' Aristotle here approaches closely to Kant's thought of a synthetic unity of apperception, though not yet a *transcendental* unity, and only operating in the sphere of sense. Only such apperception could synthesize the fleeting manifold of perception.

a *mean*, i.e. it is one, though it realizes itself in many relationships. As the point, in space or time, can be regarded as at once *terminus* and *initium*, being conceived as a mean between both, so this κοινὴ αἴσθησις (which is what is here meant by τὸ κρῖνον) while *per se* one, is in its relationships divided between the diverse objects. So far as it is *two* it applies itself to them severally : so far as it is also *one* it brings them into the conjunction required for comparison.

As Plato in the *Theaetetus* found the solution of such a difficulty in a faculty of thought transcending temporal and spatial limitations, so Aristotle finds the solution of it (as far as the comparison of *sensible* data goes) in the assumption of a *sensus communis*, which is freed from the trammels that hamper the operations of each single special sense. Each αἴσθησις—τὸ αἰσθητικὸν τοῦ ἰδίου—is a mean between the ἐναντία of its province : and τὸ αἰσθητικὸν πάντων[1] is likewise a mean between the αἰσθητά of *all* the αἰσθήσεις[2].

---

[1] Cf. 449ᵃ 17.

[2] A further explanation of the κοινὴ αἴσθησις is attempted in *de Anima* 431ᵃ 20 seqq. in which Aristotle endeavours, by the aid of the idea of a proportion between pairs of numbers or quantities, to illustrate the relation between the central sense and its objects, whether homogeneous or heterogeneous, e. g. *white* and *black*, or *white* and *sweet*. The difficulties of this passage, however, are so great that they have baffled commentators from the earliest times to the present. See Torstrik's edition of the *de Anima*, pp. 199-202 ; Trendelenburg (Belger), pp. 426-32, with the passages from Simplicius and Philoponus there quoted ; Kampe, *Erkenntnisstheorie des Arist.*, pp. 108-9 n. Also see the judicious notes of E. Wallace, ad loc. Until the disputed points of reading and interpretation are settled for this passage, we cannot venture to rely upon it for trustworthy guidance as to Aristotle's conception of the *sensus communis*. The insertion, however, of a second reference to this matter, in connexion with the psychology of reason and will, shows plainly enough that Aristotle intended to use to the full his conception of ἔν τι ἀριθμῷ, τῷ δ' εἶναι ἕτερον, which he applies (as we have seen) to explain (*a*) the individual αἰσθητήριον in relation to its function *qua* αἰσθητικόν, 424ᵃ 25 ; (*b*) the κοινὴ αἴσθησις or τὸ ἐπικρῖνον (or κρῖνον) here in its relationship to the special αἰσθήσεις ; and (*c*) in 431ᵃ 12-ᵇ 10 the διανοητικὴ ψυχή (regarded in reference to πρᾶξις) in relation to the φαντάσματα which are to it οἷον αἰσθήματα. The plan which we have followed precludes our entering any further into this last part of the subject.

In the concluding chapter of the tract *de Sensu*, we find what was perhaps chronologically Aristotle's first essay on the subject of simultaneous perception of different sensibles. The whole object of the ἀπορία, with which that chapter commences, is to lead up to the establishment of two propositions (*a*) that co-instantaneous perception of different αἰσθητά, with a single special sense, is strictly impossible ; and (*b*) that, since such perception is a fact, it must be accounted for by the agency of the one central sense there (449ᵃ 17) referred to as τὸ αἰσθητικὸν πάντων.

§ 26. The *objects* of the *sensus communis* are, chiefly, those called by Aristotle (1) the *common*[1] sensibles, and (2) the *incidental* sensibles (τὰ κοινὰ καὶ τὰ κατὰ συμβεβηκός). The κοινά variously enumerated in different passages by Aristotle consist (most fully stated) of κίνησις καὶ ἠρεμία, ἀριθμός, μέγεθος, σχῆμα, τὸ τραχὺ καὶ τὸ λεῖον, τὸ ὀξὺ καὶ τὸ ἀμβλύ (τὸ ἐν ὄγκοις). These are said[2] to be perceptions 'common to all the special senses, or if not to all, at least to sight and touch.' Wherefore (διό) with reference to these percepts errors take place (ἀπατῶνται), while with reference to the special or proper (περὶ τῶν ἰδίων) objects of each sense, such as colour, no such error occurs, or at least it occurs only in the lowest possible degree[3]. Two points are remarkable in Aristotle's statement respecting these κοινά. First, that though they are called κοινὰ πασῶν, this is corrected and their perception restricted to sight and touch ; secondly, that after declaring the above αἰσθητά to be *common*, he goes on '*wherefore* (διό) errors are possible, &c.' Why, one may ask, does the fact of these being common to several senses, render error more likely or more frequent regarding them than as regards the αἰσθητά of some special αἰσθησις ? Do the different senses which perceive any given κοινόν *contradict*, instead of *corroborating*, one another's testi-

B. The *sensus communis* as faculty of perceiving τὰ κοινά and τὰ κατὰ συμβεβηκός. Errors in such perception, scarcely at all in perception of τὰ ἴδια. The so-called κοινὰ πασῶν τῶν αἰσθήσεων really common only to sight and touch. They are *really* κοινά, because they are objects of ἡ κοινὴ αἴσθησις. All perceived in

---

[1] But see Neuhäuser, *op. cit.*, pp. 30 seqq.

[2] 418ᵃ 6–25, 425ᵃ 15, and 442ᵇ 5 where, however, κίνησις and ἀριθμός are not named.

[3] 428ᵇ 18 ἡ αἴσθησις τῶν μὲν ἰδίων ἀληθής ἐστιν ἢ ὅτι ὀλίγιστον ἔχουσα τὸ ψεῦδος.

mony? If so, why? There is an incongruity in Aristotle's position as to the relation between 'special' and 'general' sense[1].

We have here classified the κοινά as objects of the *sensus communis*. They are all perceived in virtue of one of them, viz. κίνησις[2]. But κίνησις is itself perceived by the *sensus communis*; so is χρόνος[3], and so too is μέγεθος. Though they are classed with the αἰσθητὰ ὧν καθ' αὑτά φαμεν αἰσθάνεσθαι, and distinguished from the incidental αἰσθητά[4], we find no special αἰσθητήριον dedicated to them; thus, so far as we perceive them by each αἴσθησις, we really do so only κατὰ συμβεβηκός[5]. If then they are to be really perceived καθ' αὑτά, they must be objects to some αἴσθησις, and this, being no special sense, must be the κοινὴ αἴσθησις. There could not, with profit to our experience, be any one special sense for the perception of these, e. g. of κίνησις and ἠρεμία. Were there such special sense, then when we saw an object moving or at rest, its movement or rest would, for us, be, in relation to the proper object of seeing, as sweetness is now to colour; i. e. a merely incidental percept. We *see* an object of a certain colour to be sweet. This only means that an uniform experience has taught us to connect its colour with this particular taste. We are accustomed to find the taste and the colour together in the object. There is no necessary connexion between them, however, as there is between a body and its movement or rest. Were there a special sense for the perception of movement or rest, the latter, as ἴδιον of such sense, might and no doubt would connect itself customarily, but never necessarily, with the ἴδια of other senses. We should by the assumed special sense perceive movement *per se*, not, as now, always in a moving body. Thus a gulf would be created in experience between movement and rest and bodies; and the same

[Margin note:] virtue of one of them, viz. κίνησις. There could not be one special sense for the perception of τὰ κοινά, or any of them, e. g. κίνησις, without depriving our judgments of movement and rest. of magnitude, number, and so on, of all objective necessity.

---

[1] See pp. 277, 286 n., 325-8.

[2] 425ª 16 ταῦτα γὰρ πάντα κινήσει αἰσθανόμεθα κτλ.

[3] 450ª 9 μέγεθος ἀναγκαῖον γνωρίζειν καὶ κίνησιν ᾧ καὶ χρόνον: 451ª 17 ὅτι τοῦ πρώτου αἰσθητικοῦ καὶ ᾧ χρόνου αἰσθανόμεθα: 452ᵇ 7 seqq.

[4] 418ª 8.

[5] 425ª 14 τῶν κοινῶν ... ὧν ἑκάστῃ αἰσθήσει αἰσθανόμεθα κατὰ συμβεβηκός, οἷον κινήσεως κτέ.

gulf would be created between bodies and the other κοινά, all of which are modifications of this one—movement or rest. Thus judgments of movement (mechanical science), magnitude, number, &c., would lose objective necessity. True the gulf might be bridged over by the formation of incidental customary connexions between movement, or rest, and bodies; but the necessity that a body should be either moving or at rest, would exist no longer. As things now stand, no such gulf separates bodies from the qualities called κοινά. This is so because the κοινά are κοινά, and not ἴδια of any special sense. We cannot perceive movement and rest except in necessary connexion with the perception of the qualities of body generally, i. e. by the common sense ; nor can we otherwise perceive the figure magnitude, number of bodies, than by this sense—the κοινὴ αἴσθησις[1]. Thanks to the fact that the κοινά are not proper to any one sense, but are perceptible only by the *sensus communis*, they necessarily, not merely customarily or contingently, accompany the various objects of perception[2]. Thanks to this we perceive no object in space without necessarily ascribing to it number, magnitude, motion, or rest, and so on. The κοινά are indirectly perceived by the special senses; but directly and properly by the κοινὴ

---

[1] 425ᵃ 27 τῶν δὲ κοινῶν ἤδη ἔχομεν αἴσθησιν κοινήν, οὐ κατὰ συμβεβηκός, where the seeming inconsistency with 425ᵃ 15 is easily removed, by observing that the κοινά, which to each special αἴσθησις are (ᵃ 15) κατὰ συμβεβηκός, are not so but are strictly *proper* to ἡ κοινὴ αἴσθησις.

[2] 428ᵇ 22–5 τῶν κοινῶν καὶ ἑπομένων τοῖς συμβεβηκόσιν οἷς ὑπάρχει τὰ ἴδια, λέγω δὲ οἷον κίνησις καὶ μέγεθος, ἃ συμβέβηκε τοῖς αἰσθητοῖς, i. e. the κοινά accompany the contingent objects to which the special qualities belong as qualities, as e. g. movement and magnitude accompany all contingent objects of perception. The words ἃ . . . αἰσθητοῖς may be a gloss upon τοῖς συμβεβηκόσιν οἷς ὑπάρχει τὰ ἴδια, which, however, they explain quite correctly if τοῖς αἰσθητοῖς is taken in its natural meaning. Τὰ συμβεβηκότα are here = τὰ κατὰ συμβεβηκός, i. e. objects incidentally perceived in virtue of τὰ αἰσθητά, the colours, &c., which are the proper objects of sense. All the concrete things perceived by us in space are (to the *special* senses) συμβεβηκότα in this way ; they are subjects of movement and rest, magnitude, number, &c., so far as they are objects of ἡ κοινὴ αἴσθησις.

αἴσθησις ¹. And this (not their being perceptible by all the αἰσθήσεις in common, which, indeed, according to Aristotle himself is not true) is their real title to the name κοινά.

§ 27. As already stated, all the κοινά are said by Aristotle to be perceptible κινήσει, i. e. in virtue of this one of them, κίνησις ². By this we perceive μέγεθος, and therefore σχῆμα, which is a particular mode of μέγεθος; by this we perceive also its opposite ἠρεμία, and by it we perceive ἀριθμός, which is the negation of continuity in κίνησις ³.

Aristotle, in his argument that there cannot be any one special organ for the κοινὴ αἴσθησις, is interested in the difference in point of universality and objectivity between the κοινά as they now are and as they would be if made the object of an ἴδιον αἰσθητήριον. Now, for example, we cannot perceive anything without perceiving it to have μέγεθός τι ⁴. As things stand, moreover, every αἰσθητόν has number: every visible αἰσθητόν, at least, has magnitude. If we had an ἴδιον αἰσθητήριον of number or magnitude, what Aristotle thinks is that then number would only have the incidental and occasional connexion with αἰσθητά which sweetness now has with whiteness; and this would exemplify the consequent disorganization of all experience, and the necessity for objective experience of maintaining the κοινά as κοινά.

If, however, the κοινά are perceived directly by the κοινὴ αἴσθησις, but κατὰ συμβεβηκός by each special αἴσθησις, this manifestly renders them analogous to the class of αἰσθητά

In virtue of our perception of κίνησις we perceive all the other κοινά. As the κοινά are objects not of special sense but of the sensus communis, we can perceive no object without perceiving, that it has μέγεθος and ἀριθμός. But the κοινά αἰσθητὰ κατὰ συμβεβηκός also are proper to the sensus communis, incidental only to the special senses.

---

¹ So it is called 455ᵃ 15 ἡ κοινὴ δύναμις ἀκολουθοῦσα πάσαις.

² πάντα κινήσει αἰσθανόμεθα. I cannot see what reason there is for adopting the reading κοινῇ in this passage (425ᵃ 16) for κινήσει, though Torstrik thinks he follows Simplicius in adopting it.

³ Bäumker (op. cit., p. 64 n.) explains κίνησις here as perhaps more particularly denoting 'die subjective Veränderung des Sinnes,' founding this view upon the words of Themistius, ad loc., sc. οὐδὲν γὰρ τῶν κατὰ συμβεβηκὸς αἰσθητῶν κινεῖ τὸ αἰσθητήριον κτλ. In these words, however, Themistius was not referring to the κινήσει of 425ᵃ 16, but of 418ᵃ 23 διὸ καὶ οὐδὲν πάσχει ᾗ τοιοῦτον ὑπὸ τοῦ αἰσθητοῦ (sc. τοῦ κατὰ συμβεβηκός).

⁴ 449ᵃ 20 τὸ αἰσθητὸν πᾶν ἐστι μέγεθος: where, however, he is especially thinking of perception by sight, since he goes on—ἔστι γὰρ ὅθεν μὲν οὐκ ἂν ὀφθείη, κτλ.

called τὰ κατὰ συμβεβηκός by Aristotle himself[1]. What is the αἴσθησις to which these latter are directly objective, as the κοινά are to the κοινὴ αἴσθησις? or is there any? If it is by an act of *inference* that the so-called incidental perceptions are really to be explained—an inference based on association of ideas—what prevents this explanation from being also applied to τὰ κοινά? Why does Aristotle not ascribe the incidental αἰσθητά to the operation of the κοινὴ αἴσθησις? The reason apparently lay in his feeling that this would carry him too far; such 'incidental' perception being really a matter of inference, and habitually (whether correct or incorrect) extending itself far beyond the province of comparatively simple sensation illustrated by the case of 'seeing the son of Diares.' There is here accordingly a difficulty which Aristotle apparently hid from himself. He admits—and the admission is fatal to his distinction—that error is common to our perceptions both of τὰ κοινά and of τὰ κατὰ συμβεβηκός. If we have a *sensus communis* which directly perceives τὰ κοινά as ὄψις perceives colour, there is no reason given by Aristotle to explain why we should err more easily in reference to one of the former than in regard to the latter. Our perception of magnitude or distance should be as trustworthy as that of colour. If, however, he were once to concede that magnitude and the rest of the κοινά are matter of *inference*, the whole basis of his theory of κοινὴ αἴσθησις would require reconstruction[2]. Nor must it be overlooked, that for Aristotle it is the κοινὴ αἴσθησις which really comprehends the correlated elements of the perceptions κατὰ συμβεβηκός. Such perception involves association of ideas, representation, and memory. If I see a white object and perceive 'the son of Diares' (whether I am correct in so stating my perception or not) it is the κοινὴ αἴσθησις that enables me, according to Aristotle's theory, to go beyond the *datum* of seeing to the

Margin notes:

They are really inferences. Why does Aristotle not so treat them, and ascribe them to sensus communis? The κοινά and the αἰσθητὰ κατὰ συμβεβηκός are more closely connected than Aristotle saw. Perception of these latter implies the agency of the κοινὴ αἴσθησις, as it implies association and memory.

---

[1] 418ᵃ 20.

[2] To make his theory consistent, the faculty of synthesis should be (contrary to his teaching in several places, e.g. 447ᵇ 10 seqq.) attributed to the most elementary operations of sense-perception.

mass of other sensible data already experienced by me and remembered under the name 'son of Diares.' Without this combining faculty no one sense could perceive the data of another. It is this that first gives objective reference to τὰ ἴδια. All perception, in fact—however imperfectly this is expressed by Aristotle—so far as it includes relations between the data of the same sense or of different senses, or between τὸ ἴδιον and τὸ κατὰ συμβεβηκός—is rendered possible for Aristotle by this central sense. It is by this that each sense perceives not only its object but the contrary of that object, as e.g. ὄψις perceives the visible and the invisible[1].

§ 28. The object (αἰσθητόν) of each special sense, except perhaps touch, constitutes a single *genus*; the *sensus communis* has all *genera* of αἰσθητά, not any one in particular, for its objects. That it can perceive all is due to the fact that from the first it is directed not to objects in space, as the special senses are, but rather to the αἰσθήματα, or impressions made through these senses, which abide and make re-presentation possible even after the αἰσθητά which stimulated them have departed[2]. These αἰσθήματα are to ἡ κοινὴ αἴσθησις what the φαντάσματα are to ἡ διανοητικὴ ψυχή[3]. They are what results from the process described as the apprehension by each αἴσθησις of the εἶδος, without the ὕλη, of its object. These, being without ὕλη, can present themselves to the κοινὴ αἴσθησις simultaneously, even though their perception was successive. In their detachment from their αἰσθητά, they may give rise to φαντάσματα which become sources of illusion. Even at their first occurrence, while the object is present, they may be sources of illusion, and require to be brought to order by a standard. Thus we, despite our better knowledge, continue to see the sun a foot in breadth. The controlling faculty of sense (τὸ κύριον καὶ ἐπικρῖνον)[4], however, which is that which estimates the objective reference of αἰσθήματα, may correct such illusion. The organ of this is the κύριον αἰσθητήριον.

As the special senses are directed on outward objects, so the κοινή is directed to the αἰσθήματα given by the special senses. The αἰσθήματα give rise to φαντάσματα. Even in themselves, i.e. in their first presentation, they may be sources of illusion, not merely when re-produced as φαντάσματα.

[1] The ὁρατόν and the ἀόρατον: see 422ᵃ 20, 425ᵇ 21, 426ᵇ 10.
[2] 450ᵃ 31, 460ᵇ 2.   [3] 431ᵃ 14, 432ᵃ 9.
[4] 455ᵃ 21, 461ᵇ 24 seqq.

C. *Sensus communis* as faculty of *consciousness*. It must be by sense that we perceive the fact of our perceiving; and this, too, by the same sense by which the object is perceived. For example, it is by sight that we perceive ourselves to see. We see that we see. The possibility of this lies in the fact that the faculty of sight (like that of each sense) implies two things (*a*) the primary αἴσθησις of the ὁρατόν, i. e. the apprehension of its form (εἶδος)

§ 29. 'Since we perceive (αἰσθανόμεθα) *that* we see (or hear), it must be either by the sense of seeing that we do so, or else by some other sense[1]. On the latter assumption, this "other sense" would perceive two things—both the fact of the seeing, and the object of this (the colour seen). Hence, on this assumption, there will be two senses concerned[2] with the one object. If, deterred by this, we do not make the assumption of the "other sense," it remains that the sense of seeing should perceive itself, and no such duplication would arise. But a further objection can be made against that assumption; for if the "other sense" were really different from the first, a third would be needed for consciousness of the second, and so on *ad infinitum*. To escape this we must at some point assume a sense which perceives itself in action; and, therefore, we had better do so in the case of the first perception. Let us, then, refer our consciousness of seeing to the sense of sight itself. Here, however, a fresh difficulty arises. If to perceive by the faculty of seeing is what is meant by "to see," and if the object of seeing is colour, or a coloured thing; then to "perceive by sight[3]" the seeing agent would imply[4] that this agent is something possessing colour. To this the answer is twofold. First, the expression "to perceive by sight" has more than one simple meaning[5]. That it has more is plain, if only from the fact that, even when we are not seeing anything in particular, we discern by sight between light and darkness, and such discernment is not, as an act, identical in its nature with the seeing of a particular colour at a particular time. Secondly, there is a point of view whence we can

[1] 425ᵇ 11–25: by using αἰσθανόμεθα Aristotle excludes the assumption that it is by *intelligence* that we become conscious of perceptions.

[2] Viz. the original ὄψις or ὅρασις and the ὄψις ὄψεως.

[3] εἴ τις ὄψεται τὸ ὁρῶν: 'to become conscious of seeing' means (so far as the argument has proceeded) that 'one who sees should see the seeing agent.'

[4] The point is argued as if 'to perceive *that* one perceives' were the same thing as 'to perceive the perceiving subject.'

[5] It has one meaning as expressing the act of special sense; another —and this is the point to which Aristotle is leading up—in reference to the act of the κοινὴ αἴσθησις.

even accept the assertion that "the fact of seeing is something coloured." For we have defined an organ of sense as that which is capable of receiving the *form* of its αἰσθητόν without the *matter* ; and colour, as perceived, is such form. To this capacity it is owing that even when the objects (αἰσθητά) of sense have departed, the αἰσθήσεις (or αἰσθήματα, or φαντασία, 428^b 11) which they excited remain still in our sensory organs[1].' In another passage [2] Aristotle says : ' We possess a faculty or power accompanying all the individual senses, in virtue of which power one sees *that* he sees, or hears *that* he hears, or in general perceives *that* he perceives. It is in virtue of this common power that one does so ; for assuredly it is not by the *special* sense of seeing that one sees *that* he sees.' Thus the direct objects of this *sensus communis* are not the αἰσθητά, strictly speaking, but the αἰσθήματα or impressions of the special senses. The importance of this faculty of consciousness is stated in the *Nicomachean Ethics*[3]. ' He who sees perceives *that* he sees ; he who hears perceives *that* he hears ; he who walks perceives *that* he walks. So, also, concerned in our other activities, there is something in us which perceives *that* we perform them. We perceive *that* we perceive, think *that* we think, and so on. But for us our existence consists just in this very perceiving *that* we perceive and thinking *that* we think.' Thus, so far as perception is concerned, the faculty of consciousness is the *sensus communis*. Consciousness has its empirical dawn in the emergence of this distinction between perceiving and perceiving *that* we perceive ; the distinction itself is impossible without some degree of psychical continuity—without a synthetic faculty which can bring together the present and the past. It implies elementary memory, which again implies that φαντασία, as sensory presentation, is not any longer a mere momentary appearance, but a faculty of storing up αἰσθήματα, to become

[Margin note:] without its matter ; and (b) the retention of this form—the αἴσθημα—by faculty of τὸ αἰσθητικόν in general, to which it remains as a possible object of inner vision. In this relation between the primary effect of αἴσθησις and its residual effect through its αἴσθημα in memory (i.e. in the retentive power of κοινὴ αἴσθησις) lies the dawn of empirical consciousness.

---

[1] Cf. 425^b 24. With the above cf. Plato, *Charmides* 168 D-E, οὐκοῦν (ἡ ἀκοὴ) εἴπερ αὐτὴ αὑτῆς ἀκούσεται, φωνὴν ἐχούσης ἑαυτῆς ἀκούσεται· οὐ γὰρ ἂν ἄλλως ἀκούσειεν·—καὶ ἡ ὄψις γέ που, εἴπερ ὄψεται αὐτὴ ἑαυτήν, χρῶμά τι αὐτὴν ἀνάγκη ἔχειν· ἄκρων γὰρ ὄψις οὐδὲν μή ποτε ἴδη.

[2] 455^a 15 seqq.   [3] 1170^a 29, with Prof. J. A. Stewart's note.

φαντάσματα, and on occasion also μνημονεύματα, subsidiary to the higher functions of intelligence and reason [1]. In spite of the importance assigned to consciousness in the *N.E.*, l. c., it remains in general for Aristotle a psychical πάρεργον, utterly without the importance assigned to it by modern psychologists. Science, perception, opinion, and discursive intelligence, are all concerned primarily with something other than themselves, viz. with their respective *objects*. The man of science does not as a rule think of himself as thinking; he thinks of his particular object; and of himself only indirectly, or when some interruption to the natural flow of his thought occurs [2].

II. *Sensus communis in re-presentative consciousness. Various meanings of φαντασία (1) as primary presentation, (2) as representation. A meaning of φάντασμα (as object of φαντασία) corresponds to each of these. φαντασία*

§ 30. The word φαντασία [3] often bears in Aristotle the meaning, in which Plato generally uses it, of the faculty of *presentation*, by which an object appears to the mind on the occasion of perception. Thus we read of the φαντασία of colour, i. e. the subjective impression of it upon the mind as seen [4]. Such appearance may or may not be illusory. Regarded as the source of illusion, φαντασία connects itself more with mental pathology than with psychology. Regarded on its normal side, as the faculty by which things 'appear' through sense-perception, it can be divided into two grades, according as it expresses first-hand or second-hand 'appearance.' In the one grade it is the faculty of *presentation*; in the other, the faculty of *representation*, or the reproductive imagination. Corresponding distinctions hold as to the use of the concrete φάντασμα. A φάντασμα may be illusory, or it may be the normal foundation of memory or reasoning.

---

[1] 450ᵇ 26, 449ᵇ 31 seqq. The αἰσθήματα are themselves αἰσθητά, 460ᵇ 3.

[2] Cf. *Met.* 1074ᵇ 35 φαίνεται δ' ἀεὶ ἄλλου ἡ ἐπιστήμη καὶ ἡ αἴσθησις καὶ ἡ δόξα καὶ ἡ διάνοια, αὑτῆς δ' ἐν παρέργῳ. The psychological distinction between self and its energy in thought or action, while important as revealing to us our existence, is, we may observe, as a matter of fact, one of which little use is normally made in practice ; and then chiefly either for the purposes of psychology and cognate studies, or because something abnormal occurs, which interrupts the current of objective thinking and forces the thinker in upon himself.

[3] In accordance with the use of φαίνεται, as in φαίνεται μὲν ὁ ἥλιος ποδιαῖος, 428ᵇ 3.

[4] Cf. 439ᵇ 6 ὥρισται ἡ φαντασία τῆς χρόας : 791ᵃ 17, 294ᵃ 7.

It means an individual impression made on the 'faculty' called ἡ φαντασία, or τὸ φανταστικόν. The abnormal or pathological meanings of these words are well understood by Aristotle[1], but are not to him the subject of much direct study.

The characteristic meaning of φαντασία, or τὸ φανταστικόν, in Aristotle's psychology, is that of the faculty by which φαντάσματα, mental presentations, are in the first instance formed, and in the second reproduced, in the absence of the αἰσθητά to which they are ultimately affiliated. Such reproduction is thus described. The impressions of sense, the αἰσθήματα, do not disappear or perish with the instant of their first perception. They leave traces (μοναί) of themselves[2], or persist, 'within us.' These traces are somehow stored up. This 'storing up' is effected by successive φαντασίαι, i. e. 'appearances' or presentations through immediate sense; and when a store of αἰσθήματα has been formed, the ground is prepared for φαντασία (or τὸ φανταστικόν) in the further application of this term, i. e. as the faculty of reproducing images which were once before the mind, even when the objects which gave rise to them have disappeared from perception. Thus it will be observed that an αἴσθημα and a φάντασμα are at bottom the same psychical phenomenon, which if regarded as grounded on the αἴσθησις is an αἴσθημα, but as a mere presentation or re-presentation to the 'mind's eye' is a φάντασμα. Accordingly Aristotle defines the faculty of imagination as one and the same *per se* with that of central sense, but differing from the latter in its relationships or conception[3]. The φαντάσματα, like the αἰσθήματα, are individual and concrete in their nature: they have not the universality of concepts. Until thinking takes them over they are not connected in propositions. Intrinsically the faculty of perception (τὸ αἰσθητικόν) is one with that of imagination (τὸ φανταστικόν), though they are conceived in different ways,

Margin: in either use (with the corresponding φάντασμα) as source of illusions occupies a subordinate place, and belongs rather to mental pathology. Description of the way in which φαντασία as reproductive imagination acts, and in which φαντάσματα are engendered 'in the mind.' The αἰσθήματα or impressions of αἴσθησις are 'stored up.' By this storing up the faculty of imagination is equipped for its function. Relation of the αἴσθημα to the φάντασμα. φαντάσματα are in their nature individual,

---

[1] 165ᵇ 25, 168ᵇ 19, 1114ᵃ 32, 460ᵇ 19, and 846ᵃ 37 (where φαντασία = 'apparition').

[2] 99ᵇ 34-7, 450ᵃ 27 seqq., 408ᵇ 15–18, 459ᵇ 5 seqq., 460ᵇ 2.

[3] 459ᵃ 15–17 ἔστι μὲν τὸ αὐτὸ τῷ αἰσθητικῷ τὸ φανταστικόν, τὸ δ' εἶναι φανταστικῷ καὶ αἰσθητικῷ ἕτερον.

not, like and are differently related [1]. Ἡ φαντασία as a faculty is
concepts, a process or an affection produced within the ζῷον, or
universal.
The faculty animated organism, by the exercise of sense-perception [2].
of imagina-
tion, how Thus φαντασία and ἡ κοινὴ αἴσθησις are fundamentally one :
related to and it is to be remembered that as φαντασία is rooted in the
the faculty
of general sensory faculty, so its exercise depends upon movements
sense. The continuing in the sensory organs [3], which movements serve,
imagina-
tion com- under certain conditions, from time to time, to stimulate the
paratively
idle while organ of imagination, which is that of central sense ; and
the senses thus the φαντάσματα are brought into clear consciousness by
are actively
employed. the μοναί, or traces of themselves left by the αἰσθήματα. The
organ of sense-perception is related to an external, or extra-
organic, stimulus : that of reproductive imagination receives
its stimulus from within the organism. Thus, when the senses
are not occupied with 'external objects,' the φαντασία may
be actively employed ; and, indeed, it has least to do when
the senses are engaged with the outer world energetically
and effectively. Confused and obscure, or difficult, sensory
perception is, however, apt to stimulate φαντασία to activity.
Thus, if we see a person only imperfectly at a distance, we
set about guessing who it can be : this employs φαντασία.
If we see the person well and clearly, reproductive φαντασία
has no opportunity of exercise [4]. But when the ' outer ' or
bodily eye is closed, images of many sorts crowd before
the ' inner ' or mind's eye ; and the power and activity of
φαντασία are at their maximum when the special senses are
at rest during sleep.

Differences φαντασία and αἴσθησις thus differ chronologically, the
of φαντασία former being as it were the rehearsal of the latter's work.
and αἴ-
σθησις : But they differ also in other ways. They have not the
chrono-
logical same or equal values as evidence respecting objects. The

---

[1] 459ᵃ 15–18.

[2] l. c. ἔστι δ' ἡ φαντασία ἡ ὑπὸ τῆς κατ' ἐνέργειαν αἰσθήσεως γινομένη
κίνησις : cf. 429ᵃ 1.

[3] The organ in which the κινήσεις, or μοναί, or whatever name the effect
of ἡ κατ' ἐνέργειαν αἴσθησις may be called by, persist is not the central
organ, but the particular sense-organ ; cf. 459ᵃ 3, 461ᵃ 26 ; Freudenthal,
Ueber den Begriff des Wortes φαντασία *bei Aristoteles,* p. 20.

[4] 428ᵃ 12 seqq.

evidence of αἴσθησις with respect to its *proper* object is almost always true and trustworthy. The φαντασία is a frequent cause of error, and untrustworthy in the absence of an object. They have not the same extent in the animal world. All animals have αἴσθησις: it is more than questionable whether all have φαντασία[1]. φαντασία resembles thinking in the one particular of not requiring external stimulation, as αἴσθησις does, on each occasion of its exercise. Therefore it is that φαντάσματα and νοήματα at their lowest level become somewhat difficult to distinguish[2]. But φαντάσματα are indispensable for the exercise of νόησις[3]. Indeed, in one place Aristotle goes so far as to name φαντασία as—at least according to some persons—a division of thinking[4]. φαντάσματα are distinguished, however, from νοήματα by the fact of their implicit individuality: the data of φαντασία like those of αἴσθησις are *per se* individuals, and derive their universality, so far as they possess it, from the setting in which they are placed by the activity of the thought which employs them as its material.

§ 31. The inner workings (κινήσεις) which form the basis of φαντασία are not of course *purely* corporeal: they are, like all the processes of life and mind, and in accordance with the definition of αἴσθησις given by Plato and Aristotle, movements of the soul through the body. Leaving this to be understood throughout, Aristotle gives a predominantly physiological account of the nature of φαντασία. Yet this is an activity of ψυχή. It is that on which memory and recollection depend. Without its aid sense-perception would be confined to momentary ἐνέργειαι, lacking in continuity, unassociated, incapable of forming a basis of

[Marginal notes:]
differences; evidential difference.

Difference of φαντάσματα and νοήματα.

φαντάσματα the material of νόησις; but this with its νοήματα is general or universal, not confined to individual objects as φαντασία is.

The residual movements in the organs on which φαντασία dependsare movements of body and soul together. Psychological importance of φαντασία.

---

[1] In 413ᵇ 22 there are good reasons for doubting the genuineness of the words καὶ φαντασίαν; cf. 414ᵇ 1, 415ᵃ 10, 414ᵇ 16, 428ᵃ 10. Cf. Freudenthal, *op. cit.*, p. 8.

[2] 403ᵃ 8 τὸ νοεῖν· εἰ δ' ἐστὶ καὶ τοῦτο φαντασία τις ἢ μὴ ἄνευ φαντασίας, 433ᵃ 9 εἴ τις τὴν φαντασίαν τιθείη ὡς νόησίν τινα, 432ᵃ 12 τὰ δὲ πρῶτα νοήματα τίνι διοίσει τοῦ μὴ φαντάσματα εἶναι;

[3] 449ᵇ 30 seqq.

[4] 427ᵇ 28 τοῦ νοεῖν . . . τούτου δὲ τὸ μὲν φαντασία δοκεῖ εἶναι τὸ δὲ ὑπόληψις.

ἐμπειρία. As the work of τὸ αἰσθητικὸν πάντων, it gives the αἰσθητά their first objective reference: it extends experience from τὰ ἴδια to τὰ κοινά and τὰ κατὰ συμβεβηκός. It gives their first rudimentary *meaning* to sounds, and so makes language possible [1]. It is the condition of thinking, since it is by the φαντάσματα or 'schemata' which accompany our concepts that they have the requisite clearness and distinctness, and also are capable of being *remembered*. Together with perception and thinking it forms also the basis of desire and will [2]. For the productions of art and literature its efficacy is prodigious, and quite indispensable. Who Antipheron of Oreus was we do not know: perhaps a madman, who mistook (as we learn from *de Mem.* 1) his mere φαντάσματα for μνημονεύματα; but Aristotle, as well as Shakespeare, distinguishes the poet as one who has the faculty of giving 'to airy nothing a local habitation and a name [3].'

Real nature of the residual impressions which form the physiological ground of φαντασία, unknown to Aristotle, and also to us. Correspondences between Aristotle and Hobbes, as regards this faculty.

§ 32. As to the real or physical nature of the κινήσεις in which the faculty of imagination consists, Aristotle of course can tell us nothing. We do not know whether they are regarded by him as (what would now be termed) mechanical or chemical. In this respect, modern psychologists have no great advantage as compared with him. The correspondences between his description of this faculty and that given by Hobbes (as pointed out by Freudenthal, *op. cit.*, p. 24 n.) are very well worth noticing. 'When a body' (says Hobbes) 'is once set in motion, it moveth, unless something else hinder it, eternally . . . and, as we see in the water, though the wind cease the waves give not over rolling for a long time after, so also it happeneth in that motion. . . . For after the object is removed, or the eye shut, we still retain an image of the thing seen, though more obscure than when we see it [4].' With this compare Arist. 459ᵇ 9 seqq., 460ᵇ 28 seqq. Again: 'imagination, therefore, is nothing but decaying sense'—the proposition laid down by Hobbes—might

---

[1] 420ᵇ 32.   [2] 432ᵇ 16, 433ᵃ 9–ᵇ 28.

[3] Cf. Arist. *Poet.* 1455ᵃ 32 and § 38 *infra*.

[4] *Leviathan*, pt. i. ch. 2; also *Physics*, iv. ch. 25.

be a translation of ἡ δὲ φαντασία ἐστὶν ἀσθενής τις αἴσθησις[1].
Compare also 'much memory is called experience' with
Arist. 100ᵃ 5. The words 'there be also other imaginations
. . . as from gazing upon the sun the impression leaves an
image,' remind us of Arist. 459ᵇ 7. Again : 'the phantasms
of men that sleep are dreams,' reproduces Arist. 462ᵃ 29 ;
while 'all fancies are motions within us, reliques of those
made in the sense,' might have been taken from Arist. 461ᵃ
18 αἱ ὑπόλοιποι κινήσεις αἱ συμβαίνουσαι ἀπὸ τῶν αἰσθημάτων.
'Those motions that immediately succeeded one another in
the sense continue also together after sense ' is a paraphrase
of Arist. de Mem. 2. 452ᵃ 1 ὡς γὰρ ἔχουσι τὰ πράγματα πρὸς
ἄλληλα τῷ ἐφεξῆς, οὕτω καὶ αἱ κινήσεις.

§ 33. The κινήσεις in the organs either continue latent or Latency of
propagate themselves to the central organ of perception[2]. the residual
move-
Their latency is caused by the inhibition exercised upon ments, how
caused.
them by stronger κινήσεις, in the continued use of the Their
αἰσθήσεις in external perception, or else by the activity emergence
into con-
of thinking. These stronger κινήσεις extinguish the weaker sciousness;
as a stronger light causes a weaker to pale before it[3]. But conditions
and manner
under favourable circumstances they make their way to of this.
When
the central organ and re-emerge into consciousness, i.e. latent they
either when they become strong enough to remove the are for
Aristotle
obstacles, or when the inhibiting movements become potential;
weaker, as in sleep. When latent the κινήσεις are, in in consciousness
Aristotle's phrase, potential; when they emerge into con- they
become
sciousness, they are actual[4]. They are conveyed from the actual.
special organ to the organ of central sense, and so from Their
medium
latency to consciousness, by[5] the medium of the blood[6]. In between
this organ of central sense they then produce a secondary the special
organ and
affection of consciousness with an image of the object of the central

---

[1] 1370ᵃ 28, a passage of the *Rhetoric*, of which work Hobbes made
an analysis.
[2] 459ᵇ 7, 461ᵃ 6.   [3] 460ᵇ 32, 461ᵃ 20, 464ᵇ 4.
[4] 461ᵇ 12.
[5] Or *with* the blood, by the σύμφυτον πνεῦμα, see 659ᵇ 17-20, 744ᵃ 3.
[6] 461ᵃ 25-ᵇ 18, especially ᵇ 11 κατιόντος τοῦ πλείστου αἵματος ἐπὶ τὴν
ἀρχήν κτέ.; and ᵇ 17 καὶ λυόμεναι ἐν ὀλίγῳ τῷ λοιπῷ αἵματι τῷ ἐν τοῖς
αἰσθητηρίοις κινοῦνται.

organ is the blood, or the σύμφυτον πνεῦμα which courses with the blood in the veins. Relation of φαντάσματα to hope or fear, memory, thinking, desire, and will. This is the sole guide of conduct in the lower animals, and greatly influences the conduct of men.

perception, copying this [1] as it was in its first presentation [2]. This secondary image is what Aristotle calls the φάντασμα. The *faculty*, and sometimes the *process*, by which φαντάσματα arise is called by him φαντασία, which (in the chapter expressly devoted to its explanation) is defined as 'a movement within the ζῷον produced by actualized perception [3].' Thus φαντασία is an exercise of the κοινὴ αἴσθησις, and provides the material on which this further exercises itself in memory and reminiscence, and in hope, fear, and desire [4]. We cannot think of any concrete individual thing of which we have had no previous perception [5]. Without the particular αἴσθημα we cannot have the φάντασμα, and without this we cannot have the thought—οὐδὲ νοεῖ ὁ νοῦς τὰ ἐκτὸς μὴ μετ' αἰσθήσεως ὄντα. As, if one perceives nothing he is incapable of learning anything [6], so if he has not a φάντασμα connected with the matter of scientific contemplation (θεωρία) such contemplation is impossible. Thus φαντάσματα are to ἡ νόησις what αἰσθήματα are to ἡ κοινὴ αἴσθησις. φαντασία, too, is the link which connects our thoughts with desires and impulses, and may by itself, even in defiance of scientific or any clear and accurate knowledge, guide or control the actions of men. Men, indeed, have reason (νοῦς) with which to check and control the influence of φαντασία on conduct; but to the lower animals [7] φαντασία with ὄρεξις alone presents the motives of action. All the pleasures possible to man are either *present* in perception (ἐν τῷ αἰσθάνεσθαι) or *past* in memory (ἐν τῷ μεμνῆσθαι), or *future* in expectation (ἐν τῷ ἐλπίζειν μέλλοντα). The pleasures accompanying memory and expectation are due to the φαντάσματα involved in these mental states; for the φαντάσματα are attended with

---

[1] For the *inner* stimulus is qualitatively like the *outer*; ἡ φαντασία κίνησίς τις . . . καὶ ταύτην ὁμοίαν ἀνάγκη εἶναι τῇ αἰσθήσει, 428ᵇ 10-14.

[2] 450ᵃ 10 τὸ φάντασμα τῆς κοινῆς αἰσθήσεως πάθος ἐστίν.

[3] For ἡ φαντασία generally, in itself and in its relationship to other psychical faculties, see *de An.* iii. 3. 428ᵇ 2-429ᵃ 9.

[4] Cf. *Rhet.* 1370ᵃ 28: 'When one remembers or hopes or fears (ἐλπίζοντι) a φάντασμα of the object remembered or hoped for or feared accompanies his mental act.' [5] 432ᵃ 2 seqq., 445ᵇ 16.

[6] 432ᵃ 7-10, 449ᵇ 31 seqq. [7] 429ᵃ 4-8.

αἴσθησις[1]. The pains of memory and expectation are to be explained in the same way.

§ 34. The close relation of φαντασία to intellect (τὸ νοεῖν) is most forcibly and clearly stated in de Mem. 1[2]. The intellect must have a φάντασμα to work with. This may be illustrated and in a measure proved by what we experience in geometrical reasonings. When we draw a geometrical figure, though the particular size of this figure does not matter, yet we draw it always of some particular size. In the same way generally when one thinks, even though the object of his thought be something not involving quantity, yet he envisages it (τίθεται πρὸ ὀμμάτων) as quantitative, and then proceeds in his thinking of it without any regard whatever to its quantitativeness. In the same way, too, if the object be properly quantitative but of indeterminate quantity (as when we say, e.g. 'any given circle'), in spite of this one connects it first with some determinate quantity—as if of some particular size—and then thinks of it for the purposes of his problem in abstraction from such determinateness[3]. The reason why one must do this—why we cannot exercise the intellect on any object unless under such conditions, and also why we cannot, as is likewise true[4], exercise the intellect except under the condition of time, even though dealing with conceptions not *in* time—requires separate discussion, but the fact remains[5]. After this it is not surprising that φαντάσματα and νοήματα should in Aristotle's treatment of them sometimes approach one another so closely as to appear confused. Thus we read[6]

Margin notes: Relation of φαντασία to νόησις. It is necessary for the schematism of the objects of thought, and without it these could not be remembered. Illustration from the use of geometrical figures, and the way in which they are drawn. We cannot have even objects of thought before our mind except in connexion with time-conditions. Nearness of φαντάσματα to νοήματα in Aristotle's treatment.

---

[1] *Rhet.* i. 2. 1370[a] 28–35 ; *de Mot. An.* 701[a] 4–5. The φαντάσματα are all rooted in αἰσθήματα, which if pleasurable make them pleasant.

[2] 449[b] 30–450[a] 13.   [3] νοεῖ δ' ᾗ ποσὸν μόνον.

[4] Aristotle had not before spoken of this point, yet he assumes it without hesitation, and it is the one most germane to his succeeding discussion of memory.

[5] Aristotle nowhere attempts to explain the reason of the fact thus stated and assumed here.

[6] 458[b] 23, where, however, φάντασμα appears suspicious. Simplicius does not seem to have read it: if kept, it has to bear a different sense from what it bears in the context (e.g. 458[b] 18) before and after. Without it, too, the meaning of the passage is perfect.

that dreamers sometimes have a reflection or thought which exceeds the scope of the dream, and this reflection is called a φάντασμα. But the tendency to confuse φαντάσματα and νοήματα is seen most emphatically in the unanswered query as to the point in which τὰ πρῶτα νοήματα differ from φαντάσματα [1], and in the construction given to φαντάσματα by Aristotle in relation to rational desire and will [2]. Here we find φαντασία λογιστική or βουλευτική attributed to rational beings, while only φαντασία αἰσθητική is assigned to the lower animals. Thus, from being regarded as *co-operant* with the activity of rational deliberation, φαντασία seems to have become itself invested with rationality. Yet Aristotle does not intend this. The terms λογιστική and βουλευτική need not be taken to mark powers inherent in φαντασία, but powers only belonging to it κατὰ συμβεβηκός, i. e. from its relation to the noëtic faculty. Thus φαντασία αἰσθητική would remain the only φαντασία proper [3].

§ 35. The φάντασμα may or may not be a true copy of the object, which gave rise to it through the original αἴσθημα. It is a true copy if (*a*) the κίνησις propagated from the special organ to the central organ is unmixed with alien movements also stored up in the same special organ ; and (*b*) if this organ and the medium of the movement propagated from it, viz. the blood, are not excited by some overpowering shock which would prevent each from discharging its normal function. If these conditions are fulfilled, and, of course, if the original sensory impression has been correctly taken—if the primary φαντασία is true—then the φάντασμα corresponds duly with its object, and is a true copy of it [4]. The faculty of having φαντάσματα must not

Marginal notes: φαντασία is βουλευτική or λογιστική but only κατὰ συμβεβηκός, not properly or directly. φάντασμα, a true copy of object, and truly represents it, on certain conditions. φαντασία, though the word is taken from the modality of vision, is not confined to representation of

---

[1] 432[a] 12, where, however, in the next clause Torstrik is probably right in reading ταῦτα for τἄλλα, thus denying that the πρῶτα νοήματα are φαντάσματα, and merely asserting that they are οὐκ ἄνευ φαντασμάτων— the doctrine of the *de Memoria*. [2] *De An.* iii. 433[b] 29–434[a] 10.

[3] 702[a] 19 φαντασία δὲ γίνεται ἢ διὰ νοήσεως ἢ δι' αἰσθήσεως. Here the word is used, says Bonitz, *Ind. Arist.* 811[b] 26 *latiore sensu* : the image which stimulates ὄρεξις may be suggested by a *thought* or by a *perception*. The subject is the βουλευτικὴ φαντασία, in which, as explained above, the φαντασία is allied with thinking, but not produced by it.

[4] *De An.* iii. 428[a] 15–[b] 17.

be regarded as confined to the province of vision, to which the αἰσθή-
the etymological meaning and the popular use of the word ματα of
φαντασία tend to restrict it [1]. In its definition it embraces It em-
all provinces of sensory representation. We must, therefore, provinces
suppose that to the αἰσθήματα of sounds, tastes, smells, and of sensory
of the various tangibles φαντάσματα correspond; although tion. Yet
from the associations of the word it would not be easy Aristotle,
to find φαντασία or φαντάσματα directly used of any except moderns
images derived from the sense of seeing. This requires ing with it,
to be emphasized, since Aristotle, like many modern at times
psychologists, was in the habit of treating φαντασία as if proceed as
it had no scope beyond the limits of the visual province; if it were so
just as (on the principle, ἡ ὄψις μάλιστα αἴσθησις) he also That we
habitually treats τὸ ὁρᾶν as if it were equivalent to τὸ member
αἰσθάνεσθαι in general. That, however, we must assume sounds,
φαντασία as having this wider application, and φαντάσματα smells, and
corresponding to αἰσθήματα of every αἴσθησις, follows feelings,
necessarily from the theory of memory laid down by we have
Aristotle. As we shall see memory acts by means of a φαντά-
φάντασμα, nor would it be possible for us to remember these.
the perceptions of any sense unless we had φαντάσματα of
these. The fact, therefore, that we can remember sounds,
smells, and tastes, and feelings, as well as sensations,
of every sort proves that all these as well as ὄψις leave
φαντάσματα answering to them in the mind. But, in
explaining the phenomena of dreaming (*vide infra* § 37),
Aristotle virtually asserts that the αἰσθήματα of all the
senses come under the service of φαντασία (459ᵇ 20–23).

§ 36. We have seen that ἡ κοινὴ αἴσθησις is the faculty by Sensus
which we become conscious of our waking perceptions—of as faculty
the fact *that* we perceive with any sense. Hence it might of *sleeping*
be inferred *a priori* that sleep, if it implies unconsciousness, dreaming.
is due to an affection of this faculty through its organ; Why plants
also that dreaming, which is a form of consciousness during sleep, why
sleep, is an exercise of the same faculty to which we owe why sleep
our waking consciousness. Such is the teaching of Aristotle. affects all

---

[1] 429ᵃ 2 ἐπεὶ δ' ἡ ὄψις μάλιστα αἴσθησίς ἐστιν—sight is the sense *par
excellence*—καὶ τὸ ὄνομα (sc. τῆς φαντασίας) ἀπὸ τοῦ φάους εἴληφεν.

the senses together, not some only. Formal cause of sleeping. Its final cause. The animal soul has its ἐντελέχεια in waking consciousness. Sense-perception and movement in animals have one centre in common. The efficient cause of sleep. Strange that we remember our dreams when we awake, but not the accompanying movements. This connexion of movement with perception, however, helps us to understand the exhaustion of energy which needs sleep for its repair. Sleep connects its onset normally with the nutrient

Sleeping and dreaming are affections of the κοινὴ αἴσθησις. The reason why plants do not sleep and wake is that they have no αἴσθησις; all animals, however, sleep. Sleep affects *all* the special senses: no animal sleeps with some of its senses while awake with the others. This simultaneous affection of all the senses by sleep confirms, if it does not prove, what has been asserted, viz. that sleep is due to an affection of the κοινὴ αἴσθησις; for if this were the faculty of sleep, the latter would when it occurred necessarily affect all the special senses. What affects the common sense must affect all that are dependent upon it. If sleeping were not an affection of this common sense, we should find cases of animals sleeping with some of the senses only; but we never do [1]. Sleep, formally defined, is a sort of bond which binds the general faculty of sense-perception; and wakening is as it were the loosening of this bond [2]. It implies a loss of energy, on the part of the κοινὴ αἴσθησις and its organ, due to excess in the exercise of conscious perception. Its final cause is the recuperation of this energy, and the restoration and preservation of the fitness of animals for the exercise of conscious perception. The waking state— full consciousness—exhibits the animal in its perfection [3].

Sense-perception and movement have a common centre in animals—the region of the heart, in the case of those which possess one, the analogous region or part in the case of others, such as insects, bloodless creatures, and such as do not respire atmospheric air. These show by the rise and fall, the alternate inflation and subsidence, of their bodies in the part analogous to the heart, that they have in them a 'connatural spirit' (σύμφυτον πνεῦμα) [4]. This region is the centre of motive power as well as of sensation and perception. That κίνησις and αἴσθησις should have the same seat was to be expected; for all κίνησις is normally attended with some αἴσθησις, having for its object either an external αἰσθητόν, or an internal phantasm or feeling. Thus the primary organ of sense-perception is the organ of both perception

---

[1] 455ᵃ 30–ᵇ 13.　　　　　　　　　[2] 454ᵇ 25–7.
[3] τὸ ἐγρηγορέναι is the τέλος, 455ᵇ 13–28.　　[4] Cf. 456ᵃ 2–26.

and motion. Hence the efficient cause of sleep, and the conjunction of movement with the dream consciousness. A noticeable thing about it is that though we remember our dreams when we awake, we do not remember our dream movements[1]. This connexion, however, between αἴσθησις and κίνησις shows how the ἀδυναμία διὰ ὑπερβολὴν τοῦ ἐγρηγο- ρέναι comes on: and explains the need of a period of repose.

[margin: process: the ἀναθυμίασις from food in stomach to brain where it is cooled, and returns bringing the bodily heat inwards to the heart and so cooling the outer parts. Sleep thus defined materially.]

Physiologically sleep connects its oncoming with the nutrient process. An evaporation takes place from the food in the stomach. This evaporation goes through the veins upwards to the brain, where it is cooled, and when cool returns downwards towards the heart. With its return drowsiness comes on. The outward bodily parts become cooled, and the bodily heat gathers itself in towards the region about the heart. Defined materially, from this point of view, sleep is the state consequent on the return inwards of the bodily heat and its concentration around the organ of primary perception, whither it is forced by the evaporation returning from the brain[2]. Sleep thus caused continues until the digestive process is complete, and the purer blood destined for the upper parts—the veins round the brain and connected with the sensory organs—has been secreted or separated from the coarser, which goes towards the centre and lower parts of the body.

§ 37. The faculty by which we *sleep and wake* is also that by which we *dream*[3]. Dreaming is not a function of τὸ νοητικόν, intellect, or of τὸ δοξαστικόν, the faculty of opinion; nor can it be a function of the individual senses, for these are suspended during sleep. The fact of our perceiving sensible qualities in the φαντάσματα of dreams—that we perceive colours, &c.—proves, however, that the dream-faculty is a sensory faculty, not δόξα or τὸ δοξαστικόν. We do, indeed, exercise the latter in dreams, but it cannot explain dreaming as a whole.

[margin: Sensus communis in dreaming. This not a function of understanding or of opinion, or of any special sense. Yet it is a faculty of sense, for]

[1] This observation may be paralleled by a question mentioned by Priscianus Lydus (Plotinus, p. 565, 1-6, Didot) and possibly raised by Theophrastus : why do we remember our dreams when we awake, but forget our waking life in dreams?

[2] Cf. *de Somno* 3, *passim* ; *de Part. An.* ii. 7. 653ᵃ 10-17.

[3] Cf. Arist. *de Insomn.*, *passim*.

the images seen in dreams have sensible qualities. The faculty of waking illusions is that whereby we dream: sc. τὸ φανταστικόν, freed from the control of the critical faculty.

The faculty and organ whereby we dream must be that wherewith in waking moments we are subject to illusions; for example, that whereby we seem to see the sun as only a foot in width. As in waking, so in sleeping, the presentation—the mere φάντασμα—overpowers the judgment; and in dreams this is peculiarly liable to happen, the critical faculty being then in a weak and fettered condition. In dreams, however, we sometimes become aware *that* we are dreaming. On the whole the dream state may be described as one in which there is a functional activity of the central organ or faculty of sense-perception (not, however, *qua* perceptive but *qua* representative—φανταστικόν); but in which the representations, φαντάσματα, control the critical faculty[1] owing to its weakness during sleep.

The effects of sense-perception, as has been observed, continue in the organs; exactly as local motions continue after the impact which gave rise to them has ceased. Qualitative change is propagated in the same way; and αἴσθησις is a form of such change. So heat propagates itself[2] stage by stage through a body until it has come full circuit back to its principle or source of generation (ἀρχή). Familiar instances of such persistence of sensory effects in the organs after the cessation of the stimulus are found in the phenomena of seeing. (*a*) When we look at the sun and then turn our eyes away from it, we can see nothing for a while, owing to the persistence of the light impression. (*b*) If we look steadily at some vivid colour, for example, at white (including 'bright') or green (λευκὸν ἢ χλωρόν), and then transfer our gaze to something else, the latter becomes tinged with the colour which we saw previously. (*c*) If, after looking at the sun or some other brilliant object, we close our eyes and, having adjusted our gaze, as it

Familiar instances of persistency of impressions in organs of sense: visual afterimages of light, and of colours, both negative and positive. Such persistency not confined to the sense of seeing.

---

[1] Here we come on a proposition which shows the impossibility of *finality* in a work like the present, which confines itself to the psychology of sense. What is this mysterious critical faculty, which checks and corrects illusions? A treatise on epistemology would be required to give, or attempt, the answer.

[2] 459ᵇ 3. Sc. by ἀντιπερίστασις. See Oxford Translation of *de Insomn.* with notes ad loc., and on 457ᵇ 2, 458ᵃ 27.

were, straight in the same line of vision as before, we look 'inwardly' along this line [1], we see a succession of changing colours. First we see the colour which we saw with the eyes open—the proper colour of the sun or bright object ; next, this changes to crimson (φοινικοῦν) ; and this again to violet or purple (πορφυροῦν), until the object assumes a black colour, and finally disappears [2]. (d) If we look at moving objects, e. g. a river, and then suddenly look at a body at rest, the latter, for a while, seems to be in motion. This is not, however, confined to *seeing*. Such sensory effects occur also in *hearing* and the other senses. Loud noises render us temporarily deaf ; strong odours deaden the olfactory sense for a time, and so on.

These facts go to the root of the explanation of dreaming so far as it is matter of empirical psychology.

To explain the dream phenomena, and the illusion to which we are liable in dreams, two assumptions suffice. These assumptions are :—

I. that the effects of sensation just described as persisting in the organs are capable of giving rise to after-effects in the way of perception : of becoming or furnishing objects to the central sense ; and

II. that when we are labouring under pathological conditions, e. g. strong emotions such as anger, love, or fear, we are especially liable to illusion. This can be proved by experience. Those who are in fever mistake figures on their chamber-walls for fierce animals, deceived by the resemblance. If the patients are very weak they even make bodily movements in trying to escape from the animals. So in sleep the image which comes up is strong and vivid, while the controlling faculty which should criticize its objective truth is then weak and helpless. This explains

Marginal note: These facts with two assumptions explain the illusion of dreaming. The assumptions are : I. that these persistent effects can become stimuli of the general sense-organ ; and II. that we are especially liable to illusion when labouring under pathological,

---

[1] 459^b 14 παρατηρήσασι φαίνεται κατ᾽ εὐθυωρίαν, ᾗ συμβαίνει τὴν ὄψιν ὁρᾶν. παρατηρεῖν does not here mean 'turning the gaze aside.' It gives the idea of looking *along a line*. We must keep the eyes focussed for distance as before—so Aristotle says—and look as if still gazing at the sun, but with eyes shut.

[2] As Aristotle above noticed *positive* so here he notices *negative* after-images.

the ease with which we are imposed upon by dream shapes or occurrences. Illusions of one sense, which occur even in waking moments, may be set right by the help of some other sense ; as the evidence of sight corrects the false judgment of touch respecting the apparently two marbles between the crossed fingers. But no such resource is open to us in dreaming. The central sense, whose normal tendency is to confirm and approve the reports it receives from each particular sense, unless when some one sense contradicts another, naturally inclines during sleep to affirm the objective reality of the φαντάσματα which arise before it. At such times no one particular sense is free to question another; touch, for example, is then incapable of contradicting the report of sight, or vice versa. Thus the illusion is effectual.

The residual impressions in the organs may stimulate the central sense precisely in the same kind of way as do the αἰσθήματα of which they are relics. The one κίνησις is like the other qualitatively. Whether the stimulation of the central sense is set up from without by an objective αἰσθητόν, or from within by the relic of an αἴσθημα, does not matter to a sleeping person. Hence the inevitableness of the illusion. If illusion can arise in waking moments, as already alluded to, *a fortiori* it may arise in dreams, when the critical power of the central sensory faculty is enchained by sleep. If a person sailing along the coast can be for a while deceived with his eyes open into thinking that the land is in motion, it is easy to understand how one can be deceived in sleep by fallacious sensory appearances, when the critical tests (e. g. comparison with the reports of other senses) which should detect them are not available.

Thus the residual impressions forming after-stimuli, together with the weakness of the controlling sense in sleep, account both for the φαντάσματα of dreams and for the mistake by which we in the dream regard them as realities.

Reasons
why in
sleep and
at night the
imagina-
tion is § 38. Moreover, at night, when the special senses are suspended in sleep and the atmosphere is quiet, these residual impressions have the most favourable opportunity of producing their effects on the central sense. If at such

times quiet prevails within the bodily system itself, clear most
φαντάσματα arise before the mind. If, on the contrary, active.
from any cause there is much movement going on within
the body, the images which appear are distorted, or
images do not appear at all.   Thus, too, after heavy meals
the sleep that occurs is dreamless owing to the movements
connected with nutrition then taking place.

Aristotle gives an almost wholly physiological account
of the effects which it is now customary to refer to the
productive as distinguished from the reproductive imagination[1].   Melancholia, illness of various kinds, intoxication, Conditions
all exhibit instances of the disturbing effects of pathological which are
conditions on the imagination, distorting the images, and able to the
transforming them from natural to fantastic shapes.   Such exercise
conditions affect the central organ of perception, which ' fantastic '
is also that of imagination, and, while impeding critical making
or comparative power, which it in common with every its images
sensory faculty possesses, cause the images which come nature.
before it to be untrue to nature, false copies of the αἰσθητά
whence they were derived.   The ' poetic ' imagination ' Poetic '
which moulds the forms of nature to the uses of art— imagina-
the specially so-called ' productive ' imagination—is clearly
recognized by Aristotle, but is not officially treated in his
psychology.   The ' poetic faculty ' is, he says, an attribute
which the man of genius shares with the madman.   The
plastic inventiveness of the poet or artist and the wild
aberrations of insanity are both due to cognate causes.
' Poetry implies either a happy gift of nature or a strain
of madness.   In the one case a man can take the mould
of any character ; in the other he is lifted out of his proper
self[2].'

§ 39. The general account of dreaming then is this: Summary
An image presents itself during sleep to the central faculty account of
dreaming

[1] Cf. 461ᵃ 3 seqq., 461ᵇ 17 seqq.
[2] *Poet.* 1455ᵃ 32–4 (Butcher).   Cf. Dryden :
          Great wits are sure to madness near allied,
          And thin partitions do their bounds divide.
Also Shakespeare :
          The poet's eye in a fine frenzy rolling, &c.

with its illusion. of perception—to the imagination. The latter is, as we have said, naturally disposed, in the normal course of things, to second or affirm the reports of the senses which come before it: to assume that when these forward the report of an object, the object is really there as represented. This it always does when no conflict of testimony occurs between different senses; and none ever occurs in sleep. Moreover, the critical power of the central faculty is impaired or abolished in sleep. The residual impressions which give rise to the images float inwards from the special organ to the central organ in the current of the blood, which at that time gathers towards the heart. Such impressions at such times come in a regular order of succession. The

'Association of κινήσεις' holds for dream consciousness also. Pressure of the blood around the heart during sleep is what hampers the critical faculty of central sense and makes us liable to the illusion of dreams. Efforts of the critical faculty even in dreams to penetrate the illusion: we say in our dream— 'this is only a dream.' rule of the association of ideas (κινήσεις) applies strictly to our dreaming as well as to our waking states. The ideas of the dream come in their order one after the other, just as those of reverie or memory do when we are awake. These, then, are taken by the central sense to represent outer objects, just as the αἰσθήματα of waking life do. Hence, we are deceived into supposing that we *see* what we only *dream* of. What fetters and embarrasses the critical faculty of the central sense is the pressure of the blood round the heart during sleep. If the remnant or residual impression which thus comes before the mind's eye in sleep resembles the primary impression—the αἴσθημα—we dream straightway of the *object* (αἰσθητόν) which produced this. It is, indeed, possible, and sometimes happens, that a man should be aware that he is only dreaming. In his dreams one sometimes says to himself: 'this is only a dream.' Hence to this extent he is not—in such a case—beguiled or deluded by the appearance. Generally, however, the deception is complete, and passes without detection. In waking moments we readily expose sensory illusions by the application of tests, derived also from the senses. If by inserting the finger one slightly displaces the eyeball of one eye, an object seen appears as two; but this does not cause one to believe it to be two. We know the cause of the illusory appearance, and, besides, we have the sense of touch to correct it. But during the dream no such resources are open to us. When we see the φαντάσματα we proceed just as if they were αἰσθήματα (not μοναί, or relics, of αἰσθήματα), and think and believe that we behold the actual objects (αἰσθητά) themselves.

Apart from dreams proper, we have experiences on the borderland of sleep which enable us to obtain a glimpse of the machinery by which dreams are fabricated. Often, when just sinking to sleep, we suddenly wake up, and as it were surprise a host of φαντάσματα crowding in upon our minds. Children have φαντάσματα constantly active which beset them in the dark. Such are not dreams proper, however; but they show to some extent the process of internal stimulation from which dreams come, or with which they commence. During sleep itself, too, perception of a certain sort is not uncommon, keeping us as it were in touch with waking experience [1]. We thus perceive sounds, lights, &c., in a feeble way during sleep ; especially in the moments which just precede awakening. These perceptions again are not true dreams, any more than is the corrective judgment which *does* occasionally interpose during sleep, when we dream, and, as it were, say to us— 'this is only a dream.' The dream proper results from a stimulation of the faculty of imagination by residual κινήσεις proceeding from the organs of sense ; and it consists in the φαντάσματα which then present themselves and are mistaken for objective things or events [2]. It is caused purely by the residual impressions, not by any effects of outward things conveyed through the special senses while we sleep.

Other experiences which connect themselves with our sleep or dreams, yet are not parts of the dreams, but show us the machinery of dreams at work. Objective perceptions during sleep.

§ 40. Aristotle begins his discussion of memory by distinguishing this from reminiscence or recollection, and stating that many persons with retentive memories are slow and dull at recollecting. He thinks it necessary also

Sensus communis in memory and reminiscence. Memory

---

[1] There seems to be an incongruity between this and Aristotle's repeated assertions (e. g. 455ᵃ 9-12) that the external or special senses are suspended during sleep.

[2] 462ᵃ 8, ᵃ 29-31.

(μνήμη) distinguished from perception and from expectation. Involves reference to time elapsed. φαντασία per se indifferent to time. Memory the operation of the time-sense: its organ, the organ of time-perception. This is the κοινὴ αἴσθησις with its αἰσθητή-ριον: the same with which we cognize magnitude and motion: but the

to distinguish memory from *perception* and from *expectation*. All three have to do with φαντάσματα [1] : but while those of expectation refer to the future, and those of perception to the present, those of memory refer to past time [2]. The operation of φαντασία, as *presentative* faculty, alike in expectation, memory, and perception, makes it for Aristotle more necessary than it would seem to us to distinguish them carefully. As the distinction between these three faculties—or applications of one faculty—turns altogether on the differences of time-reference (to which φαντασία *per se* is indifferent) the discussion of memory properly commences with the consideration of the time-sense. The organ or part of mind wherewith we cognize time is that wherewith we also cognize *magnitude* and *motion* ; and the φάντασμα (of time, as well as of magnitude and motion) is a product of the κοινὴ αἴσθησις, or πρῶτον αἰσθητικόν, acting as τὸ φανταστικόν [3]. Memory belongs only to creatures which possess the time-sense, and are capable of perceiving a lapse of time, and thus distinguishing the *present* from the *past*. When one remembers, he says to himself (to use Aristotle's quaint words), 'I *formerly* learned or perceived this doctrine or object.' Memory consists not in a perception or conception present to the mind,

---

[1] The αἴσθησις referred to here (*de Mem. ad init.*) includes the activity not only of the special but of the general sense.

[2] It is scarcely necessary to point out that ἐλπίς in this connexion includes fear as well as hope: expectation in general. So Plato himself states in a note on this word in the *de Legibus* 644 D. Also Aristotle below implies it in his term ἐπιστήμη ἐλπιστική which (as contra-distinguished by him from ἡ μαντική) would seem to form a parallel to our scientific induction, with resulting *power of prediction*—a genuine, if vague, anticipation of Mill's conception.

[3] 449ᵇ 25–450ᵃ 25. In other passages, e. g. 223ᵃ 25, 433ᵇ 7, it appears as if for Aristotle *reason* were a faculty which perceives time. In the former passage he says εἰ δὲ μηδὲν ἄλλο πέφυκεν ἀριθμεῖν ἢ ψυχὴ ἢ ψυχῆς νοῦς, and goes on to represent time as ἀριθμὸς κινήσεως κατὰ τὸ πρότερον καὶ ὕστερον. In the latter he says γίνεται δ' (sc. τὸ ὀρέξεις ἀλλήλαις ἐναντίας εἶναι) ἐν τοῖς χρόνου αἴσθησιν ἔχουσιν· ὁ μὲν γὰρ νοῦς διὰ τὸ μέλλον ἀνθέλκειν κελεύει, and proceeds to show that ἡ ἐπιθυμία does not see the future, as if implying that νοῦς does so. But neither really contradicts the doctrine, laid down in *de Memoria*, that time is object of αἴσθησις only.

but in the relation of one of these to time elapsed[1]; or it
is one of these as *conditioned*, or *affected*, by lapse of time.

Memory, therefore, is not a function of pure intelligence.
The latter, indeed, cannot exert itself without the help of
imagination[2]. We have already illustrated the dependence
of reasoning on imagination, by reference to the universal
and necessary procedure of the mind in connexion with
geometrical thinking and its diagrams. There our thought
is *per se* concerned with no particular figure, yet we, in
order to think, have to draw some particular figure. So,
too, in conceptions which are true irrespectively of space
or time, we find it needful, for the purpose of knowing and
discussing them, to connect them with space or time. Why
this is necessary we need not here inquire. But the fact
is so. Similarly, we cannot *remember* anything whatever
unless by the aid of a φάντασμα, through which the remembered fact may connect itself with time elapsed. This
holds of scientific and philosophic truths or theorems.
These latter, not being directly representable to imagination,
must be *schematized*, i. e. connected with φαντάσματα. Thus
only are they capable of being remembered, i. e. indirectly,
or, as Aristotle says, κατὰ συμβεβηκός. The reason why
we cannot remember except by the aid of φαντάσματα is that
we can remember directly nothing which we have not first
perceived; and only perception generates the φάντασμα,
which is the instrument of memory.

This explains how memory belongs not merely to
creatures possessing intellect, but to many of the lower
animals. These do not possess intellect, and if memory

(margin) organ of imagination is the same, only conceived in a different relation. Memory not a function of pure intellect, which cannot, indeed, act without the support of a schematizing imagination. Proofs and illustrations.

[1] 449ᵇ 24 ἡ μνήμη οὔτε αἴσθησις οὔτε ὑπόληψις ἀλλὰ τούτων τινὸς ἕξις ἢ πάθος ὅταν γένηται χρόνος. See p. 313. By πάθος is suggested the *genesis* of the ἕξις. The αἴσθησις or ὑπόληψις is *affected* by the lapse of time: from this affection arises the *relative* character of the μονή, its ἕξις, in which consists the time-perspective of memory. There are some places in which ἕξις = 'having,' but this is certainly not one of them.

[2] This passage (449ᵇ 30 seqq.) more clearly than any other exhibits the relation of dependence on the lower in which the higher mental faculties are placed by Aristotle, in accordance with his theory of the gradual evolution of scientific knowledge from individual sensible experience.

were a function of pure intellect, none of them would be able to remember [1]. However, many of them manifestly do remember. Those which cannot remember are those which lack the sense of time. If memory were a function of *pure* intelligence, even man could not remember [2]; for our intellectual acts are not capable of being remembered *per se*, but only indirectly, in virtue of their sense-derived φαντάσματα. Memory, therefore, is a function of the same part of the soul to which imagination belongs. All facts capable of being presented to imagination can be directly remembered; all others can be remembered only so far as they link themselves with φαντάσματα, i.e. only indirectly.

How do we, with only a present image to help us, remember the *past*? The memory-image is always relative to, and representative of, an object; related to it as a picture

§ 41. How then do we, by the help of φαντάσματα, remember, i.e. *know the past*? Our sole datum is the image present to the mind. This, however, is not past but present, whereas the past is absent: it is gone. How then is it known [3]? We must try to conceive the answer to this question as follows. The foundation of memory is laid in perception. When, therefore, we perceive, a sort of picture (ζωγράφημα, γραφή) is painted in the soul, or in the part of the body which contains the perceptive organ concerned in the perception; or else the sensory κίνησις stamps an impression as it were of the particular sense datum upon the organ, as a person with a seal ring stamps its impression on

---

[1] This assumes Rassow's correction θηρίων for θνητῶν, 450ᵃ 18.

[2] This explains the traditional θνητῶν, the difficulty of which is that it forces us to press the word 'pure,' which is not really in the text.

[3] As regards the physical character of the *impression* which generates the φάντασμα Aristotle gives no clear statements, but expresses himself in a variety of metaphors. It is 'imprinted' by a κίνησις ὑπὸ τῆς κατ' ἐνέργειαν αἰσθήσεως γιγνομένη, and is ὅμοιον ὥσπερ τύπος ἢ γραφή (450ᵃ 30, ᵇ 15). Freudenthal (*op. cit.*, pp. 20 seqq.) examines minutely into Aristotle's statements to discover, if possible, an exact account of his conception of this memory image, but to little purpose. He concludes, with every appearance of truth, that the τύποι were, for Aristotle, not really like seal-impressions, but rather qualitative or 'chemical' changes of tissue, not involving mechanical movement. The question of agreement on this point between Aristotle and Hobbes is merely a question how far Hobbes followed Aristotle.

a piece of wax [1]. The question now arises: is this impression, thus taken, what we remember? Do we not remember rather that of which it is an impression—the object, or event, which produced it in the mind? For if what we remember is this impression, we do not remember the past at all: it is a mere mistake to think we do. But if we really remember the past object or event (as experience proves that we do), how is it possible to do so through an impression which is not past but present? This Aristotle proceeds to treat as the real question to be answered. He imagines an objector to say that it would be as easy to suppose a person seeing some colour, or hearing some sound, which was not present to sense, as to suppose him knowing the past, which is now gone. To this he replies: do we not as a matter of fact, in a certain way, see and hear the non-present? Do we not in pictures see absent persons? Now this will illustrate what takes place in remembering by means of a φάντασμα. A picture is not merely a painted object: it is more than this. It is a likeness of some person or thing. While *per se* numerically one and the same thing, it may be viewed in two relations. In the same way, the φάντασμα before the mind in memory—the impression bequeathed by sense to imagination—may be regarded purely and simply as a φάντασμα, or it may over and above this be regarded as a likeness, a representation of something else. Taken in

to its original, or connected with it by association in some way.

The memorial φάντασμα can be regarded either (1) as a mere appearance, or (2) as a

---

[1] 450ᵃ 27–32 δεῖ νοῆσαι τοιοῦτον τὸ γινόμενον διὰ τῆς αἰσθήσεως ἐν τῇ ψυχῇ καὶ τῷ μορίῳ τοῦ σώματος τῷ ἔχοντι αὐτήν, οἷον ζωγράφημά τι [τὸ πάθος οὗ φαμὲν τὴν ἕξιν μνήμην εἶναι: I suspect this of being a gloss on τὸ γινόμενον]. ἡ γὰρ γινομένη κίνησις ἐνσημαίνεται οἷον τύπον τινὰ τοῦ αἰσθήματος, καθάπερ οἱ σφραγιζόμενοι τοῖς δακτυλίοις. Cf. Plato, *Rep.* 377 B ἐνδύεται τύπον (so Adam) ὃν ἄν τις βούληται ἐνσημήνασθαι ἑκάστῳ: also especially *Theaet.* 191 D. For the ζωγράφημα, cf. *Phaedrus* 276 D. Aristotle 450ᵇ 5–11 introduces some observations on the causes of defective memory. Persons in whom, like those very old or very young, a great deal of movement exists are bad subjects for mnemonic impressions: it is as difficult to impress a durable mark on their organs as on running water. If the surface is too hard, no impression is taken by it; whereas if it is too easily impressed—too soft—the impression is taken but not retained long.

representa-
tive ap-
pearance.
As the
latter, it is
a μνημό-
νευμα. But
besides this
reference
to an
original,
the μνημό-
νευμα refers
always also
to time
elapsed.

the latter way it is a *memorial* or *reminder* (μνημόνευμα), no longer a mere φάντασμα. Thus regarded, it explains how we remember by its means. It is like a picture which is a portrait of a friend, by which, when I look at it, I can have my absent friend present to my mind. Two marks distinguish the μνημόνευμα from the mere φάντασμα; viz. (*a*) the conscious reference to past time involved in having a μνημόνευμα, and (*b*) the relationship of the μνημόνευμα to an object which it resembles, or otherwise represents, and so recalls to mind.

Confusion
of memory
with
imagina-
tion, and
of imagina-
tion with
memory.
Antipheron
of Oreus.
Mnemonics
aim at
confirm-
ing the
represen-
tative
character
of an 'ap-
pearance.'

Certain ordinary experiences partly confirm, partly illustrate, what has here been said. Sometimes, when men have a φάντασμα before the mind, they ask themselves— for they are not sure—whether they are or are not then remembering; whether, that is, the phantasma which they contemplate is a likeness or not of a past experience. In such cases, indeed, we often discover that it is a likeness; the original flashes upon our minds, and we remember. We pass from regarding it in its individual character to regarding it as related to its original. The contrary also occurs in occasional experience. Men mistake their mere φαντάσματα for μνημονεύματα; they confound their fancies with past experiences. Such was the mental condition of Antipheron of Oreus, and certain other deranged persons; they recounted the events or objects which merely presented themselves to their imaginations as though these were facts of their past experience which they remembered [1].

The practical value of the mnemonic art rests on the truth of what has been above stated. Mnemonics aim at training a person to regard certain presentations not merely as single or unrelated, but as in connexion with, or as likenesses of, certain objects. Thus the former become *reminders* (μνημονεύματα) for the latter.

Remi-
niscence

§ 42. Memory, in general, can accordingly be defined as *the relationship which a* φάντασμα (or *mental presenta-*

---

[1] In discussing the subject of dreams Aristotle refers to the way in which φαντάσματα can be mistaken for αἰσθήματα, and how certain forms of hallucination arise; cf. 460ᵇ 3-27.

tion), *as a likeness, bears to that of which it is a* φάντασμα¹. This general faculty of retention (μνήμη) is the presupposition of reminiscence or recollection (ἀνάμνησις). If one does not remember—if the already described conditions are not fulfilled—he cannot recollect. But he may remember without being able to recollect, i. e. without being able to *recall* at the moment the ideas which represent fully to consciousness the past object or event. Often there is a difficulty felt in doing this. Some persons succeed better than others in doing it, and all persons do it better in some cases than in others. This is the faculty whose nature and procedure Aristotle next undertakes to explain.

We must not, he says, hastily define recollection as the mere *recovery* of memory. It is no more this than it is the *inception* of memory². Memory may exist without reminiscence, i. e. there may be no need of the latter. No breach may have occurred in the continuity of our memory of an experience. Reminiscence or recollection has no place until after such a breach of continuity has intervened.

(marginal notes: (ἀνάμνησις). Definition of memory, and distinction of memory from reminiscence. Memory is the general faculty of retention: reminiscence the particular faculty of recollection. One may remember without then being able to recollect; he cannot recollect if)

¹ 451ᵃ 15 φαντάσματος, ὡς εἰκόνος οὗ φάντασμα, ἕξις. The obvious rendering of ἕξις here (approved by Zeller) as 'having,' introduces a superfluous notion. The more Aristotelean interpretation, though less easy to work into a translation, as 'relation' or 'relative state' alone gives the sense required. So taken, this definition sums up with force and brevity the preceding account of the mnemonic φάντασμα. It might be paraphrased τὸ εἶναι ἐν ἡμῖν φάντασμά τι οὕτως ἔχον πρὸς ἐκεῖνο οὗ φάντασμά ἐστι, ὡς εἰκὼν ἔχει πρὸς ἄλλο τι οὗ εἰκών, which use of οὕτως ἔχον . . . ὡς ἔχει would explain ἕξις. Freudenthal accordingly supports the view that ἕξις here comes from the intransitive ἔχειν, but finds it hard to get a German equivalent. He likes the word 'Stand,' but thinks it unidiomatic. His own rendering p. 36 n. is: *die Andauer einer Vorstellung als eines Abbildes von dem, dessen Vorstellung sie ist*. I prefer to use '*relative* state,' or 'relationship,' rather than 'state,' as its equivalent, and base my right to do so on Aristotle's definition 1022ᵇ 10 ἄλλον δὲ τρόπον ἕξις λέγεται διάθεσις καθ᾽ ἣν ἢ εὖ ἢ κακῶς διάκειται τὸ διακείμενον, καὶ ἢ καθ᾽ αὑτὸ ἢ πρὸς ἄλλο.

² 451ᵃ 20–ᵇ10, Aristotle here seems to criticize (unfairly, as Plato's αὐτὴ ἐν ἑαυτῇ shows) the definition (accepted by Plato, *Philebus* 34 B) of ἀνάμνησις as = μνήμης ἀνάληψις. He points out that this is possible by a fresh exercise of αἴσθησις or μάθησις, and that these, though they lay the basis of memory, cannot synchronize with it, for memory implies that *time has elapsed* since the αἴσθησις or μάθησις took place.

he does not remember.

Definition of reminiscence. Distinction between it and re-experiencing or re-learning.

But when the chain of memory has been temporarily broken, we may re-unite its parts in either of two ways. We may by an effort of recollection recall the vanished ideas required for knowledge of the past experience— whether αἴσθησις or μάθησις. But it is also possible for us to repeat this experience itself. Such repetition, however, would not be reminiscence. It would, indeed, be our sole resource if the ideas had absolutely vanished : if we *no longer remembered*. Reminiscence, however, properly takes place only when the vanished ideas are recalled by the activity of an internal impulse or spring, over and above any external means of recalling them. When a man recollects, this implies that he was able somehow of himself, and without appealing to anything outside himself, to proceed onwards to the goal of his effort; to recover the wished-for idea. When he is unable to do this, he simply has no memory of the fact or experience. He no longer remembers. When he can do this, i. e. when, proceeding by internal activity, he reaches the missing idea, he recollects in the proper sense, and his full memory of the experience ensues, or is revived[1]. If I have to see a face again in order to form an idea of it, I do not remember it, and therefore cannot, try as I will, recollect it. If I can recollect it, then the idea of it recurs after the effort of reminiscence, and so I again remember it[2]. So if I have to relearn a lesson by having recourse to my book or my teacher; or if I have to go through the forms of calculation by which I first made a discovery, in order to recall the discovery to mind, I do not thereby recollect. I recover my memory of the

---

[1] 451ᵇ 4 τοῦτ᾽ ἔστι καὶ τότε τὸ ἀναμιμνῄσκεσθαι τῶν εἰρημένων τι· τὸ δὲ μνημονεύειν συμβαίνει καὶ ἡ (so Biehl) μνήμη ἀκολουθεῖ. These last words, which have perplexed some persons, merely convey the idea of the revival of memory as contingent on the act of successful reminiscence. It must be borne in mind that memory is not only the *prius* but also the *posterius* of reminiscence.

[2] The terms μεμνῆσθαι and μνήμη have a tendency to ambiguity, since each may be used of its object either δυνάμει or ἐνεργείᾳ. Potential μνήμη is the presupposition of successful ἀνάμνησις; actual μνήμη is its result or *sequel*; cf. ἀκολουθεῖ, last note.

lesson indeed ; but not according to the conditions of recollection : not by means of the 'further internal spring ¹.'

§ 43. Given the internal spring, however, acts of reminiscence are facilitated by the natural law that the κινήσεις
left in our organs by sense-perception (in which the ideas
which we wish to recall, or the φαντάσματα with which they
are associated, must have originated) tend to reproduce
themselves in a regular order of succession whenever they
return to consciousness. The order in which they do so
depends mainly on the objective order of the sensible
experiences by which they were generated. There are
movements in nature which are followed by others according to necessary mechanical law. Such, however, is not
the case with the mnemonic movements. These follow the
law of custom ; i. e. they *tend to* succeed one another in
a certain order, and do so succeed *as a general rule.* If the
connexion between antecedent and consequent among our
κινήσεις were necessary, then whenever the antecedent came
up the consequent would follow invariably, and *efforts* of
recollection would be superfluous ². It is with the movements whose succession is customary that reminiscence has to
do, and with these, therefore, we are here chiefly concerned.

The effects of habituation or custom vary with the
various types of mind. Some are impressed by κινήσεις
in a single experience more firmly than others by several

Margin: So-called law of association of ideas. All κινήσεις naturally follow one the other in regular order. This order is either necessary or habitual. The κινήσεις on which memory depends follow the latter order. It is with customary connexion of ideas that we in treating of reminiscence have to do. Effects of habituation in fixing such connexion

---

¹ 451ᵇ 8 δεῖ οὖν διαφέρειν τὸ ἀναμιμνῄσκεσθαι τούτων, καὶ ἐνούσης πλείονος ἀρχῆς ἢ ἐξ ἧς μανθάνουσιν ἀναμιμνῄσκεσθαι.

² Themistius (Sophonias), who illustrates the 'necessary connexion' by the relation of the idea of *heat* to that of *fire*, &c., seems to miss the purpose of the distinction made here by Aristotle. What the latter really means is to deprecate the notion that we can expect in the succession of internal κινήσεις that invariableness which we find in many of the movements of nature. Therefore, in 451ᵇ 11, πέφυκεν ἡ κίνησις ἥδε γενέσθαι μετὰ τήνδε seems to express a general law applying to merely physical as well as to psychical κινήσεις ; only that while in the former it is often true ἐξ ἀνάγκης, in the latter it holds merely ἔθει (see 452ᵇ 1–3). Reminiscence for Aristotle implies voluntary effort. Taking the passage as Themistius does, I fail to understand how the succession of κινήσεις ἐξ ἀνάγκης could be relevant to the explanation of *efforts* at reminiscence. If ἀνάγκη operated, voluntary efforts would be needless.

repeated experiences. The effects of custom vary also with the nature of the experience. There are experiences which we never forget when once they have occurred to us, one single occurrence sufficing to produce a firm connexion between the successive κινήσεις. Other experiences require to be frequently repeated before a firm connexion is produced. The rule is that the connexion is strengthened in proportion to the frequency of the experience. What we often rehearse in our minds we easily and quickly recollect, custom becoming as it were a second nature.

When a person sets himself to recollect something he may for a while fail, but afterwards succeed. His procedure is like that of one searching for something lost. After exciting many trains of movements he at last rouses that particular train in which the idea which he desires to recall is to be found. Recollection depends upon our exciting some κίνησις which has a customary connexion with that one which we want to revive. When it succeeds, it reinstates *in consciousness* the required sequence of ideas.

When we make the voluntary attempt to recollect we act upon these principles; but even when we recover ideas involuntarily (as we may do) the process is similar: the κινήσεις and ideas following the order which the objective events of which they are the representatives pursued. In our voluntary efforts, therefore, availing ourselves of this known fact, we deliberately 'hunt up' (θηρεύομεν) the order of succession, endeavouring to come as near as we can to what this was in objective experience. We start the train of reminiscence either from a present intuition [1], or from some other, which promises to carry us whither we wish to go. We may begin with a κίνησις (representative movement) *like* the one we seek, or *contrary* to it, or *contiguous* to it [2]. The κινήσεις of its *like* are specifically identical with those of

Margin notes:
vary with persons and experiences. As a rule, frequency of experience confirms custom, and custom becomes second nature.

Process of voluntary efforts at recollection described.

The case of involuntary revival of ideas involves the same laws. Reminiscence is the 'hunting up' of an idea. Need of a 'good start.' Connexion of ideas by *similarity, contrariety, contiguity* (in space or time).

---

[1] For what follows *vide* 451ᵇ 18–23.

[2] The contiguity directly referred to here is. probably that of space: yet contiguity in the time order is not excluded. For though we have been told that in this order the former κίνησις recalls the latter, yet we are not debarred from reversing the process. We can even start as has just been said ἀπὸ τοῦ νῦν, which would necessarily imply 'hunting' backwards.

that which we seek to revive; those of its *contrary* are concomitant with them; while those of the *contiguous* idea form part of a whole of movements set up by both, so that but a portion of this whole remains to be revived[1]. Whether we recollect by voluntary effort, or the idea comes back to us without our making or after we have ceased to make[2] the effort, the psychical process is just the same. The succession of ideas is generally determined in one of these three ways. In order to illustrate the psychical process there is no need to refer to remote cases, or those in which the links in the series of κινήσεις are very numerous. The simplest cases will serve for illustration. The cardinal fact is that the κινήσεις have a regular order which they tend to follow, corresponding to the order in which the αἰσθήματα, or sensible impressions, on which they are based took place.

Therefore, in trying to revive a vanished idea[3], one should choose as his starting-point the *beginning* of the train of ideas in which it is likely to be found. When this is done reminiscence proceeds most easily and quickly. As the sequence of the κινήσεις corresponds to the objective sequence of events to which they refer, we should try to think of some event in this latter series. Thus a κίνησις representing the forgotten event is likely to be aroused. Well arranged facts like those of mathematics are, owing to the regularity of their sequence, easily remembered, and as they are easily remembered, so they are easily recollected. On the contrary, confused ill-digested experiences are difficult to remember, and once forgotten equally difficult to recollect,

Facts logically well-arranged, as those of mathematics, easily recalled to mind; ill-arranged matters difficult to recall or recollect.

[1] Thus the picture of Socrates with its specifically identical 'movements' calls up the idea of Socrates himself; the idea of black recalls that of white, the κινήσεις of the one being habitually concurrent in the mind with those of the other. The idea of a thing seen in a certain place together with something else recalls the latter to mind; as also the idea of one of two events synchronously perceived recalls that of the other event.

[2] For this case, see 453[a] 18.

[3] i.e. one which has disappeared from the field of consciousness, not one which has absolutely passed away and which we no longer remember.

or bring back to memory. But the chief thing is to select a good starting-point.

§ 44. Such a starting-point may be anything whatever which has a customary connexion with the idea to be recalled. Hence the surprisingly strange suggestiveness of some things in reviving in our minds ideas with which at first they seem to have nothing to do[1]. But the connexion is always real nevertheless. Thus from the thought of *milk* one's mind passes to the thought of *white*, from this to that of *mist*[2], from which it goes on to *moist* (ὑγρόν), upon which it recalls *autumn*, if this happens to be a season which one seeks to recollect[3]. The central point in a series also forms a good beginning for the attempt at recollection. If one who starts from this does not succeed, he probably has no further chance. He has totally forgotten what he wishes to remember.

It happens, however, that starting from the same initial point one sometimes succeeds and at other times fails in the effort to recollect. A reason (a) of this may be that from

---

[1] I am inclined to read, after Sir William Hamilton, ἀπ' ἀτόπων, 452ᵃ 13, instead of ἀπὸ τόπων which makes δοκοῦσι unintelligible.

[2] ἐπ' ἀέρα. The colour of ἀήρ (misty air, fog) is distinctively white for Aristotle: the ἀήρ in them is what causes the whiteness of foam and snow. Cf. 786ᵃ 6; Prantl, *Arist.* Περὶ Χρωμάτων, p. 105.

[3] Cf. Keats, *Autumn*, 'Season of *mists* and mellow fruitfulness.' With this illustration may well be compared that given by Hobbes for a similar purpose. The passage occurs in his *Leviathan*, i. 3, and is quoted by Sir W. Hamilton in his excellent note on the history of mental association printed at the end of his edition of the works of Reid (Edinburgh, 1849) : 'And yet in this wild ranging of the mind, a man may oft-times perceive the way of it, and the dependence of one thought upon another. For in a discourse of our present civil war, what could seem more impertinent, than to ask, as one did, what was the value of a Roman penny? Yet the coherence to me was manifest enough. For the thought of the war introduced the thought of the delivering up of the King to his enemies; the thought of that brought in the thought of the delivering up of Christ; and that again the thought of the thirty pence, which was the price of that treason; and thence easily followed that malicious question ; and all this in a moment of time; for thought is quick.' Sir W. Hamilton's observation that in this whole doctrine of association of ideas and reminiscence Hobbes is an *alter ego* of Aristotle is literally true.

one and the same point his mind may chance to move in any one of several trains of κινήσεις. One may make sure of his point of departure, but cannot always be certain of the direction in which he shall subsequently move. When one starts, intending to reach a certain terminus, if his mind chances not to move in the former or old[1] path leading thither, it is borne by custom to some more familiar terminus. For, as we have said before, custom in these matters is a second nature; and frequency of repetition produces 'naturalness' of sequence in *our* κινήσεις. But as in objective nature events occur which are unnatural or due to chance, we can easily see how in the sphere of custom irregularities are to be expected. Indeed they should occur *a fortiori* in the latter sphere, since in this natural law has less control[2]. Such is a true explanation (sc. by reference to τύχη) of facts like that above-mentioned. If, however, (*b*) there happens to be some intervening cause which diverts our thoughts from their true direction, and, as it were, switches them off towards itself, such failure to recollect is more easily and obviously accounted for. So when we wish to recollect a name, it often happens that some other name beginning with the same sounds carries our thoughts off to itself, and we either pronounce this wrong name, or blunder upon some compound which is a jumble of both together[3].

§ 45. But, in trying to recollect an experience (object or event), nothing is of so much importance[4] as knowing the *time* of the experience, either determinately or inde-

[margin: fortuitousness which is even more pronounced in the realm of custom than in the realm of nature; influence of distracting associations, which tend to draw one's thoughts out of the train or track.]

[margin: Importance of knowing the time]

---

[1] 452ᵃ 24–30. ἐὰν οὖν μὴ διὰ παλαιοῦ (Bekker) gives the correct sense. The same three or four initial notes may form the commencement of a variety of tunes. Thus I have heard a person sing a few notes and then ask—'What song am I thinking of?' The different answers given show how easily one's 'mental ear' may go off in a wrong series of notes, before hitting upon the right series in which a few notes more would infallibly recall the required tune.

[2] 452ᵃ 29 seqq. ἐπεὶ δ' ἐν τοῖς φύσει γίνεται καὶ παρὰ φύσιν καὶ ἀπὸ τύχης, ἔτι μᾶλλον ἐν τοῖς δι' ἔθος, οἷς ἡ φύσις γε μὴ ὁμοίως ὑπάρχει. Imperfect as was Aristotle's conception of 'natural law,' yet, for the above interpretation of φύσις, cf. *N. E.* 1103ᵃ 19–23 (Stewart).

[3] Themistius (Sophonias) gives as examples of such words Πλευρωνία (in Aetolia) and πλευρῖτις, Λεωφάνης and Λεωσθένης. [4] 452ᵇ 7–453ᵃ 4.

of what we wish to recollect.

Distance in time is marked in our imaginations like distance in space. Memory is 'vision in time.'

Function of the time-mark in discriminating between φαντά-σματα intrinsically alike, and so giving them their correct respective relations (to objects) as μνημο-νεύματα.

terminately. For the faculty whereby we remember is that by which we perceive and estimate lapse of time. It is also that by which[1] we cognize distances in space, and magnitudes in general[2]. The mode in which we perceive distances in time is analogous to that in which we perceive distances in space: i. e. by representative κινήσεις within us. We have 'within our minds' a distance-κίνησις[3], i.e. one which represents or stands for the objective distance; and so, too, we have a time-κίνησις similarly related to the objective time elapsed. As several objective space or time distances are to one another, so are the subjective space or time κινήσεις, which represent them, to one another. But besides these κινήσεις, which symbolize the time and space *distances*, we have 'in our minds' κινήσεις corresponding to the forms[4] (εἴδη) of the *objective experiences* themselves which are projected at such distances. Now, if these experiences are to be properly and fully recollected, it is of cardinal importance that the κινήσεις which 'formally' represent them should be duly connected in consciousness with their time-κινήσεις. By the aid of the latter we not only recall the experiences themselves but also distinguish experiences which may be intrinsically similar. If two non-synchronous experiences have been in themselves exactly alike, the κινήσεις which survive the apprehension of their forms are exactly alike. For recollection, therefore, these experiences would be indistinguishable, were it not that they have annexed to them different time-κινήσεις, by which they are respectively assigned to their separate positions in the series of past experiences. They are 'dated' and thus saved from being confounded with one another in memory. The time-κίνησις, therefore, is most fruitful for reminiscence if we have it to start with when we make the effort to remember an experience. By its close association with the εἶδος of the object or event it is of the utmost service

---

[1] Probably ᾧπερ should be read for ὥσπερ 452ᵇ 9.

[2] In what here follows memory is for Aristotle, what it is for Ribot, *vision in time.*

[3] This is all that had been suggested by Aristotle or his predecessors for explaining the perception of distance.

[4] εἴδη : sc. τὰ ἄνευ ὕλης.

for reviving this εἶδος in consciousness, and recalling the event itself to mind. Nor can we remember a past experience in the full sense until, besides envisaging it, we likewise connect it with its date, i. e. fix its true place in the objective time series [1].

Note on Aristotle's diagram-matic illustration of the function of the time-κίνησις.

[1] The passage in which Aristotle tries exactly to explain his assertion of the importance of 'knowing the time' is 452[b] 17–24. Biehl prints it thus: ὥσπερ οὖν εἰ τὴν ΑΒ ΒΕ κινεῖται, ποιεῖ τὴν ΓΔ· ἀνάλογον γὰρ ἡ ΑΓ καὶ ἡ ΓΔ. τί οὖν μᾶλλον τὴν ΓΔ ἢ τὴν ΖΗ ποιεῖ; ἢ ὡς ἡ ΑΖ πρὸς τὴν ΑΒ ἔχει, οὕτως ἡ [τὸ] Θ πρὸς τὴν Μ ἔχει. ταύτας οὖν ἅμα κινεῖται. ἂν δὲ τὴν ΖΗ βούληται νοῆσαι, τὴν μὲν ΒΕ ὁμοίως νοεῖ, ἀντὶ δὲ τῶν ΘΙ τὰς ΚΛ νοεῖ· αὗται γὰρ ἔχουσιν ὡς ΖΑ πρὸς ΒΑ.

ὅταν οὖν ἅμα ᾖ τε τοῦ πράγματος γίνηται κίνησις καὶ ἡ τοῦ χρόνου, τότε τῇ μνήμῃ ἐνεργεῖ.

The last sentence gives the clue to the meaning of this passage as a whole. Here no doubt Aristotle had introduced a diagram with letters of the alphabet to illustrate his argument. This diagram perished. To suppose (with Wendland, p. 13) that the diagram given by Themistius (Sophonias) may be the one given by Aristotle himself is impossible, for the simple reason that it would have committed Aristotle to a geometrical blunder. The diagram, however, having been lost, the letters were easily corrupted. The MSS. differ widely in recording them. To reconstruct Aristotle's figure we must divine his meaning first from the remainder of the context. The hazards of this are apparent. Yet it is indispensable, and needs no apology. There would be some satisfaction in introducing tolerable sense (even if merely hypothetical) into a passage which as it stands has for ages baffled commentators. The cardinal thought in our passage is that of mnemonic *representation*. As usual Aristotle thinks of one sense in particular—the sense of sight—while speaking of the procedure of reminiscence in reference to all sensible experiences. Like Ribot he holds that memory is (primarily and chiefly) *vision in time*.

Having asserted that we distinguish longer and shorter times by the organ whereby we cognize different μεγέθη, he briefly indicates how this is done, and restates his theory of perception, as basis of his theory of memory, by representative analogy or similarity.

That which in the 'outer world' consists of spatial objects in spatial relations (τὰ μεγάλα καὶ πόρρω) is, as perceived, represented 'internally' by κινήσεις—psychical affections—which are (a) similar, i.e. 'analogous' to the objective experiences, and (b) related to one another as the latter are to one another. Between the outer or objective sphere and the inner or subjective which thus represents it the parallelism is complete. Therefore, says Aristotle, what difference does it make whether the mind *moves* in the inner or *knows* in the outer sphere? In virtue of the identical proportions, the 'moving' in the one *is* the 'knowing' in the other. Applying what is thus said of perception to the ex-

Illusions of memory.

§ 46. A person may erroneously think that he remembers, fancying that there is a time-mark or date affixed planation of memory and recollection, he proceeds: In the inner world of memory events and objects no longer perceived have their εἴδη and ἀποστήματα (distances in time or space) depicted in imagination. There are within us κινήσεις representing *events* and others also representing the *times* of these events. If the 'same' event has occurred twice in our experience distinct memory would require that its inner εἴδος should be connected with different time-κινήσεις, respectively analogous to the real time-ἀποστήματα. Thus the same εἴδος of an event may, by being associated with different time-κινήσεις, be capable of recalling different portions of past experience; whose difference, however, would not be remembered but for the distinct time-κινήσεις conjoined with it in relation to each portion. In accordance with these preconceptions of Aristotle's meaning I write the passage as follows : ὥσπερ οὖν εἰ τὴν ΑΒ ΒΕ κινεῖται, ποιεῖ [? νοεῖ] τὴν ⟨ΑΓ⟩ ΓΔ· ἀνάλογον γὰρ ἡ ΑΓ ΓΔ—τί οὖν μᾶλλον τὴν ΑΓ ΓΔ ἢ τὴν ΑΖ ΖΗ ποιεῖ [? νοεῖ] ; ἢ ⟨ὅτι⟩ ὡς ἡ ΑΒ ⟨ΒΕ⟩ πρὸς τὴν ΑΓ ΓΔ, οὕτως ἡ Θ πρὸς τὴν Ι· ταύτας οὖν ἅμα κινεῖται. ἂν δὲ τὴν ⟨ΑΖ⟩ ΖΗ βούληται νοῆσαι, τὴν μὲν ⟨ΑΒ⟩ ΒΕ ὁμοίως νοεῖ, ἀντὶ δὲ τῶν Θ, Ι, τὰς Κ, Λ, νοεῖ· αὗται γὰρ ἔχουσιν ὡς ΑΒ ⟨ΒΕ⟩ πρὸς ΑΖ ΖΗ. ὅταν οὖν ἅμα κτέ.

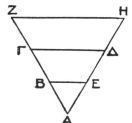

The figure was, as I take it, somewhat like this. In this triangle, divided 'similarly,' AB BE stands for the εἴδος representing either the objective event ΑΓ ΓΔ, *or* the similar event ΑΖ ΖΗ. But $\frac{ΑΓ}{ΓΔ} = \frac{ΑΖ}{ΖΗ}$; therefore the two are distinguished by the different time-marks associated with their common εἴδος. When, therefore, AB BE stands for ΑΓ ΓΔ it has the time-κίνησις Θ, corresponding to the objective time Ι; when it stands for ΑΖ ΖΗ, it has the time-mark Κ corresponding to the objective time Λ. The time-marks and objective times cannot be represented in the *same* geometrical diagram with the εἴδος and the objective events; because their *distinguishing* functions would thus be lost, and the question τί οὖν μᾶλλον would remain unanswerable. Premising this, I translate: 'As, therefore, the mind, if it *moves* subjectively through AB BE, *knows* (the objective event) ΑΓ ΓΔ, since AB is to ΒΕ as ΑΓ is to ΓΔ, why does it in fact know ΑΓ ΓΔ rather than ΑΖ ΖΗ? ⟨The answer is⟩: because as AB ⟨ΒΕ⟩ is to ΑΓ ⟨ΓΔ⟩, so is Θ (the subjective time-mark of the former) to Ι (the objective time of the latter). Hence the mind moves in *these* lines (viz. AB BE, ΑΓ ΓΔ) simultaneously (i. e. it moves subjectively in the former, objectively in the latter ; or while *moving* in the one it *knows* the other, according to the principle laid down in 452[b] 13 τίνι οὖν διοίσει κτλ.). But if a person wishes to think (not of ΑΓ ΓΔ, but) of ΑΖ ΖΗ, his mind moves as before (ὁμοίως) in the representative εἴδος-

SENSUS COMMUNIS 323

to the φάντασμα before his mind. The contrary error is
impossible. A person who really remembers something,
cannot delude himself into thinking that he does not remember this. One cannot remember without being clearly
conscious of doing so, and indeed remembering consists
essentially in such consciousness, i. e. the recognition of the
image of a past experience *as* an image of the experience
which it represents and which was therefore ours. The time-
κίνησις may be definite or indefinite; but even the latter
is sufficient for genuine memory. By its help a person is
able to think and say that he *remembers* something as
having taken place, though he cannot tell *when* it did so.

Conditions of genuine memory. Memory and reminiscence their differences. Reminiscence involves a process of reasoning: nature of this process.

Such is the account of recollection or reminiscence. It
differs, we must observe, from memory in two respects.
First, the latter is chronologically the *prius*, and logically
the presupposition of the former. Secondly, while memory
belongs to many of the lower animals, recollection belongs
to man alone. The reason of this is that it is, or involves,
a sort of inference. In recollecting a person proceeds from
a φάντασμα before his mind to some other which he wishes
to recall. That which he has presents a problem to be
solved. He first reasons that it has conditions—viz. the
circumstances under which it was generated. The major
premiss in such inferences is that every φάντασμα of a certain sort is to be connected with, and explained by, a past
experience. The minor is: this is such a φάντασμα. Having
concluded thus, he proceeds to seek for the experience from
which the φάντασμα is derived—to trace the history of the
φάντασμα and determine its date, or the circumstances when
it first arose[1]. This mental process belongs only to those

lines AB BE, with this difference, however, that instead of also moving
as before in ΘΙ it moves in ΚΛ (i. e. κινεῖται μὲν τὴν Κ, νοεῖ δὲ τὴν Λ).
For these (Κ, Λ) are to one another as AB BE to AZ ZH. When,
therefore, in this way the subjective κινήσεις of the experience and of
its time concur, then, and only then, one actually and fully remembers.'
See *Hermathena*, No. xxv. pp. 459-66; Oxford Trans. of *de Mem.*,
notes *ad loc.*

[1] τὸ ἀναμιμνῄσκεσθαί ἐστιν οἷον συλλογισμός τις· ὅτι γὰρ πρότερον
εἶδεν ἢ ἤκουσεν ἤ τι τοιοῦτον ἔπαθε, συλλογίζεται ὁ ἀναμιμνησκόμενος, καὶ

Y 2

who are capable of rational deliberation ; for such deliberation also is or involves a sort of inference [1].

That memory and reminiscence involve a corporeal, and not merely a psychical, process, shown. (a) We continue involuntarily

§ 47. Memory, like every function of the κοινὴ αἴσθησις and of αἴσθησις generally, involves a corporeal as well as a psychical process [2]. Recollection, too, the search for a missing idea, involves a corporeal process. This is proved by (a) the bodily discomfort caused by fruitless and persistent efforts at recollection ; and (b) by the fact that sometimes even *after giving up* the attempt to recollect a person suddenly remembers what he failed to recall when he tried. The explanation of this can only be that, after the voluntary effort has been given over, the process which

ἔστιν οἷον ζήτησίς τις. τοῦτο δ' οἷς καὶ τὸ βουλευτικὸν ὑπάρχει, φύσει μόνοις συμβέβηκεν (453ᵃ 10-13).

[1] συλλογισμός is a term wide enough to include not only deductive reasoning—the element of which involved in ἀνάμνησις, though fundamental, is slight—but, also inductive with the process of reasoning from particulars to particulars. This last is especially what takes place in the ζήτησις of recollection, when we proceed 'discursively,' turning our minds, so to speak, hither and thither, from point to point, until we have covered the area within which we think the missing idea is to be found. That it *is* somewhere in this area we deduce from the nature of the φάντασμα or idea which prompts the attempt to recollect. If we did not make this deductive step at first : if i. e. we did not feel that we remember and can, if we try, perhaps recollect, we should not make the effort at all. Sir William Hamilton errs by taking συλλογισμός here as merely= syllogism or deductive reasoning (ἀπόδειξις). Aristotle by referring ἀνάμνησις to the deliberative faculty, τὸ βουλευτικόν, shows what he means. The function of the latter faculty is to analyse the conditions of a τέλος (believed possible, and regarded as desirable) until τὰ πρὸς τὸ τέλος, the means, are discovered ; whereupon, if we are satisfied with them, we proceed to πρᾶξις. Cf. *E. N.* 1112ᵇ 12-21 βουλευόμεθα δ' οὐ περὶ τῶν τελῶν, ἀλλὰ περὶ τῶν πρὸς τὰ τέλη . . . Ἀλλὰ θέμενοι τέλος τι, πῶς καὶ διὰ τίνων ἔσται σκοποῦσι . . . ἕως ἂν ἔλθωσιν ἐπὶ τὸ πρῶτον αἴτιον, ὃ ἐν τῇ εὑρέσει ἔσχατόν ἐστιν· ὁ γὰρ βουλευόμενος ἔοικε ζητεῖν καὶ ἀναλύειν τὸν εἰρημένον τρόπον ὥσπερ διάγραμμα. Thus the ζήτησις, which from the end analyses the means in the case of βούλευσις, proceeds, in that of ἀνάμνησις, to analyse from the φάντασμα (whatever starts us off thinking) the conditions in which it originated, i. e. to remember the event which is related to our φάντασμα. The explanations given by Themistius (Soph.) and other old commentators may be disregarded.

[2] It may be mentioned here and should have been stated earlier, that all κινήσεις properly belong to body, and only metaphorically, or κατὰ συμβεβηκός, to ψυχή. Cf. *de Anima*, i. 3. 406ᵃ 11 seqq.

it set up still continues, and that this process is one which ^{trying to} goes on in the body. Such persistence of a corporeal process independently of, or in spite of, the will is not uncommon in persons of the 'melancholic' temperament. Just as one who throws a stone cannot by a mere effort of will stop its course when once it has left his hand, so one who sets the process of recollection going excites, in the part of the body which (as will be seen) is the seat of memory (as of κοινὴ αἴσθησις), a corporeal process consisting of a train of κινήσεις among which somewhere the idea to be recalled has its own place. The discomfort above alluded to is felt particularly by those who have much moisture around or in the region or seat of sense-perception[1]. When this moisture has been set moving, it is not easily restored to rest. It keeps on until the missing idea is found, whereupon or in which event it 'finds a straight path' for itself, and lapses into quiescence[2]. So when strong excitement such as fear or anger has stirred a person, he may struggle to subdue his emotions, but they refuse to be allayed, and continue for a while to resist all the efforts of his will. So, too, it is with us when some popular air or cant expression has become inveterate on our lips. We endeavour to forgo the air or the expression, but in vain. It returns again and again, and we find ourselves humming the forbidden tune or uttering the prohibited phrase before we have time to check ourselves.

[margin: trying to recollect even after we have made up our minds to cease trying. (b) Such involuntary efforts sometimes succeed, and we are surprised by the emergence of the idea when we did not expect it. Illustrations of this involuntary process from other mental phenomena.]

§ 48. What—in Aristotle's[3] theory—is the relation of the so-called 'outer' senses to the 'inner,' or *sensus communis*? Do processes of sense complete themselves in the special senses? Or is each affection of the latter something merely inchoate and requiring to be completed in the central office of the *sensus communis*? There are advocates of both views. In favour of the *second* it may be said that the more

[margin: Relation of sensus communis to the special senses. Never really cleared up by Aristotle himself.]

[1] περὶ τὸν αἰσθητικὸν τόπον: is this the seat of *special* or of *general* sense?

[2] ἕως ἂν ἐπανέλθῃ τὸ ζητούμενον καὶ εὐθυπορήσῃ ἡ κίνησις.

[3] For what follows in this paragraph, cf. C. Bäumker, *op. cit.*, pp. 78-82, and J. Neuhäuser, *Aristoteles' Lehre von dem sinnlichen Erkenntnissvermögen und seinen Organen*, pp. 60-70.

narrowly we scrutinize the details of special perception the more we find it dependent on the activity of the *sensus communis*. The different species of the genus which falls under each outer sense must, in order to be distinguished and compared, come under the ken of the inner sense. This is plain from the argument of *de Sensu* vii (447ᵇ 6–21), where it is urged that each sensory δύναμις is capable only of one ἐνέργεια at one time, and that, therefore, no one sense can perceive more than one even of its proper objects at one time. The aid of the 'common sense' has to be invoked, if any two objects, even the ἐναντία of a single sense, such as white and black, are to be perceived together.

In favour of the *first* may be quoted the many passages in which *each* αἴσθησις is defined as a δύναμις κριτική, having under it (like each ἐπιστήμη) a province of its own, whose content forms one *genus*, consisting of a plurality of *species*. Such passages seem to negative the view that each special αἴσθησις is incapable of perceiving its object without the aid of the common or central sense. Other passages may be added bearing rather on the physiological relation between the inner and outer senses. Thus we read[1] that the objects of sense produce a sensation in each sensory organ, and the affection generated by the object remains in this organ even after the object that produced it has departed. We read[2] that the affection is in the sensory organs not only at first while they are perceiving, but even when they have ceased to do so—in them both deep down and at the surface of the organ; that[3] there are presentative movements (κινήσεις φανταστικαί) in the sensory organs (ἐν τοῖς αἰσθητηρίοις). It may be urged that the affections thus referred to are only physiological facts which do not attain to their psychological meaning until they reach the central organ and are 'informed' by the κοινὴ αἴσθησις. Or we may expect it to be said, according to a passage of Aristotle[4], that the soul has to 'move outwards' to them, as in recollection, in order to impart to them their meaning. Yet this will not get rid of such assertions as that[5]

---

[1] 459ᵃ 24–7.   [2] 459ᵇ 5.   [3] 462ᵃ 8.   [4] 408ᵇ 15–18.   [5] 426ᵇ 8.

'each αἴσθησις has its own αἰσθητόν subjected to it, while it (the αἴσθησις) subsists in its organ *qua* organ'; and that [1] 'αἴσθησις in all animals is engendered in the homogeneous parts' (i. e. the αἰσθητήρια). Moreover, when Aristotle argues that σάρξ is not the true organ of touching, but is related to the latter (the heart), as the external translucent medium is to the organ of vision (κόρη), the analogy would lose its whole point if the pupil itself were not the organ of vision. Again [2], Aristotle describes the stimulation of the eye *qua* diaphanous as being ὅρασις—actual seeing, which would seem to prove that in his opinion seeing has its seat *in* the pupil, not merely that it is effected *through* it. The passage [3] in which he draws a parallel between ὁ ὀφθαλμός and τὸ ζῷον, making the ὄψις of the former answer to ψυχή in the latter, while the eyeball corresponds to the σῶμα, seems to point to the same conclusion; especially when he adds the remark that as the eye is the κόρη *plus* visual power (ὄψις), so the ψυχή and the σῶμα make up the ζῷον [4]. Thus it would seem that seeing completes itself in the eye, not in the central organ; from which it is of course permissible to reason by analogy that the other senses do likewise.

If, therefore, the special senses (with the exception of touching) have separate peripheral seats, each must have a kind of independent office. This, however, can only be a qualified and relative sort of independence. For the consciousness of one's sense-perceptions and the distinction and comparison of the data of the different senses can only take place by means of the central sense, the head-office of the special senses, to which these are related as its contributors [5]. When, however, we inquire more closely into the nature of this relationship of outer and inner sense, to discover how they are united while yet divided, we can receive from Aristotle no assurance that he had ever cleared up this matter even for himself. A psychology completed

---

[1] 647ᵃ 2 seqq.   [2] 780ᵃ 3.
[3] 412ᵇ 18 seqq.
[4] 413ᵃ 2 ὥσπερ ὁ ὀφθαλμὸς ἡ κόρη καὶ ἡ ὄψις, κἀκεῖ ἡ ψυχὴ καὶ τὸ σῶμα τὸ ζῷον.   [5] 469ᵃ 4–12.

on his lines might provide the answer to the question ; but he has not supplied it.

§ 49. The clue to the organ of the central sense seems to lie in Aristotle's treatment of the organ of the sense of touching. For this sense can exist without any of the other senses (even without its modification, tasting); while none of the others can exist apart from it [1]. Now the organ of touching is not what it seems to most at first sight to be, viz. the flesh of the body. The πρῶτον αἰσθητήριον of touch is something in the interior [2]. The superiority which man enjoys over the other animals he owes to the fineness of his sense of touch [3]. This testifies implicitly to the connexion between the organ of touch and that of the central sense. But the connexion is directly stated. The organ by whose function we distinguish white from sweet is a bodily part connected with all the special organs of sense, but especially with that of touch, on which all depend for their existence [4]. Thus what we were led to expect from the fact that touching is the primary sense, by which animal is distinguished from infra-animal life [5], turns out to be true, to a considerable

[1] 415ᵃ 3.

[2] 422ᵇ 21–423ᵇ 23, 426ᵇ 15 ἡ σὰρξ οὐκ ἔστι τὸ ἔσχατον αἰσθητήριον: 656ᵇ 35 οὐκ ἔστι τὸ πρῶτον αἰσθητήριον ἡ σὰρξ καὶ τὸ τοιοῦτον μόριον, ἀλλ' ἐντός. The πρῶτον αἰσθητήριον and the ἔσχατον are the same thing looked at from different standpoints.

[3] 421ᵃ 22, 494ᵇ 12–18.

[4] 455ᵃ 22 τοῦτο δ' ἅμα τῷ ἁπτικῷ μάλισθ' ὑπάρχει.

[5] With this *dictum* of Aristotle that touch is the primary sense, Dr. Ogle compares the words of John Hunter: 'Touch is the first sense, because no animal that has a sense (as far as I know) is without it, while there are many animals without the others'; and again, 'Touch I call the first sense ; it is the simplest mode of receiving impressions; for all the other senses have this of touch in common with the peculiar or specific; and most probably there is not any part of the body but what is susceptible of simple feeling or touch' (J. H., *Museum Cat.* iii. 53, 51). Dr. Ogle resists the temptation to find in this view of Aristotle the theory that the higher sensibilities have been 'evolved by gradual differentiations of parts, originally endowed in common with the rest of the body with sensibility to resistance and temperature, both of which are included by Aristotle under touch ; in other words, that the remaining special senses are but modifications of touch or general sensibility.' He resists this natural temptation be-

extent. For even if Aristotle nowhere *expressly* identifies
the organ of touch with the κοινὸν (or πρῶτον, or κύριον)
αἰσθητήριον of perception, they are certainly for him most
intimately associated. This central organ was the heart or
the region of the heart.

§ 50. Plato and Alcmaeon had taught that the *brain* was The heart,
the organ of intelligence[1]. Aristotle deliberately rejects not the
this view[2]. Plato looked upon the brain as an enlarged for Aris-
portion of the spinal marrow; Aristotle declared it to be organ of
something quite different[3]. The brain, says Aristotle[4], is central
itself as much without sensibility as the blood or any of intelli-
the secretions (ὥσπερ ὁτιοῦν τῶν περιττωμάτων); and there- gence (at
fore cannot be the cause of sensations. The connexion as the
which the brain has, or seems to have, with the eyes or dependent
ears proves nothing to the contrary. The πόροι from brain on φαν-
to eye conduct not sensory currents, but only the moisture Why
which, as internal diaphanous medium, is essential to the κόρη. Aristotle
Though he says[5] that a vein leads from the brain to the the brain
ear, yet he does so with a certain looseness of expression; as central
for in the previous line[6] he had stated that there is no
πόρος from the inner ear to the brain, but that there is one
from it to the roof of the mouth or palate. Hence in the next
line he must be understood to refer to what he elsewhere

cause in *de Sens.* ch. 4 this latter view which was held by Democritus is repudiated by Aristotle. Touch, thinks Dr. Ogle, was for
Aristotle the primary sense; *first*, because it is the most universally
distributed, no animal being without it; *secondly*, because by it we are
able to recognize the four primary qualities of matter, *hot, cold, solid,
fluid*—θερμόν, ψυχρόν, ξηρόν, ὑγρόν. What Dr. Ogle says is most true;
yet it is hard to suppose that Aristotle—the pioneer, in *general* terms, of
the theory of evolution not only physical, but physiological and psychological—should in this particular application of his theory, have failed
to recognize it, or have denied its truth simply because it was a doctrine
of Democritus. However, we have only to do with the facts as
Aristotle himself states them. Cf. Dr. Ogle, Trans. of Arist. *de Part.
An.*, notes, pp. 169-70, and SENSATION IN GENERAL, § 23.
    [1] All doubt on this question had vanished for Galen, thanks to the
anatomical discoveries of Herophilus and Erasistratus. Cf. Galen. *de
Placit. Hipp. et Plat.* § 644 seqq.
    [2] 656[a] 17 seqq.   [3] 652[a] 24 seqq.   [4] 656[a] 23 seqq.
    [5] 492[a] 20.   [6] 492[a] 19.

speaks of as a vein not extending to the brain, but to the membrane (μῆνιγξ) surrounding this[1]. In this membrane there is a network of veins with fine and pure blood running through them ; while there is no blood in the brain itself. Dr. Ogle sums up (substantially, and almost verbally) as follows Aristotle's reasons for rejecting the brain theory. He did so—

' (a) Because the brain is insensible to external mechanical stimulation[2]. If the brain of a living animal be laid bare, the hemispheres may be cut without any signs of pain whatever, and without any struggling on the part of the animal—a difficulty which was impenetrable to Aristotle.

(b) Because he could find no brain or anything apparently analogous to a brain in any of the invertebrata except in the cephalopods[3], the cephalic ganglia in the other animals having, owing to their minute size, escaped his unaided vision. Yet sensation was the special characteristic of an animal. The absence of a brain, then, from numerous sentient creatures, was quite incompatible for him with the notion that the brain was the central organ of sensation.

(c) Because he erroneously regarded the brain as bloodless, as also did Hippocrates; and all experience taught him that those parts alone were sensitive that contained blood[4].

(d) Because he thought it manifest to inspection that there is no anatomical connexion between the brain and sense-organs[5].

(e) Because he believed himself to have good grounds for supposing another part, viz. the heart, to be the sensory centre.'

§ 51. The same author summarizes also the reasons for which Aristotle held the heart to be the sensory centre :—

' (a) He thought he discovered connecting links between the sense-organs and the heart. This he took to be obviously the sense-organ of touch and taste ; while the other organs were connected by ducts with the blood-vessels, and therefore ultimately with the heart[6].

Why Aristotle adopted the alternative theory of the heart as the organ of central sense and intelligence.

---

[1] 495ª 7.   [2] 656ª 23 seqq., 520ᵇ 16.   [3] 652ᵇ 23-6.
[4] 514ª 18, 656ᵇ 20.   [5] 514ª 19.   [6] 781ª 20 seqq., 469ª 4-23.

(*b*) The heart is the centre of the vascular system and of the vital heat[1].

(*c*) The heart is the first part to enter into activity, and the last to stop work (*primum vivens ultimum moriens*); therefore, probably the seat of sensibility—the essential characteristic of animal life[2].

(*d*) The heart's action is augmented or diminished when intense pleasure or pain is felt.

(*e*) Loss of blood causes insensibility.

(*f*) The heart has the central position in the body[3], which seemed to fit it to be the organ of central sense[4].'

For these reasons then Aristotle satisfied himself that the heart is the central sense-organ. He held that, in all sanguineous animals, the centre of control over the sensory operations is situated in this organ (sc. the heart). The κοινὸν αἰσθητήριον, to which all the particular αἰσθητήρια are subordinated, must be in the heart. Two particular senses we plainly see to converge towards it: those of touching and tasting. Hence we may infer that the others likewise do so. . . . Apart from these considerations, if in all animals the life-process is centred in this organ, it follows clearly that the origin of sense-perception is there also[5]. The heart is the principle of motion *qua* consisting of *heterogeneous parts*; and of sensation, *qua* consisting of *simple* (=homogeneous) *parts*[6].

§ 52. The heart being thus the κοινὸν αἰσθητήριον, the Physiological connexion of the special organs of blood, though itself without sensation, plays a most important part in connexion with sensation. Its vessels are the channels whereby sensory κινήσεις are conveyed from

[1] 478[a] 29, 458[a] 14.   [2] 479[a] 1.
[3] 666[a] 14 seqq., 467[b] 28 seqq.
[4] *Vide* Dr. Ogle's translation of the work *On the parts of Animals*, with his notes thereto, pp. 168–9, 172–3. His commentaries on the physiological portions of this work, and on the latter half of the *Parva Naturalia*, are of the greatest service to 'mere scholars,' whose confidence in his scientific authority is not diminished by his evidently thorough acquaintance with the language and writings of Aristotle.
[5] 469[a] 4–23.
[6] 647[a] 27 ἀναγκαῖον ᾗ μέν ἐστι δεκτικὸν πάντων τῶν αἰσθητῶν, τῶν ἁπλῶν εἶναι μορίων, ᾗ δὲ κινητικὸν καὶ πρακτικόν, τῶν ἀνομοιομερῶν.

sense with the special or peripheral to the central or general sense-organ. The principal passages containing information respecting this function of the blood-vessels are found in the third chapter of the tract de Insomn., which deals with the way in which, from residuary movements continuing in the sensory organs after αἴσθησις, 'appearances' arise in consciousness, not only in waking moments but in time of sleep. The residuary movements are conveyed inwards from the special organ—their origin and home, when not actualized or 'in consciousness'—to the central organ. 'We must suppose,' he says, 'that, like the little eddies which are for ever being formed in rivers, the sensory movements are processes continuous but distinct from one another . . . When one is asleep, according as the blood subsides [1] and retires inwards towards its fountain, these residual movements whether potential or actual accompany it inwards [2]. They are so related that, if anything has caused some particular movement in the blood, some given psychic movement comes to the surface, emerging from it [3], while, if this fails, another takes its place. They are to one another like certain toys consisting of artificial frogs [4] submerged in water, which rise in a fixed succession to the surface, according as the various quantities of salt, which keep them severally submerged, become successively dissolved, and so release them [5] from their submersion.' The movement of heat in the blood, however, interrupts the course of the sensory movement [6]. Hence the more exact kinds

the general organ for the media-tion of the sensory processes between them. The agency of the blood in this connexion. Is it the actual vehicle of sense impressions? Or is it only a concomitant, which may impede as well as further their progress? At all events to favour sensory processes the blood must be cool and pure.

[1] 461ᵃ 8, 464ᵇ 8 seqq.

[2] The potential are those which have been already in consciousness, but have sunk into latency, the actual are, we must suppose, the waking perceptions which accompany us into the land of sleep : those which have not yet ceased to affect consciousness, or keep occurring up to the moment when sleep supervenes.

[3] 461ᵇ 14 ἐξ αὐτοῦ, SC. τοῦ αἵματος.

[4] ὥσπερ οἱ πεπλασμένοι βάτραχοι οἱ ἀνιόντες ἐν τῷ ὕδατι τηκομένου τοῦ ἁλός. Some well-known invention—possibly for the amusement of chil-dren—of the time is referred to. So Kant refers to Vaucanson's 'duck.'

[5] For the function of the blood in disseminating κινήσεις, cf. Plato, Tim. 70 A seqq. and § 18, p. 271 supra.

[6] 656ᵇ 5 ἐκκόπτει γὰρ ἡ τῆς ἐν τῷ αἵματι θερμότητος κίνησις τὴν αἰσθητικὴν ἐνέργειαν.

of sensation are necessarily conveyed through the parts which have in them the purer and cooler blood[1]. These, therefore, are in the head near the brain which cools the blood in the small vessels that traverse the membrane surrounding it. Unconsciousness results from compression of the 'veins of the neck[2].' Probably Aristotle would have accounted for this by the interruption of the course of the αἰσθητικὴ ἐνέργεια through these veins towards the heart.

§ 53. But in the conveyance of sensory effects from the outer organs, besides the blood, another agency has to be taken into account, namely the 'connatural spirit' (σύμφυτον πνεῦμα). 'The organ of smelling and that of hearing are πόροι which are in connexion with the outer air, and are full of connatural spirit[3].' The πόρος of the organ of hearing terminates in the region where in some animals the pulsation of the connatural spirit, in others the process of respiration, is located[4], i.e. in the heart or the 'part analogous[5].' For Aristotle's curious explanation of the process of learning from dictation, based on the connexion of ἀκοή with the σύμφυτον πνεῦμα (or at least with the πνεῦμα), see HEARING, § 26, p. 120. This connatural spirit is found in all animals. The vital heat resides in it; and its ἀρχή is in the heart.

The question is how we are to understand the relation between this connatural spirit and the blood in the vessels with regard to the conveyance of sensory effects from the outer organs to the heart. We may understand the πόροι by which the organs of seeing, hearing, and smelling are connected with the heart to be the veins; for of the nerves or their sensory function Aristotle was ignorant. But these

The real agency in the transmission of sensory impressions from the special to the central organ is probably the σύμφυτον πνεῦμα. The πόροι connected with the senses of hearing and smelling (and probably also those connected with seeing) contain this πνεῦμα. If by these πόροι Aristotle meant veins (i.e. blood-vessels of some sort),

[1] He refers to the sensations of sight, hearing and smelling: ἔτι δὲ τὰς ἀκριβεστέρας τῶν αἰσθήσεων διὰ τῶν καθαρώτερον ἐχόντων τὸ αἷμα μορίων ἀναγκαῖον ἀκριβεστέρας γίγνεσθαι, 656ᵇ 3.

[2] 455ᵇ 7. Such unconsciousness is to be distinguished, says Aristotle, from that of sleep.

[3] 744ᵃ 1 ἥ δ' ὄσφρησις καὶ ἡ ἀκοή . . . πλήρεις συμφύτου πνεύματος.

[4] 781ᵃ 23–5 ὁ μὲν οὖν τῆς ἀκοῆς (πόρος) . . . ᾗ τὸ πνεῦμα τὸ σύμφυτον . . . ταύτῃ περαίνει.

[5] 456ᵃ 7 seqq.

πόροι, whatever they were, conveyed in Aristotle's opinion more than the blood[1]. We are told expressly that those of hearing and smelling are full of σύμφυτον πνεῦμα, and this in such a connexion as to lead us to think that the πνεῦμα is the sensory agency in them. On the other hand Aristotle often refers to the blood in a manner which leads one to suppose that he regarded it—at all events in its grosser form—as a mere impediment to the transmission of sensory impressions. It is this that, when it gathers around the heart in sleep, fetters τὸ κύριον—the faculty of judgment[2]. The residual movements in the outer sense-organs are liberated successively[3] in sleep as the blood in these organs is diminished. The senses that are most exact—ἀκριβέσταται—are found in the parts where the blood-vessels are finest and thinnest, and where the blood is coolest and purest, i.e. near the brain[4]. Thus on the whole it would appear—though Aristotle has not worked his conception out clearly—as if he conceived the sensory effects to be conveyed *with* the blood, in the same vessels, but not to be affections of the blood itself or primarily connected with it, but rather with the σύμφυτον πνεῦμα. This view seems decisively confirmed by one clause of a passage already quoted, κατιόντος τοῦ αἵματος ἐπὶ τὴν ἀρχὴν συγκατέρχονται αἱ ἐνοῦσαι κινήσεις[5]. He had before illustrated the nature of the κινήσεις as like eddies in a stream—ὥσπερ τὰς μικρὰς δίνας τὰς ἐν τοῖς ποταμοῖς γινομένας. Thus it might seem fairly as if the κινήσεις of sensation were small 'purls' in the blood, produced by the πνεῦμα, as an interfering force; dependent on the blood, and furthered or restrained by it according to its temperature and quantity, but preserving a form and direction derived from and sustained by

---

[1] In the *History of Animals*, 496ᵃ 30, we read ἐπάνω δ' εἰσὶν οἱ ἀπὸ τῆς καρδίας πόροι· οὐδεὶς δ' ἐστὶ κοινὸς πόρος, ἀλλὰ διὰ τὴν σύναψιν δέχονται τὸ πνεῦμα καὶ τῇ καρδίᾳ διαπέμπουσιν. Plato, too, held that air passes through the blood-vessels. See *Tim.* 82 E.

[2] 461ᵇ 27 and several other passages.

[3] So I take λυόμεναι, not with Neuhäuser (*op. cit.*, p. 131) as 'losing their determinateness.'

[4] 461ᵇ 18.   [5] 461ᵃ 8 seqq.

the πνεῦμα. A similar doubt affects us as to what Plato conceived to be the exact agency in the conveyance of sensory impressions. Are the φλέβια, by which in the *Timaeus* he represents these impressions as distributed through the body, agents of such distribution in virtue of the blood contained in them, or in virtue of the air which (according to Plato) they also contain? The former is the assumption made by Zeller [1]. Our difficulty with respect to Aristotle largely arises from his use of the ambiguous word πόροι to designate the vessels, or connexions generally, of the sensory organs. In some cases this possibly means nerves [2]. In others it certainly means blood-vessels. We are unable to say always which it is in any given case [3]. At all events the σύμφυτον πνεῦμα was conceived by him as having its ἀρχή in the heart, where also that of the blood lies. From this ἀρχή the σύμφυτον πνεῦμα diffuses vital heat throughout the body. The σύμφυτον πνεῦμα is different, of course, from the πνεῦμα of respiration, but takes the place of the latter in creatures which do not respire. It was certainly, on the other hand, the opinion of Aristotle that the blood-vessels are channels of sensory processes. On the whole it seems probable that, while the blood in these vessels was (as Aristotle himself might say) συναίτιον, or a joint agent in the conveyance of such processes from the organs of outer to the organs of inner sense, the σύμφυτον πνεῦμα held rather the office of αἴτιον or principal agent. This becomes more probable the more we reflect on the importance of such πνεῦμα in Aristotle's biology. The 'energetic' factor in the generation of living creatures consists of πνεῦμα. We

[1] *Plato* (E. Tr.), p. 429 n., cf. Plato, *Tim.* 65 C, 67 B, 70 A seqq., 77 E.

[2] The theory of 'animal spirits,' coursing along the nerves, which persisted so long even in modern psychology, dates from the connexion of πόροι in this sense (which after the discovery of the function of nerves was natural enough) with Aristotle's σύμφυτον πνεῦμα. Cf. p. 86, n. 1 *supra*.

[3] We must avoid the common error of supposing that Aristotle regarded the arteries as conveying only air. This arises from ignorance of the meaning of ἀρτηρία in Aristotle, for whom it was the τραχεῖα (ἀρτηρία) or windpipe. Besides he did not even know of the difference between veins and arteries in the modern use of these terms.

are told by Aristotle that what makes seeds fruitful is τὸ θερμόν—the 'caloric' which they contain. This caloric, however, is not ordinary fire, but a πνεῦμα, or rather a natural substance (φύσις) inherent in this πνεῦμα ; a substance like or analogous to the element of which the celestial bodies consist. The blood is thus a comparatively late formation in the animal economy. The πνεῦμα is at the very origin of the life process ; and for Aristotle the origin of life must contain *potentially* (in the case of animals) that of sense. Therefore if we could discover all the properties and functions of the σύμφυτον πνεῦμα, we should (from Aristotle's point of view) have penetrated to the inmost secrets of sense-perception, not merely as regards the origin of the μεσότης or λόγος which essentially characterizes a sensory organ, but also as regards the means provided by nature for the distribution of sensory messages within the organism, and the conveyance of sensory impressions, from the eye and ear and other external senses, to the organ governing them all[1]. The σύμφυτον πνεῦμα had, for him, a primordial and subtle efficacy operative throughout the origin and development of animal existence. It was the profoundest cause and the most intimate sustaining agency from beginning to end of life and sensory power.

---

[1] Cf. 736ᵇ 33–737ᵃ 1 πάντων μὲν γὰρ ἐν τῷ σπέρματι ἐνυπάρχει, ὅπερ ποιεῖ γόνιμα εἶναι τὰ σπέρματα, τὸ καλούμενον θερμόν. τοῦτο δ' οὐ πῦρ οὐδὲ τοιαύτη δύναμίς ἐστιν, ἀλλὰ τὸ ἐμπεριλαμβανόμενον ἐν τῷ σπέρματι καὶ ἐν τῷ ἀφρώδει πνεῦμα καὶ ἐν τῷ πνεύματι φύσις, ἀνάλογον οὖσα τῷ τῶν ἄστρων στοιχείῳ.

INDICES

I. ENGLISH

primary, 52; of rainbow, 53, 66; three not producible artificially, 53; of the diaphanous, how produced, 57; Aristotle's definitions of, 57, 59–60; visible only in light, 58; not = χροιά, 59–60; Aristotle's two definitions of, 60; its species limited, 61; six, seven, or eight chief species of, 61, 69; a ποιότης or πάθος, 61; not purely subjective for Aristotle, 63; objects of vision other than, 64; of the four elements, 65; due to reflexion, 66; determined by diaphanous in body, 68; confounded with luminosity, 69; ἐναντία of, 69; generation of, from primitive black and white, 69; compound, analogous to chords, 70; intermediate, how produced, 70; the pleasing and displeasing, 70–1; three possible theories of formation of intermediate, 70–4; list of particular species of colour, 75–6; contrast, colour effects of, 76, 77; in clouds, 76; complementary, 76; illusions as to, by lamp-light, 77; colour-blindness unknown to Aristotle and his predecessors, 90; only externally mediated αἰσθητόν which takes no time in transit, 153; changes of, in after-images, 303.

Communion of substances, 19.

Comparing and distinguishing, faculty of, 7.

Complementary colours, 76.

Concha, of ear, 95.

Concords pleasing, why, 117; formed of opposites, 126; perceptible by one sensory ἐνέργεια, 126.

Confluence of rays, 18.

Connexion of κινήσεις, customary or necessary, 283–4.

Consciousness, 8, 252; of perception, explained = perceiving the subject which perceives, explained, 288; not due to intellect, 288; faculty of, 288–9; empirical dawn of, 289; neglected in general by Aristotle, 290.

Consonance, 126, 127.

Consonant and non-consonant vibrations, ratios of, 128.

Contact, between organ and object defeats perception, 150; supposed, really only close proximity, 193.

Contiguity, 316–17 (*see* Association).

Continuity of substrates with discreteness of αἰσθητά, 61.

Contraries, 61; perception by, 208, 237.

Contrariety, 316–17 (*see* Association).

Copernican thought, the, 244.

Cranium, rational soul seated in, 270–3.

Crimson, 61, 67, 75.

Critical faculty hampered in sleep, 306.

Cupping-glass, 110.

Curtain (or lid) on olfactory organ, 151.

Custom, law of, in reminiscence, 315–16.

Darkness, 57–8; darkness a στέρησις, 58, 59.

Data of sense for Democritus, 25.

Date of φαντάσματα, 325.

Day, vision by, 20, 22, 23.

Dazzling, sensation of, 51.

Deaf, the congenitally, less intelligent than the congenitally blind, 89, 123.

Deliberation, 324.

Demiourgos, Plato's: arrangement of tripartite soul, 269–73.

DEMOCRITUS, 1, 7, 17, 18; on vision, 23–37; hearing, 99–102; smelling, 136–7; tasting, 163–7; touching, 181–4; sensation in general, 205–8; sensus communis, 254–6; made all senses modes of touching, 24, 200, 230; exact impressions of things impossible for sense, 24; his physical theory, 24; not named by Plato, 25; visual images, necessarily imperfect, 25; visual organ, of water, 25, 82; inconsistently implies a φύσις χρώματος, 25; colour non-objective, 25; vision is ἔμφασις, 25; atoms and void alone objective, 25; distinguished between 'primary' and 'secondary' qualities, 25; his visual theory criticized by Aristotle, 25; vision by contrariety of colour, 26; ignorance on subject of ἀνάκλασις, 26, 82; conditions of perfect vision, 26; peculiarity of his visual theory, 26; visual theory criticized by Theophrastus, 27–9; περὶ εἰδῶν, 27; cognate things see cognates, 29; whole body participant in visual perception, 29; theory of colours, 30–4; four primary colours, 34; colours infinite, 34; colour theory criticized by Theophrastus, 34–6; colour non-objective, 36, 49; on production of leek-green, 53, 54, 61; he and Plato wrong in holding kinds of colour infinite, 62; wrong in thinking colour purely subjective, 63, 72; wrong in thinking vision would succeed best *in vacuo*, 78; vision not (as he held) due to ἔμφασις, 82; peculiarities of his theory of hearing,

line lens, 10 ; like perceives like, 14 ; doctrine of four elements, 14 ; πόροι and ἀπόρροιαι, 14 ; primary colours (two or four?), 15 ; lantern simile, 14–15 ; like Alcmaeon, a physician, 15 ; his theory of vision and Plato's, 18, 46–8, 49, 54, 57 ; held that light travels, 58, 59 ; Empedocles, Anaxagoras, and Aristotle, views on colour, 65 ; Aristotle rejects his theory of light travelling, 77, 80, 81 ; on vision criticized, 83 ; his explanation of γλαυκότης, 85 ; agrees with Alcmaeon on hearing, 94 ; the κώδων within the ear, 95 ; differs from Alcmaeon on hearing, 97 ; Theophrastus asks, ' How do we hear the κώδων itself ?' 97 ; theory of smelling criticized by Theophrastus, 134 ; his theory of ἀπόρροιαι as to touching and tasting unsatisfactory, 161 ; on tastes, criticized by Aristotle, 174 ; his theory of touching criticized by Theophrastus, 180–1, 201 ; his theory of ξυμμετρία, 233 ; forced to recognize λόγος as true φύσις of bodies, 220 ; theory of temperaments and genius, 253 ; no doctrine of synthesis, 253, 260, 269.

Empirical psychology, 1, 3, 8.

Energy, exhaustion and repair of, 300.

Engelmann, 104.

Enthusiasmus, divination by, 273.

Epicurus, 7, 17, 18.

Epistemology, 214.

Equal, the, a branch of the *one*, 127.

Erasistratus, 5, 329.

Error, 4 ; of sight and of inference or judgment, 90.

Euripides, 12, 256.

Eustachian tubes, 95, 121.

Evaporation, fumid, 243 ; from food, 301.

Expectation, 264.

Experiments, 4.

Eye, as optical system, 9 ; a mirror, 10 ; outgrowth from brain, 12, 86 ; constitution of, 19 ; differences of, 19 ; gleaming, 21 ; best constitution of, 23 ; its essential feature for Democritus, 24 ; ' duplicates itself' when moved, 64 ; compared by Empedocles to lantern, 15–16, 83 ; the embryonic, over-moist and over-large, 85–6.

Eye-ball, displacement of, causes double vision, 306.

Faculties, higher, depend on lower, 309.

Faculty, comparing and distinguishing,

7 ; judging and controlling, 303 ; the central, normally seconds reports of special senses when uncontradicted, 306.

Falsehood, 4.

Farbenlehre, Aristotle's, 69.

Feeling, 270 ; no single term for, in Greek, confused with cognitive αἴσθησις, 273–4.

Fenestra ovalis, 96.

Fever patients, their hallucinations, 303.

Fiery element, not *our* fire, 64.

Fifth, in music, 129.

Figure, of atoms, 36, 182 ; geometrical, 297, 309.

Fire, intra-ocular, 10, 11, 13, 18; smaller destroyed by greater, 22 ; its atoms spherical, 32 ; three fires concerned in vision, for Plato, 46, 48 ; kinds of, for Plato, 65 ; visible in darkness, 57, 64 ; visual organ, not of, 82–3 ; extinguishable, not so light, 83 ; by it in organ of touch we discern hot and cold, 240 ; how far contained in αἰσθητήρια, 248.

Fishes, in Acheloüs, 118 ; voiceless, 119.

Five senses, 1, 2, 207.

Flame colour, 34.

Flesh, need of, as medium of sensation, 192.

Fluid and solid, 190.

Forgetfulness, total, 318.

Forgetting, Plato's definition of, 259, 264.

Form, ranks higher than matter, 219 ; implicitly universal even in perception, 224.

Forward and backward, meanings of, 90.

Foster, Sir M., on olfactory function, 133; on odours, 143; on taste, 160.

Four elements, 18.

Fourth, in music, 129.

Freudenthal, J., 292, 293, 294, 310, 313.

Frogs, artificial, illustration from, 332.

Galen, 5, 25 ; agrees with Aristotle that light does not travel, 59, 95 ; approves Plato's three ἀρχαί of ψυχή, 275, 329.

Gas, our idea of, represented by ἀήρ or καπνός, 149.

Gellius, A., 102.

Generation, 335.

Genus, divisible only into species, which are finite, 61 ; a discrete quantity, 61, 217.

Geometrical qualities of atoms, 37.

II. GREEK

PASSAGES OF GREEK AUTHORS EXPLAINED OR DISCUSSED

Oxford : Printed at the Clarendon Press by HORACE HART, M.A.